Lecture Notes in Computer Science 10171

Commenced Publication in 1973
Founding and Former Series Editors:
Gerhard Goos, Juris Hartmanis, and Jan van Leeuwen

More information about this series at http://www.springer.com/series/7409

Peter W. de Vries · Harri Oinas-Kukkonen
Liseth Siemons · Nienke Beerlage-de Jong
Lisette van Gemert-Pijnen (Eds.)

Persuasive Technology

Development and Implementation of Personalized Technologies to Change Attitudes and Behaviors

12th International Conference, PERSUASIVE 2017
Amsterdam, The Netherlands, April 4–6, 2017
Proceedings

Springer

Editors
Peter W. de Vries
University of Twente
Enschede
The Netherlands

Harri Oinas-Kukkonen
University of Oulu
Oulu
Finland

Liseth Siemons
University of Twente
Enschede
The Netherlands

Nienke Beerlage-de Jong
University de Twente
Enschede
The Netherlands

Lisette van Gemert-Pijnen
University of Twente
Enschede
The Netherlands

ISSN 0302-9743 ISSN 1611-3349 (electronic)
Lecture Notes in Computer Science
ISBN 978-3-319-55133-3 ISBN 978-3-319-55134-0 (eBook)
DOI 10.1007/978-3-319-55134-0

Library of Congress Control Number: 2017933539

LNCS Sublibrary: SL3 – Information Systems and Applications, incl. Internet/Web, and HCI

Printed on acid-free paper

This Springer imprint is published by Springer Nature
The registered company is Springer International Publishing AG
The registered company address is: Gewerbestrasse 11, 6330 Cham, Switzerland

Preface

Persuasive Technology (PT) is a vibrant interdisciplinary research field, focusing on the design, development, and evaluation of technologies aimed at changing people's attitudes or behaviors through persuasion and social influence, but not through coercion or deception. This field is serviced by a series of international conferences on PT, bringing together academic researchers, designers, and practitioners from social scientific and technological disciplines, as well as from the fields of health, health care, safety, sustainability, and ICT. This community aims to facilitate healthier lifestyles, make people feel or behave more safely, and reduce consumption of renewable resources, e.g., by using big data, sensor technology for monitoring, personalized feedback and coaching, mHealth, data visualization techniques, serious gaming, or social media.

The 12th International Conference on Persuasive Technologies was hosted by the Centre for eHealth and Wellbeing Research, Persuasive Health Technology Lab, University of Twente (UT), the Netherlands. Rather than in the UT's hometown Enschede, however, it took place on the other side of the country, in the picturesque city of Amsterdam. Its special theme was "Smart Monitoring and Persuasive Coaching," building bridges between persuasion and personalized health care via real-time data collection and smart, empathic, user-adaptive engaging technology. The previous successful conferences were organized in Salzburg, Chicago, Padua, Sydney, Linköping, Columbus, Copenhagen, Claremont, Oulu, Palo Alto, and Eindhoven.

We feel that the field of PT is reaching maturity. Not only does the number of papers submitted to PT conferences continue to increase – perhaps testifying to a growing shared belief of the importance of PT – but the same can be said for the quality and breadth of the submitted papers. The papers that we received focused on PT in different domains (e.g., health, safety, coaching), examined specific psychological mechanisms that positively or negatively influence PT effectiveness, emphasized methodology for design, evaluation, and implementation, and synthesized theory from PT and adjacent fields.

This volume contains the papers that were presented at the 12th edition of the conference series in April 2017. A total of 85 papers were submitted, which fell into three categories: Design Research, Conceptual, Theoretical and Review, and Empirical. The latter category encompassed a variety of papers, from user studies and user surveys to experimental studies, and could be either qualitative or quantitative. Boiling this number down to the 23 papers published in this volume (27% acceptance) required the dedication of our reviewers (on average three per paper) and some tough decisions.

These papers were subsequently grouped based on content. Grouped under the heading "Health(care), Monitoring, and Coaching" are the papers that tie in with the special theme of this conference. A number of papers focused on the link between persuasion attempts and personality, for instance, attempting to motivate healthy eating habits by tailoring messages to specific personality characteristics. Others examined persuasion as the result of situational factors such as motivation, awareness, facilitators, and barriers. The former papers were grouped under the heading "Personality, Personalization, and Persuasion", whereas the latter were labeled as "Motivations, Facilitators, and Barriers." The largest group of papers focused on identifying or testing specific persuasion strategies from a design perspective. These were grouped under "Design Principles and Strategies". These categories should be seen as an attempt to reflect the diversity of the submitted papers.

Posters, demos, symposia, doctoral consortium papers, and workshop and tutorial contributions were printed in the adjunct proceedings.

We would like to thank all authors for their high-quality contributions, and the reviewers for their valuable feedback. Furthermore, thank all of our colleagues and students involved in the overall organization, the workshops, tutorials, doctoral consortium, posters, demos, symposia, and the conference itself.

April 2017

Peter W. de Vries
Harri Oinas-Kukkonen
Liseth Siemons
Nienke Beerlage-de Jong
Lisette van Gemert-Pijnen

Organization

General Chair

Lisette van Gemert-Pijnen	University of Twente, The Netherlands

Organizing Chair

Nienke Beerlage-de Jong	University of Twente, The Netherlands
Liseth Siemons	University of Twente, The Netherlands

Program Chair

Peter W. de Vries	University of Twente, The Netherlands
Harri Oinas-Kukkonen	University of Oulu, Finland

Tutorial/Doctoral Chair

Jaap Ham	Eindhoven University of Technology, The Netherlands
Cees Midden	Eindhoven University of Technology, The Netherlands
Luciano Gamberini	University of Padova, Italy

Workshop Chair

Saskia Kelders	University of Twente, The Netherlands
Geke Ludden	University of Twente, The Netherlands

Poster Chair

Thomas van Rompay	University of Twente, The Netherlands

Demos

Thomas van Rompay	University of Twente, The Netherlands

Public Relations

Hanneke Kip	University of Twente, The Netherlands
Floor Sieverink	University of Twente, The Netherlands

Social Media Committee

Geke Ludden	University of Twente, The Netherlands
Agnis Stibe	MIT Media Lab, MA, USA

Administration

Marieke Smellink-Kleisman	University of Twente, The Netherlands

Program Committee

Ali Rajan	Bournemouth University, UK
Nienke Beerlage-de Jong	University of Twente, The Netherlands
Shlomo Berkovsky	CSIRO, Australia
Robbert Jan Beun	Utrecht University, The Netherlands
Samir Chatterjee	Claremont Graduate University, USA
Luca Chittaro	University of Udine, Italy
Jacqueline Corbett	Smithsonian Institution, USA
Janet Davis	Whitman College, USA
Johannes de Boer	Saxion University of Applied Sciences, The Netherlands
Boris de Ruyter	Philips Research, The Netherlands
Peter de Vries	University of Twente, The Netherlands
Alexander Felfernig	Graz University of Technology, Austria
Jill Freyne	CSIRO, Australia
Luciano Gamberini	University of Padua, Italy
Sandra Burri Gram-Hansen	Aalborg University, Denmark
Ulrike Gretzel	University of Southern California, USA
Jaap Ham	Eindhoven University of Technology, The Netherlands
Marja Harjumaa	VTT, Finland
Stephen Intille	Northeastern University, Massachusetts, USA
Giulio Jacucci	University of Helsinki, Finland
Anthony Jameson	German Research Center for Artificial Intelligence (DFKI), Germany
Maurits Kaptein	Tilburg University, The Netherlands
Sarvnaz Karimi	CSIRO, Australia
Pasi Karppinen	University of Oulu, Finland
Saskia Kelders	University of Twente, The Netherlands
Sitwat Langrial	Sur University College, Oman
Thomas MacTavish	Illinois Institute of Technology, USA
Alexander Meschtscherjakov	University of Salzburg, Austria
Cees Midden	Eindhoven University of Technology, The Netherlands
Alexandra Millonig	AIT Austrian Institute of Technology, Austria
Harri Oinas-Kukkonen	University of Oulu, Finland
Rita Orji	University of Waterloo, Canada

Peter Ruijten	Eindhoven University of Technology, The Netherlands
Liseth Siemons	University of Twente, The Netherlands
Anna Spagnolli	University of Padua, Italy
Agnis Stibe	MIT Media Lab, USA
Piiastiina Tikka	University of Oulu, Finland
Kristian Tørning	Danish School of Media and Journalism, Denmark
Manfred Tscheligi	University of Salzburg and AIT, Austria
Lisette van Gemert-Pijnen	University of Twente, The Netherlands
Thomas Van Rompay	University of Twente, The Netherlands
Julita Vassileva	University of Saskatchewan, Canada
Vance Wilson	Worcester Polytechnic Institute, USA
Khin Than Win	University of Wollongong, Australia

Sponsoring Institutions

Exxellence
www.exxellence.nl

ConnectedCare
www.connectedcare.nl

Coolminds
www.coolminds.nl

Centre for eHealth & Wellbeing Research
www.cewr.nl

Sponsors

We would like to thank our sponsors for their support:

Exxellence Labbs is the lab for innovative IT concepts. Through means of innovation we want to realize the best service for citizens, customers, patients and students. Anytime, anywhere.

In collaboration with The University of Twente, Saxion University and several other research institutes and partners, we set up innovative projects that will potentially lead to successful start-ups. Exxellence Labbs aides in marketing new ideas as Minimal

Viable Products (MVPs) and giving these ideas a chance to take root in the current market. We then look for third parties that see the potential in these projects and invite them to contribute to them.

Exxellence Labbs is a sandbox within an inspirational environment. We flesh out concepts and hold lectures on innovation aimed at businesses and governments.

COOLMINDS

Coolminds distinguishes itself by providing a diverse range of internet services and the virtual aspect on which we focus. Passion for healthcare, continuous innovation, creativity and social involvement makes Coolminds the partner in the leading position for VR in healthcare.

connectedcare

ConnectedCare develops engaging digital communication and collaboration applications for independent living and ageing, and brings innovations to market. Our team of ICT experts, designers, and business developers are experienced in design methodologies, user-centered design, service design, care collaboration, user interface development, and business development in the context of care collaboration and independent living. ConnectedCare – your flexible partner in EU care innovations.

Centre for eHealth & Wellbeing Research

A leading research centre for personalized health care. The centre captures the available scientific expertise within the Department of Psychology, Health and Technology (University of Twente). Our mission is to apply psychological knowledge in the design and evaluation of technological innovations that contribute to well-being, health, and personalized health care.

Contents

Health(care), Monitoring, and Coaching

Design Decisions for a Real Time, Alcohol Craving Study Using Physio- and Psychological Measures

Hendrika G. van Lier[1(✉)], Mira Oberhagemann[1], Jessica D. Stroes[1], Niklas M. Enewoldsen[1], Marcel E. Pieterse[1], Jan Maarten C. Schraagen[1,2], Marloes G. Postel[1,3], Miriam M.R. Vollenbroek-Hutten[1,4], Hein A. de Haan[3], and Matthijs L. Noordzij[1]

[1] University of Twente, Postbus 217, 7500 AE Enschede, The Netherlands
h.g.vanlier@utwente.nl

[2] Netherlands Organization for Applied Scientific Research (TNO), Soesterberg, The Netherlands

[3] Tactus Addiction Treatment, Deventer, The Netherlands

[4] Ziekenhuis Groep Twente, ZGT Academy, Almelo, The Netherlands

Abstract. The current study was a pilot for an alcohol craving monitoring study with a biosensor (E4 wristband) and ecological momentary assessment (EMA) smartphone app. The E4 wristband was evaluated on compliance rates, usability, comfort and stigmatization. Two EMA methodologies (signal- and interval-contingent design) were compared on data variability, compliance and perceived burden. Results show that both EMA methodologies captured variability of craving and compliance rates were between medium to low. The perceived burden of the designs was high, in particular for the signal-contingent design. Participants wore the wristband ranging from occasionally to often and the usability was rated good. Many participants reported frequent questioning about the bracelet, which they indicated as positive. However, addicted individuals are expected not to appreciate this attention, we therefore propose to provide them with coping strategies. Efforts should be made to increase compliance, we therefore propose the interval contingent design with micro incentives.

Keywords: Biosensor · Ecological momentary assessment · Data variability · Compliance rates · Perceived burden · Usability · Wearing comfort · Stigmatization

1 Introduction

Alcohol abuse is currently the fourth leading unhealthy lifestyle behavior contributing to morbidity and mortality. Oinas-Kukkonen and Harjumaa [1] argue that in the case of (alcohol) addiction, persuasive systems should aim at reinforcing proper attitudes and making them easier to stick with even in challenging,

P.W. de Vries et al. (Eds.): PERSUASIVE 2017, LNCS 10171, pp. 3–15, 2017.
DOI: 10.1007/978-3-319-55134-0_1

spontaneous situations. In a similar vein, Rohsenow and Monti [2] state that prompting patients to mobilize their coping resources by making them aware of their craving in challenging situations may protect them from relapse.

Innovations in wearable technology offer new possibilities to determine these challenging craving situations, by continuous measurement of fluctuations in physiological parameters, not only in the lab setting, but also in the real world (e.g. [3]). Craving responses have been proven multiple times in lab settings. According to a meta-analysis [4], alcoholics evidently show heightened physiological responses (heart rate, electrodermal activity) and psychological response (self-reported craving) to alcohol cues (e.g., pictures of beer). Yet, whether these effects are transferable from the lab setting to continuously measuring in the real world, the "wild", has not been investigated, therefore this pilot and eventually a monitoring study is performed. The lack of prior studies impedes making informed study design decisions. This study pilots the design decisions of an alcohol craving monitoring study using a smartphone app and a wearable biosensor. Using an app to administer questions is known to increase the compliance and to reduce the perceived burden [5]. The questionnaires were administered according to a Ecological Momentary Assessment design (EMA, [3]). EMA studies are repeated measurements of participants experiences and behavior in real time and in their natural environment [6].

A possible limitation of measuring in the "wild" is the completeness of both the EMA and wearable and data. Since the completeness of the data relies on the compliance of respondents and the functioning of the technology, missing values are likely to occur [6]. Additionally, if the wearable biosensor is not perceived as usable and comfortable, this could be a reason to stop wearing the sensor over time leading to an increase in the number of missing values. Furthermore, because the target population are alcoholics who are known to value anonymity [7], wearing a biosensor for an alcohol craving study might establish a feeling of stigmatization. A possible drawback of using a repeated intensive assessment like EMA is the high burden which might discourage participation. This can result in a sampling bias where only participants with certain personality factors and high motivation enter or complete the assessment [8]. This might also impose an compliance problem for the sample in general, since addicts are not known for their motivation, responsibility and compliance to schedules and instructions [6].

Two design decisions, the EMA cue type and the user acceptance of the E4 wristband as biosensor, are evaluated in this study. For the self-reported measures the perceived burden, compliance rates and craving variability of two types of EMA cue type measures were assessed; (1) interval-contingent measures and (2) signal-contingent measures [9]. In interval-contingent designs, a fixed number of measures are taken according to a standard schedule of intervals. With signal-contingent designs, a fixed number of measures are taken at randomly scheduled intervals, possibly within specified periods of time. Regarding the E4 wristband three factors were evaluated to determine the user acceptance of the device: (1) usability [10], (2) wearing comfort [11], and (3) perceived stigmatization [11] of the wearable.

1.1 Perceived Burden of the EMA Design

The perceived burden, variability and compliance of self-reported craving will be explored to determine which cue type (interval or signal) is most feasible for a monitoring study.

Interval-Contingent Design. The strength of interval-contingent is that the measures can be taken at certain times of the day at which it is likely that construct of interest will occur [12], in this case craving. A risk is that a at unexpected moments at which craving could occur could be missed if the interval time is not selected properly, which causes a systematic bias [13]. Another weakness is that an interval-contingent design can result in high levels of predictability: patients can predict the timing of interval assessments. This may alter their behavior in preparation of the recording time [13].

Signal-Contingent Design. The main advantage of a signal-contingent measures is that the occurrence of craving during multiple different time frames of the day can be examined [7]. A weakness is that signal-contingent recordings can sometimes be perceived obtrusive and disruptive or a recording can be missed because the signal was unexpected or could not be completed at the moment of the signal [13].

Compliance Rate EMA. Compliance rates are associated with the persuasiveness of a system, since the system will be used more when it is highly persuasive [1]. Compliance rates in EMA studies vary from 90% to 50%, but many studies have compliance rate around 75–80% [6]. It is not clear what causes this variation, but the contingent design is likely to affect these compliance rates.

Variability EMA. Researchers of EMA studies found low levels of self-reported craving in alcoholics [6]. It is not clear whether this is a genuine finding or that alcoholics are unwilling to admit their craving. This pilot study explores the variability of self-reported in two contingent designs.

1.2 User Acceptance E4

To determine user acceptance of the E4, usability, wearing comfort and perceived stigmatization are tested in the pilot. The E4 has the size of a large watch and is designed for continuous, real-time data acquisition in daily life.

Usability E4. It is important that the wearable biosensor has high usability, so that the participants use the device as intended by the researcher. Usability is described as the effectiveness, efficiency and satisfaction with which specified users can achieve specified goals in a particular environment [10].

Wearing Comfort E4. An important aspect in integrating and accepting the E4 sensor in the respondents daily life is the wearing comfort. Fensli and Boisen [11] found several factors that play a role in this perceived comfort of wearable technology, for example perceived burden during daily activities like bathing or sleeping. Users preferred small discrete sensors that are as much as possible integrated in everyday objects [14] and did not affect daily behavior. Furthermore, they highly valued a comfortable, compact, reliable and easy to operate device. However, the E4 has some specifications that can not be adjusted, such as the device has to be charged every few days, cannot be worn during showering or extreme rain and the size and comfort of the device cannot be modified. It is explored to what extent this influences the user acceptance of the wearable.

Stigmatization E4. Fensli en Boisen [11] also found that people can experience stigmatization when wearing a sensor. Bergmann, Chandaria and McGregor [14] showed that users prefer sensors that cannot be seen by other people. This preference could be even more important for alcoholics, who often want to keep their addiction hidden from their social environment, including family and friends [15]. Stigmatization might cause respondents not to wear the wristband at certain moments (for example at work) or stop wearing the wristband altogether, posing a problem with compliance.

2 Methods and Materials

This study was a mixed methods feasibility pilot, in which both physiological and self-reported EMA data were collected. To explore reasons for possible (non)compliance, multiple questionnaires and an exit interview regarding perceived burden and user acceptance were performed.

2.1 Participants

In total, eight participants (3 male) between 19 and 24 years (M = 21.5, SD = 1.77) took part in the experiment over a period of eight days. Four (2 male) of these participants completed the whole experiment, the other four were only included for a part of the study, namely wearing comfort and stigmatization, since they did not perform the EMA part of the study. Participants were included if they met at least 2 of the 11 diagnostic criteria of the DSM-V for an alcohol use disorder, if they drank more than 14 glasses on average of alcohol per week and did not have a diagnosis of dependence of another substance than alcohol (except nicotine). Participants also had to own a smartphone and have access to a laptop or computer. The participants were college students from the University of Twente and received course credits to participate in the study. The Ethical Commission of the University of Twente approved this study and participants signed an informed consent before entering into the study.

2.2 Materials

EMA Recording. For recording the daily questionnaires a smartphone app named 'UTSurvey' [16] was used. The app issues an notification when a questionnaire had to be administered.

Physiological Data. For uploading the data of the E4 wristband, the program 'Empatica Manager' was used.

2.3 Procedure

Data-acquisition started with a meeting during which informed consent was obtained and the DSM-V criteria were assessed. Participants were then trained to use the EMA app, use the E4 sensor, charge the wristband and upload the data. During the week, the participants had to wear the E4 during the day, and charge it at night when they were sleeping. Additionally, they had to upload the recorded data at least every two days. In the EMA app, participants had to fill in an alcohol registry every morning, and complete a brief questionnaire four times a day. The first four days, phase one of the experiment, an interval-contingent design was used for the EMA app. The time slots at which questions had to be administered were predefined and the same for every day (11 am, 3 pm, 7 pm and 12 midnight). This final time slot was not mandatory since some participants might already be asleep. The second set of four days, the second phase, a signal-contingent design was employed. The time slots were randomized; between 9:00 am and 12:45 pm, between 1:00 pm and 4:45 pm and between 5:00 pm and 9:00 pm. The final time slot of the day was again at midnight, again not mandatory. After the experiment the participants administered online questionnaire about the perceived burden of the EMA designs, a questionnaire with the 'System Usability Scale' (SUS) about the usability of the E4 wristband and an exit interview on the wearing comfort and the perceived stigmatization of the E4 wristband.

2.4 EMA Measures

Morning Report. Upon awakening, the app questioned the participants about the time and number of standard drinks consumed on the previous day.

Prompts. Participants had to answer three questions after a prompt: alcohol craving, mood and coping ability. Since craving was the only construct of interest for this pilot and the two other questions were merely included to represent a realistic EMA burden, only craving will be further explained. Craving for alcohol was measured on a 10-point Likert Scale ranging from 1 (not at all) to 10 (very much). A single item measure of craving is a straightforward and time effective manner for assessing the level of subjective craving of a participant [17].

2.5 Measures

Perceived Burden of the EMA Design. For measuring the perceived burden of the participants for the EMA designs, an online open ended questionnaire was taken (See Table 1.) (translated from the original Dutch items).

Table 1. 11 Questions on perceived burden of the two EMA design

1.	What did you think of answering a questionnaire every day? Please explain.
2.	Would you be able to integrate answering (daily) questionnaires in your day-to-day life? Why could or couldn't you?
3.	What is your attitude towards daily surveys? Please explain.
4.	How did the continuous answering of questions go? Please explain.
5.	How difficult was answering the questions for you? Please explain.
6.	How burdensome did you find it to answer questions every day? Please explain.
7.	Did anything irritate you while answering the questions? If yes, what and why?
8.	Did anything go wrong with answering the questions? If yes, what and why?
9.	What did you think of answering questions on an app? Please explain.
10.	Did you think the app was easy to use? Please explain.
11.	Did you miss many administration moments? If yes, why?

Usability E4. Usability of the E4 wristband was measured with the System Usability Scale [10] consisting of ten 5-point Likert scale items. The SUS gives a global view of subjective assessments of usability [10].

Wearing Comfort E4. A semi-structured interview was performed, starting with a non-directive question "How did you experience last week?" and becoming more and more directive, leading to the closing question "How would you describe wearing the sensor", when a participant did not give any information on wearing comfort in earlier questions.

Stigmatization E4. A semi-structured interview was performed with the same non-directive to directive structure. The opening question was "Did you have any conversations about your participation in the experiment in the last week?" leading to "What was the impact of the sensor on your feeling of anonymity?".

2.6 Data Analysis

EMA Design

Compliance Rate and Variability EMA. The EMA data was extracted from the program 'Limesurvey' [18]. The compliance rate and variability in craving for the self-reported measures per EMA design were determined. Because of the exploratory nature and small sample in this research no statistical test was performed. Instead, multiple graphs for every participant per design were made with craving measurements for every time slot in order to make the variability and distribution of the craving scores visible.

Perceived Burden EMA. The questionnaire for the perceived burden was analyzed by filtering out the main statements relevant to the perceived burden of participants. After which the statements were labeled. The labeling continued until no new labels could be given. The inter-rater reliability was determined using Cohens Kappa, by labeling all data by a different researcher.

User Acceptance E4 Wristband

Compliance Rate E4. The physiological data was extracted from the Empatica Manager. The times on which data was recorded were visualized in a graph in order to evaluate the compliance rates.

Usability E4. The SUS scores were calculated according to [10]. With the use of the adjective rating scale [19], the SUS score was interpreted.

Wearing Comfort and Stigmatization E4. Wearing comfort and stigmatization were measured using a semi-structured interview. The text was divided in relevant fragments and labeled, as close as possible to the text of the fragment. The labeling continued until no new labels could be given. The inter-rater reliability was determined using Cohens Kappa, by labeling 12.5% of the data by a different researcher.

3 Results

The female participants reported drinking an average of 12.8 glasses of alcohol a week (sd = 6.53 glasses) divided over four evenings during the week (sd = 1.45 days). The male participants reported drinking 14 glasses (sd = 3.46 glasses) on average divided over three evenings during the week (sd = 0.69 days).

3.1 Self-reported EMA data

This research explored the variability and the compliance of alcohol craving measurements when using interval- and signal-contingent designs. During one time slot no prompt occurred and during another it was not possible for the participants to administer the questions due to technical problems. However, these minor anomalies are not expected to have influenced the results, as they constituted two out of the total of 81 completed time slots.

Compliance Rate and Variability EMA. Participants reported craving greater than 1 relatively often: out of 81 completed time slots, noticeable urge to drink was reported 42 times. The average craving was 3.68 (SD = 3.36) with the interval-contingent design and 3.05 (SD = 2.90) with the signal-contingent design. Figure 1 visualizes the compliance and variability of alcohol craving per participant. It seems that the compliance and variability of the craving measurements between the interval- and signal-contingent design do not differ greatly.

The average compliance in the interval-contingent design was 67% and in the signal-contingent design 61%. The craving scores were higher in the evening compared to the morning and afternoon. As expected, the number of completed time slots was much lower at the non-mandatory time slot at midnight (time slot 4, 8, 12 & 16). Participant four mentioned she missed a lot of time-slots, due to not having an Internet connection or beeing occupied at work.

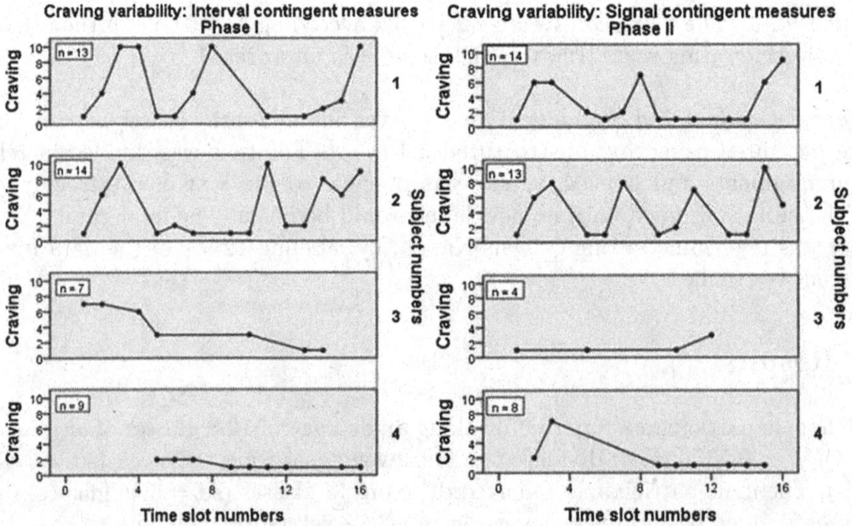

Fig. 1. Craving scores per participant over time. The missing values are left out (n = the number of administered data points). The time slot numbers are displayed on the horizontal axes. Every four numbers represent one day: First time slot = morning, second time slot = afternoon, third and fourth time slots are evening.

Perceived Burden EMA. The statements about the perceived burden had an inter-rater reliability of 0.71 which corresponds with a substantial strength of agreement. The effort to complete the questions at a single time slot was generally experienced as low. All participants thought the questions were easy to answer: "Answering the questions went well, the questions were easy and you could answer in a detailed, but easy manner, how you felt at the moment." (Participant 4, q.4). The questionnaire at midnight was sometimes perceived as an higher burden: "(...) the questionnaire at midnight was sometimes unpractical (I was at the cafe without my phone then or was sleeping already)." (Subject 2, q.1). The burden of filling in the questionnaires with a frequency of four times a day was perceived as relatively high by two participants. They reported feeling bad or stressed because of the questionnaires: "I don't like the feeling of always having to answer something. I felt very pressured. When I was not able to answer questions I felt bad because I did not finish the assignment" (Participant 4, q.6). However, others mentioned that they did not feel it as a burden to answer the questions multiple times a day: "Not really a burden, you have your phone often near you anyway." (Participant 1, q.6). The burden was heightened with the signal-contingent design, compared to the interval-contingent design: "It is easier if the time slots are at a set time, in that manner you can anticipate to react and after a while you don't forget to fill it in anymore." (Participant 1, q.2).

3.2 User Acceptance E4 Wristband

Compliance Rate E4. The compliance rate of wearing the E4 differed over participants. An overview of the times each participant wore the E4 can be seen in Fig. 2. Note that participant four only wore the E4 on 4 of the 7 days and participant three wore the E4 every day, however sometimes only for a few hours (see for example Monday). The other two participants all had a higher compliance.

Usability E4. The usability of the E4 wristband when wearing it on a daily basis for a week, with use of the System Usability Scale was on average 65 points on a 100-point scale, which is considered to be slightly below average [10]. Scores on individual questions could range between 1 and 4. The participants thought the E4 was easy to use (3.25) and did not feel like they needed a lot of technical support (1). However, all but participant 4 did not really wanted to use the E4 frequently (1.25) and found the E4 sometimes frustrating (3).

Wearing Comfort E4. The agreement rate regarding the qualitative analysis had a Cohens kappa of 0.77 which corresponds with a substantial strength of agreement. The general impression of the participants on wearing the sensor was neutral to positive and was described by all participants as "fine". All participants indicated that they had to get used to the sensor, but this happened quickly and with ease. Participant one remarked that it was fine to wear

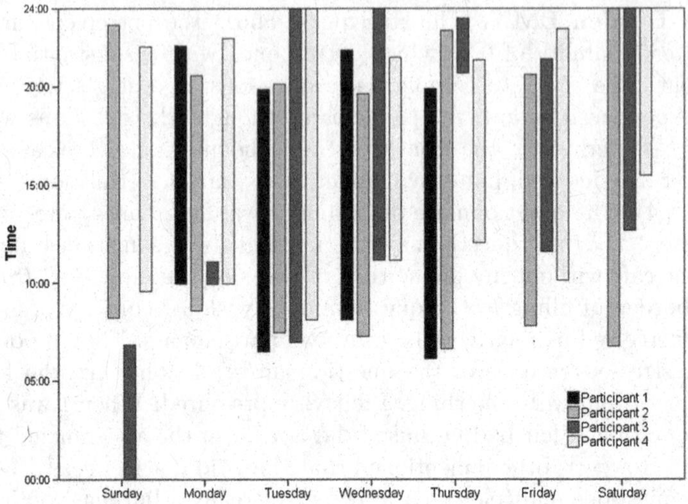

Fig. 2. Time each participant wore the E4 over a course of a week.

the sensor for a week, but would not have wanted to wear the wristband any longer. The participants mentioned that it is burdensome that the sensor is only splash-waterproof. For many activities, like doing the dishes, showering or even cycling through the rain, it was not possible to wear the wristband. One participant indicated that he found it annoying to have to remember to bring the charger, charge the wristband and upload the data. Multiple participants did mention that the sensor sometimes pressed painfully on the bone and was either too tight or too loose. One of the men, as well as both females found the sensor "ugly" and considered it to be too big to be worn with tight clothes. A final negative point about the comfort that was mentioned by the participants was that the sensor is not easy to wear during exercise, since in some sports wearing anything around your wrists is prohibited and some are too "rough" for gathering good data.

Stigmatization E4. All participants indicated that they were approached by many people about the sensor. The majority of the participants therefore became more aware of the sensor. Especially family, friends and acquaintances addressed the participants about the sensor. All participants indicated that they had a positive feeling about the conversations and explaining why they had to wear the sensor. One participant mentioned that if he should wear the sensor for longer, it would become irritating that his environment would continue to question the sensor. The participants indicated that the sensor was often thought of as a watch or sports and fitness tracking device. Three participants also said they were asked if they were wearing a house arrest curfew band due to a criminal offense.

4 Conclusion and Discussion

In order to make more informed design decisions when carrying out a full-scale monitoring study in the field of alcohol craving, the current study pilot tested two EMA designs and the validity of both physiological and self-report measures. The current study showed that both the interval- and signal-contingent EMA study designs can capture relevant and meaningful variability of self-reported alcohol craving. This is in contrast with prior mentioned findings [6], however students might be more willingly to admit their craving then alcoholics. Therefore, it is not definite that these results will transfer to the target population, but a relevant finding is that no significant difference between the two designs was found. The overall compliance of the self-reported measures was lower (i.e. 64%) then found by multiple EMA studies [6], but highly different among participants. The type of EMA design did not seem to explain the differences in compliance rates. Further research should focus on explaining the variations in compliance rates.

The burden of filling in multiple questionnaires a day was perceived as quite high, the interval-contingent was preffered over the signal-contingent design, as it was perceived as less obtrusive. It should be noted, though, that this preference for interval-contingent EMA may also be explained by an order effect, as all participants started with this design and the perceived burden may have increased in course of time. Nevertheless, all participants experienced both designs, allowing a direct comparison of perceived burden within subjects.

All the participants found the E4 easy to use, rated the usability slightly below average and described the wristband as fine. Women experienced less wearing comfort, due to the size of the E4. Participants found it annoying that the wristband is only splash water proof. Despite the positive rating of the E4 wristband, the compliance rate of the physiological data was quite low for two of the participants. As mentioned, addicts are notoriously non-compliant to schedules and instructions [6]. An effort should be made to increase this compliance, we therefore propose for the monitoring study to give participants a micro incentive to stimulate compliance to both the self-reported and physiological data. Musthag, Raij, Ganesan, Kumar, and Shiffman [20] showed these incentive studies to be low cost, while ensuring high compliance, good data quality, and lower retention issues.

A potential threat to the user acceptance of the E4 may be the apparent visibility and the attracted attention of the wearable biosensor, which may particularly apply to the target population of the monitoring study, persons with addiction, might not prefer this scrutiny, and consider this to be a stigmatization. Unfortunately no other research device is available that unobtrusively measures heart rate and electrodermal activity. Therefore, we propose to prepare the participants by providing coping strategies. For example, the participant can learn to respond that the sensor is helping to improve their health. It should be explored whether this lowers feelings of stigmatization.

In conclusion, the present feasibility study showed that an interval-contingent EMA design with micro incentives and the E4 wristband are feasible and low burden design decisions for a real time, alcohol craving study. This finding has relevance for similar studies on other addictive behaviors, but may also contribute to less related field of research (e.g. occupational stress). For the latter types of studies the present work can also be seen as a blueprint on how to do a thorough, relatively quick formative evaluation for a subsequent daily life study. Daily life studies are often rather time consuming, costly and put a heavy burden on participants [3]. It is therefore essential to make well substantiated choices for the study design based on both literature and a feasibility pilot, such as the one presented here.

References

1. Oinas-kukkonen, H., Harjumaa, M.: Key issues, process model, and system features persuasive systems design: key issues, process model, and system features. Commun. Assoc. Inf. Syst. **24**(28), 485–500 (2009)
2. Rohsenow, D.J., Monti, P.M.: Does urge to drink predict relapse after treatment? Alcohol Res. Health **23**(3), 225–232 (1999). The Journal of the National Institute on Alcohol Abuse and Alcoholism
3. Csikszentmihalyi, M., Mehl, M.R., Conner, T.S.: Handbook of Research Methods for Studying Daily Life. Guilford Publications, New York (2013)
4. Carter, B.L., Tiffany, S.T.: Meta-analysis of cue-reactivity in addiction research. Addiction **94**(3), 327–340 (1999)
5. Takarangi, M.K.T., Garry, M., Loftus, E.F.: Dear diary, is plastic better than paper? I can't remember: comment on Green, Rafaeli, Bolger, Shrout, and Reis. Psychol. Methods **11**(1), 119–122 (2006). Discussion 123–125
6. Shiffman, S.: Ecological momentary assessment (EMA) in studies of substance use. Psychol. Assess. **21**(4), 486–497 (2009)
7. Hufford, M.R., Shiffman, S.: Methodological issues affecting the value of patient-reported outcomes data. Expert Rev. Pharmacoeconomics Outcomes Res. **2**(2), 119–128 (2002)
8. Scollon, C.N., Kim-prieto, C., Diener, E.: Experience sampling: promises and pitfalls, strengths and weaknesses. J. Happiness Stud. **2003**(4), 5–34 (1925)
9. Serre, F., Fatseas, M., Swendsen, J., Auriacombe, M.: Ecological momentary assessment in the investigation of craving and substance use in daily life: a systematic review. Elsevier Ireland Ltd. (2015)
10. Brooke, J.: System usability scale (SUS): a quick-and-dirty method of system evaluation user information. In: Usability Evaluation Industry, vol. 189, pp. 4–7 (1996)
11. Fensli, R., Boisen, E.: Human factors affecting the patient's acceptance of wireless biomedical sensors. In: Fred, A., Filipe, J., Gamboa, H. (eds.) Biomedical Engineering Systems and Technologies, BIOSTEC 2008. Communications in Computer and Information Science, vol. 25, pp. 402–412. Springer, Heidelberg (2008)
12. Moskowitz, D.S., Russell, J.J., Sadikaj, G., Sutton, R.: Measuring people intensively. Can. Psychol. Psychol. Can. **50**(3), 131–140 (2009)
13. Smyth, J., Wonderlich, S., Crosby, R., Miltenberger, R., Mitchell, J., Rorty, M.: The use of ecological momentary assessment approaches in eating disorder research. Int. J. Eat. Disord. **30**, 83–95 (2001)
14. Bergmann, J.H.M., Chandaria, V., McGregor, A.: Wearable and implantable sensors: the patient's perspective. Sensors (Switzerland) **12**(12), 16695–16709 (2012)

15. Howard, M., McMillen, C., Nower, L., Elze, D., Edmond, T., Bricout, J.: Denial in addiction: toward an integrated stage and process modelqualitative findings. J. Psychoactive Drugs **34**(4), 371–382 (2002)
16. Trompetter, H.R., Borgonjen, F., Zwart, M., van Tongeren, S.: UTSurvey Manual v1 (2015). http://www.utwente.nl/igs/datalab
17. Drobes, D.J., Thomas, S.E.: Assessing craving for alcohol. Alcohol Res. Health **23**(3), 179–186 (1999). The Journal of the National Institute on Alcohol Abuse and Alcoholism
18. Schmitz, C.: Limesurvey: an open source survey tool (2012). http://www.limesurvey.org
19. Bangor, A., Kortum, P.T., Miller, J.T.: An empirical evaluation of the system usability scale. Int. J. Hum. Comput. Interac. **24**(6), 574–594 (2015)
20. Musthag, M., Raij, A., Ganesan, D., Kumar, S., Shiffman, S.: Exploring micro-incentive strategies for participant compensation in high-burden studies. In: Proceedings of the 13th International Conference on Ubiquitous Computing - UbiComp 2011, p. 435 (2011)

Argumentation Schemes for Events Suggestion in an e-Health Platform

Angelo Costa[1], Stella Heras[2(✉)], Javier Palanca[2], Jaume Jordán[2], Paulo Novais[1], and Vicente Julián[2]

[1] Centro ALGORITMI, Escola de Engenharia, Universidade Do Minho, Guimarães, Portugal
{acosta,pjon}@di.uminho.pt

[2] D. Sistemas Informáticos y Computación, Universitat Politècnica de València, Valencia, Spain
{sheras,jpalanca,jjordan,vinglada}@dsic.upv.es

Abstract. In this work, we propose the introduction of persuasion techniques that guide the users into interacting with the Ambient Assisted Living framework iGenda. It is a cognitive assistant that manages active daily living activities, monitors user's health condition, and creates a social network between users via mobile devices. The objective is to be inserted in a healthcare environment and to provide features like adaptive interfaces, user profiling and machine learning processes that enhance the usage experience. The inclusion of a persuasive architecture (based on argumentation schemes) enables the system to provide recommendations to the users that fit their profile and interests, thus increases the chance of a positive interaction.

1 Introduction

e-Health has become an important area in the latest years. Devices and technologies that compose an e-Health environment are more accessible and big entities, such as the European Commission and the World Health Organization, are supporting the development of new technological solutions to old problems.

One of the main focus in terms of care is the elderly community. Studies [5,6] show that this community is the most affected by health problems and, in overall, represents a higher cost in terms of care services. To respond to this issue the scientific community presented a solution in form of two areas, the Ambient Intelligence (AmI) and the Ambient Assisted Living (AAL) [2].

The objective of these areas is to provide technological solutions through devices and software that help the elderly or disabled population to overcome their limitations and have an active life. Currently the focus is directed at helping to perform activities of daily living (ADL). To provide this type of assistance an ecosystem of participants has to be established, namely: the caregivers, the family carers and stakeholders (in the form of technicians or the company in charge of the computer systems). These participants play a major role, as they verify and assure that the system provides the expected service and that it is in

P.W. de Vries et al. (Eds.): PERSUASIVE 2017, LNCS 10171, pp. 17–30, 2017.
DOI: 10.1007/978-3-319-55134-0_2

accordance to the care-receiver demands. Furthermore, technological devices may increase the possibility of human interaction and the creation of social bonds.

The introduction of new technologies has also downsides like elderly people describing as being troubling to use technology that they have no experience or that it is difficult to learn or understand [6]; or that the caregivers receive too much or too little information from these type of systems. A reasonable way to deal with this issue is to endow the systems with decision support procedures and persuasion procedures that in combination provide more and better information to the users of these systems.

For instance, using intelligence decision support systems in medical diagnosis can result in a better supported and assured diagnostic. The justifications can be inferred from different sources of knowledge, e.g. clinical practice guidelines or previous experiences (clinical cases). Case-Based Reasoning (CBR) is one of the most suitable Artificial Intelligence (AI) techniques for building clinical decision-support systems [4]. The reasoning from examples simulates a physician's way of thinking, where new patients can be diagnosed in view of the experience gathered from previous similar clinical cases. In addition, a CBR system presents the advantage of being able to justify its conclusions, by referring to similar cases where a certain solution was found to be successful. In this sense, previous cases stored in a case-base can be used as a *knowledge resource* to generate arguments to explain the decisions (or recommendations) provided by the system.

How these arguments can be generated and be interchanged in an argumentative discourse and what are the relations that underlies from these argumentations are core research topics of the argumentation theory. Nowadays, several well-known concepts of the argumentation theory have been adopted for the AI community to manage argumentation dialogues in computational settings and digital systems. Among them, the theory of argumentation schemes is one of the most widely applied. Argumentation schemes represent stereotyped patterns of common reasoning whose instantiation provides an alleged justification for the conclusion drawn from the scheme. The arguments inferred from argumentation schemes adopt the form of a set of general rules by which, given a set of premises, a conclusion can be derived. Many authors have proposed different sets of these argumentation schemes, but the work of Walton [11], who presented a set of 25 different argumentation schemes, has been the most widely used by the AI community. Walton's schemes have associated a set of *critical questions, (CQ)* that, if instantiated, questions the elements of the scheme and can represent potential attacks to the conclusion drawn from it. This characteristic of Walton's argumentation schemes makes them very suitable to reflect reasoning patterns that the system can follow to bring about conclusions and, what is more important, to devise ways of attack any other alternative conclusions.

What is missing is a true effort to provide the users of AAL systems with a truly adaptive and responsive system that fits the user's needs and disabilities. Our proposal is to use an e-Health platform (iGenda) coupled with a persuasive module that has a set of argumentation schemes that map the reasoning procedures that physicians and caregivers follow to recommend activities to patients.

These schemes are used to generate arguments to support the recommendation of activities or to attack other potential alternatives.

The rest of the paper is structured as follows: Sect. 2 shows an overview of the iGenda system; Sect. 3 presents the argumentation schemes that we have used in this version of the iGenda system; Sect. 4 explains the structure and operation of the persuasion module proposed; Sect. 5 provides a running example; and finally, Sect. 6 summarises the contributions of this paper and proposes future work.

2 The iGenda Framework

iGenda is a cognitive assistant platform [3] with the aim of assisting all the players in the elderly's sphere of people, e.g., family, relatives, health assistants, caregivers. The main feature is a time management service that schedules events and manages time conflicts between events, automatically promoting ADL's according to the user's profile. The main social goal is to increase the happiness levels of the users, by maintaining them active and facilitating social connections and human contact.

Following the current trends, iGenda's main way of access is through mobile devices, having several features like: create, delete, update, and accept events. Moreover, most mobile devices have a set of sensors, like GPS or accelerometer, that may help the iGenda by giving it more information about the current location and environment status, which could be useful in suggestions.

There are two visual interfaces: (i) care-receivers - directed for the elderly, friends and relatives; (ii) caregivers - directed to health assistants, like registered nurses and physicians. The reason behind this divide is the different needs of each group. The care-receivers will receive activities and perform them, creating the

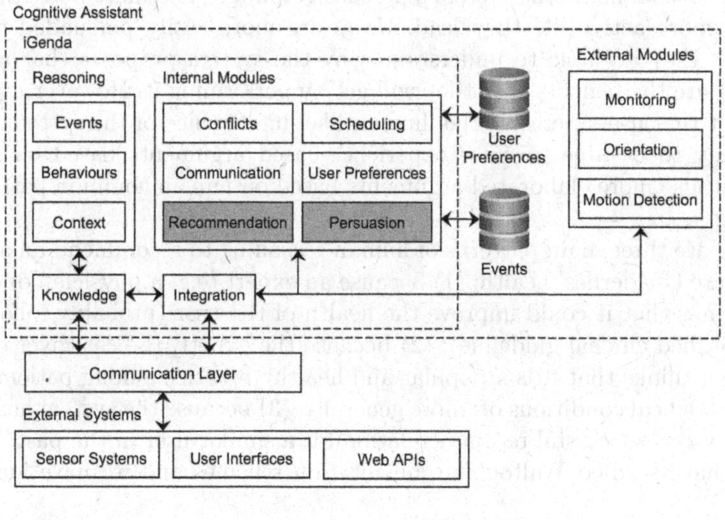

Fig. 1. iGenda architecture

expected social network with other users, while the caregivers will attend to their assigned care-receiver's health status and assure that they are well and secure.

iGenda periodically schedules activities that promote active living, selected from the *free time events database* through the use of the recommendation module. The events are filtered according to the users' medical condition, the weather condition, and the available free time. However, the potential willingness of the user to accept a specific action (based on his/her current social context or similar past experiences) varies greatly. Our proposal enables more information and more specific to each user about why each event is recommended. Figure 1 shows the architecture of the iGenda system established over a multi-agent system, which facilitates the deployment of new features and the addition of new modules.

In the next section, how the argumentation schemes improve the operation of the iGenda is explained.

3 The iGenda Argumentation Schemes

AAL platforms tend to automatise every aspect of the operation, thus taking little consideration to the opinion of the users that they are caring. Holzinger et al. [8] and Lindley et al. [9] have shown that it is important that the users feel included and part of the decision process; it builds trust and promotes the usage of the platform, which in turn promotes an active and healthy lifestyle. To provide a reason or justification of why the events should be performed may lead the users to concede the usage of the system. Therefore, the inclusion of persuasive methods that may provide the motives to the suggestion of events may compel the users to attend to it.

In this work, we aim to improve the acceptance levels of the activities by the users by enhancing the system's persuasive power. The more activities they accept a more active life they lead. Users are more easily persuaded to take activity if they are able to understand why the system proposes that activity and what are the benefits that they will get by performing it. However, it is also important that users perceive the 'human-like' intelligence of the system, which is not only to be able to show experience-based arguments based on similar cases, but also more elaborated arguments based on human common patterns of reasoning.

There are three main patterns of human reasoning to recommend an activity to take care of elderlies' health: (1) because an expert (e.g. a physician or a caregiver) thinks that it could improve the health of the user (probably following a well-stablished clinical guideline); (2) because the expert, the caregiver or even the system thinks that it is a popular and healthy practice among patients with the same medical conditions or; more generally, (3) because the system has found that this was a successful recommendation for a similar user in the past. Therefore, we have studied Walton's argumentation schemes and we have identified those of them that directly apply to our domain of justifying such activities[1]: the *Argument From Analogy*, the *Argument From Popular Practice*, and the *Argument From Expert Opinion.* In this section, we provide an adaptation of these argumentation schemes for the iGenda application domain. We refer the reader to [11, Chap. 9] for the original version of these schemes.

The **Argument From Analogy** is the foundation of all case-based reasoning [11, Chap. 2]. In this sense, the basic experience-based arguments provided by the persuasive module of iGenda follow this pattern of reasoning.

Similarity Premise: Domain-case(s) X is(are) similar to the current situation
Base Premise: Activity A was listed in domain-case(s) X
Conclusion: Activity A should be recommended in the current situation
CQ1: Are there any distinguishing attributes between domain-cases X and the current situation?
CQ2: Was activity A recorded in the domain-case(s) with a suitability degree higher than a minimum acceptable threshold?
CQ3: Are there any domain-cases Y that represent the same situation but that propose a different activity B?

The **Argument From Popular Practice** is the practical form of the most general Argument From Popularity [11]. This scheme represents the pattern of reasoning that humans follow when the group's opinions related to what decision to take when deliberating a course of action are considered as acceptable recommendations. Obviously, this scheme is highly used as a fallacy of reasoning, since popular opinions cannot be always taken as valid. However, it still captures a common line of reasoning that people follow when they are looking for recommendations in the medical domain, and hence we have decided to implement it in the iGenda persuasion module.

Major Premise: Activity A is a common recommendation among patients with medical conditions X
Minor Premise: Activity A should be considered as an acceptable recommendation for those patients with medical conditions X
Conclusion: Activity A should be recommended in the current situation
CQ1: What data shows that a large majority of patients with medical conditions X accept activity A?
CQ2: Even if the majority accepts activity A, why activity A should be considered as suitable?

The **Argument From Expert Opinion** is probably the most commonly used argumentation scheme in the recommendation domain. It captures the pattern of reasoning that humans follow when an expert on a specific field provides

[1] We acknowledge that there are more Walton's argumentation schemes that could apply, but we found that these three capture the most common ways of reasoning in our application domain. New schemes could easily be added to the persuasion module if required.

an opinion regarding, in a domain, the best recommendation to provide in a specific situation given its expertise in such domain. In this sense, this scheme can be considered as a specialisation of the line of reasoning that the Argument From Position to Know [11]. Note that critical questions 3 and 6 are assumed to be true by the same nature of this recommendation domain, since all activities recorded in the iGenda database have a proposer by default (the doctor, caregiver or at least the system that created the activity). Thus, they cannot be instantiated as potential attacks for this argumentation scheme.

Major Premise: Expert E (doctor, caregiver or expert system) is an expert on the area of expertise X where activity A belongs to
Minor Premise: Activity A is proposed by expert E
Conclusion: Activity A should be recommended in the current situation
CQ1: How credible is E as an expert source?
CQ2: Is E an expert on the area of expertise X where activity A belongs to?
CQ3: Did expert E recommend activity A?
CQ4: How personally trusted is E as an expert source?
CQ5: Is A consistent with what other experts have recommended?
CQ6: Is E's recommendation based on evidence?

4 The iGenda Persuasion Module

In this section, we provide an overview of the persuasion module of the iGenda tool, focusing on the operation of the module and the new argumentation schemes knowledge resource. When iGenda calls the recommendation module to recommend activities, the system tries to create one argument (or more) to support each activity and decide which one would be preferred by the user. Then, an internal argumentation process takes part to decide the activity that is better supported by its arguments.

4.1 Argumentation Framework

The persuasive module of iGenda implements the agent-based argumentation framework for agent societies proposed in [7]. This framework takes into account the values that arguments promote (the preferences of the users over the activities' *motion* characteristics, *location*, *social* requirements, *environmental* conditions, or *health conditions*), the users' preference relations (preference orderings over values (*Valpref*)), and the dependency relations between agents (the relations that emerge from agent interactions or are predefined by the system) to evaluate arguments and to decide which ones defeat others.

In our system, agents can play the role of *patients*, *caregivers* (e.g. relatives, personal health assistants, friends), and *doctors*. We also consider the following dependency relations: (i) *Power*: when an agent has to accept a request from another agent because of some pre-defined domination relationship between them. For instance, in our agent society S, $Patient <^S_{Pow} Doctor$, and *Caregiver* $<^S_{Pow}$ *Doctor* since patients and caregivers must follow the guidelines recommended by their doctors; (ii) *Authorisation*: when an agent has committed itself to another agent for a certain service and a request from the latter leads to an obligation when the conditions are met. For instance, in S, *Patient* $<^S_{Auth}$ *Caregiver*, if the patient has contracted the health assistant service that a caregiver offers; and (iii) *Charity*: when an agent is willing to accepts a request from another agent without being obliged to do so. For instance, in S, by default *Patient* $<^S_{Ch}$ *Patient*, *Caregiver* $<^S_{Ch}$ *Caregiver* and *Doctor* $<^S_{Ch}$ *Doctor*.

In this work, we have adapted the knowledge resources of this framework to cope with the requirements of the iGenda domain: a case-base with *domain-cases* that represent previous problems and their solutions and a database of *argumentation-schemes* with a set of schemes that represent stereotyped patterns of common reasoning in our application domain.

Figure 2 shows an example of the structure of a specific domain-case in our system. This domain-case is the representation of a set of previous activities that have been successfully recommended to the same kind of user. Each case has a set of attribute-value pairs (variables of any value type) that describe the characteristics of the user, the environmental context where the recommendation was provided, and the list of activities recommended. The characteristics of a user are a representation of users with the same attributes. These are their medical status (moderate, severe, mild, ...), their role (elderly, family, medical, ...), the medical term that defines them (psychological, physical, both, ...) and whether or not the user is allowed to go outside his/her house or just the perimeter. Besides the above, these characteristics also define if the user is physically constrained, semi-constrained or unconstrained and if the user is allowed to practice high-intensity activities or not. The environmental context where the recommendation was provided is useful to be aware of the suitability of an activity regarding the environment. It's easy to conclude that an outdoor activity is directly dependent on the weather. The characteristics that are stored in the environmental context are: the weather, which is usually only important for outdoor activities, the time range when the activity was done, the season (there are activities that are more desirable than others regarding the season), whether the day was a holiday or not and, finally, if the user is at home or at another residence (hospital, holidays residence, ...). Finally, the list of activities includes the activity that was recommended (Id), the proposer of that activity (ProposerId) and a degree of suitability that represents if the activity was good or not for that case (Suitability).

Arguments that iGenda generates are tuples of the form:

Definition 1 (Argument). *$Arg = \{\phi, p, \langle SS \rangle\}$, where ϕ is the conclusion of the argument (e.g. the activity to recommend), p is the value that the argument promotes and $\langle SS \rangle$ is a set of elements that justify the argument (the support set).*

The support set $\langle SS \rangle$ is the set of features (*premises*) that represent the context of the domain where the argument has been put forward (those premises

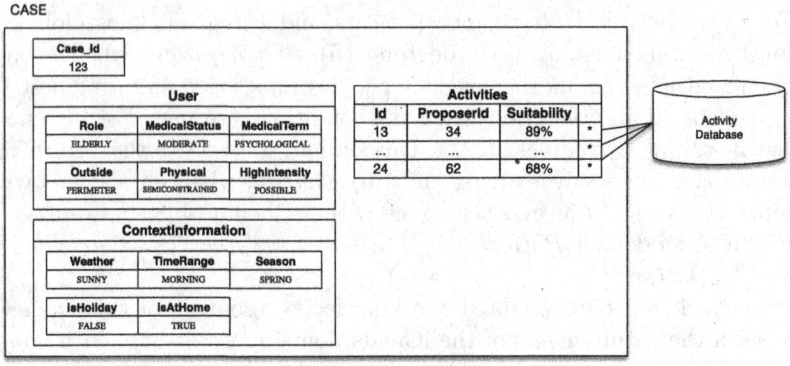

Fig. 2. Structure of a domain-case

that match the problem to solve and other extra premises that do not appear in the description of this problem but that have been also considered to draw the conclusion of the argument) and any knowledge resource used by the proponent to generate the argument (domain-cases and argumentation schemes).

Now, the concept of conflict between arguments defines in which way arguments can attack each other. There are two typical attacks studied in argumentation: *rebut* and *undercut*. In an abstract definition, rebuttals occur when two arguments have contradictory conclusions (i.e. if an argument Arg_1 supports a different conclusion for a problem description that includes the problem description of an argument Arg_2). Similarly, an argument undercuts another argument if its conclusion is inconsistent with one of the elements of the support set of the latter argument or its associated conclusion (i.e. if the conclusion drawn from the argument Arg_1 makes one of the elements of the support set of the argument Arg_2 or its conclusion non-applicable in the current recommendation situation). Thus, in our framework we can define the defeat relation as:

Definition 2 (Defeat). *An agent's ag_1 argument Arg_1 put forward in the context of a society S $defeats_{ag_1}$ another agent's ag_2 argument Arg_2 iff $attack(Arg_1, Arg_2) \wedge (val(ag_1, Arg_1) <^S_{ag_1} val(ag_1, Arg_2) \notin Valpref_{ag_1}) \wedge (Role(ag_1) <^S_{Pow} Role(ag_2) \vee Role(ag_1) <^S_{Auth} Role(ag_2) \notin Dependency_S)$*

That is, an argument Arg_1 of an agent ag_1 defeats from ag_1's point of view another argument Arg_2 of an agent ag_2, if Arg_1 attacks Arg_2, the value that promotes Arg_1 is preferred to the value that promotes Arg_2, and the role of ag_2 is not dominant (the role of ag_1 has higher preference).

4.2 Recommendation Process

When iGenda has to schedule a new activity for a user it starts its recommendation process. Then, a list of possible candidate activities that match the requirements of the current situation is retrieved from the database. Next, our

Table 1. Argumentation schemes instantiation

Argument From Analogy	
Elements of the scheme	Related data
Similarity Premise	Similar domain-cases
Base Premise	Activities recorded in the domain-cases
CQ1	Any attributes distinguishing between the case recovered and the current situation (distinguishing attributes)
CQ2	Activity proposed with a suitability degree higher than the threshold specified
CQ3	Any domain-cases that represent the same situation but that propose a different activity
Argument From Popular Practice	
Elements of the scheme	Related data
Major Premise	Activity proposed, Medical conditions, Number of times (higher than a threshold) that the activity has been recommended to similar users (computed either from the iGenda database and/or from the retrieved domain-cases)
Minor Premise	Activity proposed
CQ1	Number of times (higher than a threshold) that the activity has been accepted and actually was not finally executed by similar users (computed either from the iGenda database and/or from the retrieved domain-cases)
CQ2	Low degree of suitability/satisfaction (lower than a threshold) experienced by similar users when performing the activity (computed from the retrieved domain-cases)
Argument From Expert Opinion	
Elements of the scheme	Related data
Major Premise	Proposer, area of expertise, activity proposed
Minor Premise	Activity proposed
CQ1	Proposer reputation lower than a threshold or or less preferably (computed from all recommendations provided by this proposer)
CQ2	Proposer area of expertise does not exactly match the required in this situation
CQ4	Trust degree between the user and the proposer lower than a threshold or less preferably (computed from previous interactions between them)
CQ5	Other different proposers that recommend different activities for this same situation (computed either from the iGenda database and/or from the retrieved domain-cases)

persuasion module is in charge of selecting from this list the best activities to recommend in view of past similar experiences. This is done by means of a case-based reasoning cycle [1] (the Retrieve, Reuse, Revise, and Retain phases).

With this information, the persuasion module tries to generate scheme-based arguments for each of the activities selected by the recommendation module. The iGenda database[2] provides the pieces of information that support the instantiation of each reasoning pattern that each argumentation-scheme represents. Furthermore, the information stored in the domain-cases can also be useful to instantiate argumentation schemes. These related data is shown in Table 1. Thus, if any scheme can be instantiated, the module generates new scheme-based arguments to support the activity under consideration. Also, if a scheme is instantiated, the system also tries to retrieve data to instantiate their associated critical questions. In this way, attack arguments to the argument generated from the scheme and hence to the activity that it supports can be also created.

Once all possible arguments have been generated to support or attack each potential activity to recommend, we start an evaluation of the arguments to decide which are rebutted and which hold. The formal specification of this process is out of the scope of this paper. We refer the reader to [3,7] for details. At this point, the recommender proposes the activity that it is deemed to be more suitable and persuasive for the user. This activity is the one supported by more arguments and/or with higher weights (in the case of experience-based arguments). Finally, when an activity is scheduled, the system receives a feedback from the user to indicate whether the activity was actually performed and his/her degree of satisfaction with it. Then, the recommender executes the *retention* phase in order to learn from the recommendation experience and store the degree of suitability of its recommendations. This degree of suitability is taken from the user feedback. Then, if the system was able to retrieve a domain-case that matches the current situation and the activity was in the list of activities associated with this case, the suitability degree of this activity is increased; otherwise, the activity is added to the list or, if no matching cases were found, a new domain-case is created to store the new knowledge acquired.

5 Example

In this section, we present an example of how the iGenda framework, with the new persuasion module using argumentation schemes, suggests several healthcare activities to an elderly patient. Let us assume that iGenda is requested to schedule activities for a patient *Patient*1 with a psychological disease. The patient has a moderate medical status (i.e. not too severe), he is allowed to practice high-intensity activities but in semi-constrained way, and he is only allowed to leave his house in a small perimeter. Also, the activities have to be scheduled in the morning of a spring weekday, with good weather. Furthermore, the system

[2] The database includes different tables to store information about patients, activities, doctors, caregivers, etc. The full specification of the database is not provided due to space restrictions.

has established a minimum suitability threshold of 75% for the activities, and a preference relation that grants the higher reputation to doctors, followed by caregivers.

Firstly, iGenda framework uses its recommendation module and selects some activities (*Activity*13: 'Music listening (alone)', *Activity*24: 'Reading a book or magazine', and *Activity*33: 'Home gardening') from its database taking into account the current weather forecast, the medical and contextual condition of the patient, and his preferences. According to the iGenda database, *Activity*13 and *Activity*24 were prescribed by $CaregiverC$, whereas *DoctorD* prescribed *Activity*33. After that, the persuasion module tries to generate all possible support and attack arguments for each activity. Then, it first searches its case-base of domain-cases looking for any cases that match the current situation and represent past recommendations provided for similar users. In doing so, the module follows the *Argument From Analogy* pattern of reasoning and tries to generate experience-based arguments. Let us assume that iGenda is only able to retrieve one domain-case $DC1$ that matches the current situation (user characteristics and context), as shown in the example of Fig. 2, and includes *Activity*13 and *Activity*24 in its recommended activities list, but not *Activity*33. Thus, the persuasion module can generate the arguments[3] $SA1 = \langle Activity13, v1, \langle \text{DC1,AFA} \rangle \rangle$ and $SA2 = \langle Activity24, v2, \langle \text{DC1,AFA} \rangle \rangle$ to support *Activity*13 and *Activity*24 respectively. Note that the support set of both also includes the *Argument From Analogy* (AFA), since they instantiate this scheme.

After that, the module tries to generate more support arguments by following the patterns of reasoning represented by its *Argument From Popular Practice* (APP), and the *Argument From Expert Opinion* (AEO) argumentation schemes. Now, for instance, let us assume that by searching again its case-base, iGenda finds that there are a number of domain-cases that match the characteristics of $Patient1$ (higher than the threshold specified in the system to be considered a 'common practice') and that include *Activity*33 in their activities list. This means that *Activity*33 can be a common practice among patients that are similar to $Patient1$, but maybe it has been recommended in other contexts (e.g. different season or weather conditions). Note that despite these differences between the current context and the context represented by these cases, it does not necessarily mean that *Activity*33 is not suitable, but maybe iGenda still has not faced a situation like the current one, so it has not yet been able to record a similar past experience. Thus, argument $SA3 = \langle Activity33, v3, \langle \{\text{DC}\}, \text{APP} \rangle \rangle$ can be generated to support *Activity*33. Finally, following its *Argument From Expert Opinion* scheme and taking into account who prescribed each activity, arguments $SA1b = \langle Activity13, v3, \langle \text{AEO} \rangle \rangle$, $SA2b = \langle Activity24, v2, \langle \text{AEO} \rangle \rangle$, and $SA3b =$

[3] In this example, we do not specify the values v that arguments promote for simplicity purposes. Also, all arguments are mathematical abstracts. A textual interpretation for $SA1$, for instance, may be "You should'Listening music alone' since it suits your profile and needs and this activity has been successful for people similar to you."

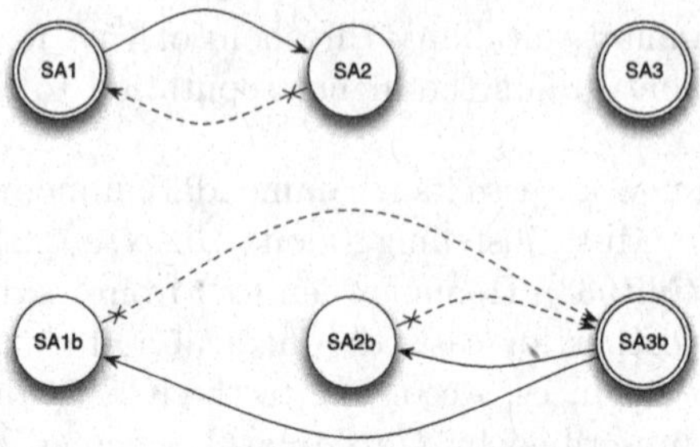

Fig. 3. Example argumentation graph. Arrows: successful attacks; Crossed dotted arrows: unsuccessful attacks; Nodes: defeated arguments; Double circled nodes: prevailing arguments.

$\langle Activity33, v3, \langle \text{AEO} \rangle \rangle$ can be generated to support *Activity*13, *Activity*24, and *Activity*33, respectively.

Once all possible support arguments have been generated, the iGenda persuasion module tries to generate other attack arguments and determine the attack relations between arguments. For simplicity purposes, in this example we assume that no extra arguments can be generated, but attack relations are established between the support arguments, and that for the simple fact that an argument supports a different conclusion to another argument, this does not mean that both arguments are attacking each other. Therefore, the system can compute the following attacks, as shown in the argumentation graph of Fig. 3: (1) by definition, $SA1$ and $SA2$ are counter-examples and attack each other. They are arguments that share their support set, $DC1$ and AFA, but that support different conclusions. However, (2) argument $SA1$ can defeat $SA2$ since the suitability degree of *Activity*13 in $DC1$ is greater than the one of *Activity*24 and also greater than the minimum suitability threshold established by the system (instantiating $CQ2$ of the AFA scheme); (3) $SA1b$ and $SA2b$ attack $SA3b$ and vice versa instantiating $CQ5$, since these arguments support activities prescribed by *CaregiverC* and $SA3b$ supports an activity prescribed by *DoctorD*. However, (4) $SA3b$ defeats $SA1b$ and $SA2b$ instantiating $CQ1$ of the AEO scheme, since the reputation of caregivers is lower than the reputation of doctors in the system. Argument $SA3$ does not receive any attack. At the end of the argumentation process, arguments $SA2$, $SA1b$ and $SA2b$ are defeated, and arguments $SA1$ (supporting *Activity*13), and $SA3$ and $SA3b$ (both supporting *Activity*33) prevail. Therefore, iGenda has more reasons to believe that *Activity*33 is potentially more persuasive, and will recommend and schedule it for *Patient*1.

6 Conclusions

This work has presented an extension of the persuasive module included into the iGenda Ambient Assisted Living framework. This extension improves user engagement through the selection of activities that are supported by the creation of arguments. These arguments are generated using argumentation schemes that allow to capture the way of reasoning that physicians and caregivers follow to recommend activities to patients. With this new process, the persuasive power of iGenda is enhanced because the selected action is presented with reasons that support it and people tend to trust recommendations more when the system can justify them.

In its current version, the argumentation process is an internal process that helps iGenda to decide the activity that is better supported by its arguments. This method has the advantage of allowing the system to provide justifications and explanations for its decisions, which adds value in comparison with other recommendation techniques that act as a black-box for the user (e.g. collaborative-filtering or knowledge-based [10]). Furthermore, we are currently developing a new user interface that allows discourse between the system, patients, and caregivers. For future work, we are testing the iGenda framework with these new features to support the recommendation of activities, both from the perspectives of providing appropriate recommendations (efficiency) and of providing convincing recommendations (persuasiveness). This will be done in a mobile application specially designed for elderly people. Also, the collected data about the users' experience will be very valuable to include new argumentation schemes into iGenda to create more powerful justifications to the activities that this framework recommends.

Acknowledgements. A. Costa thanks the Fundação para a Ciência e a Tecnologia (FCT) the Post-Doc scholarship with the Ref. SFRH/BPD/102696/2014. This work has been supported by COMPETE: POCI-01-0145-FEDER-007043 and FCT – Fundação para a Ciência e a Tecnologia within the Project Scope: UID/CEC/00319/2013. It was also supported by the by the projects TIN2015-65515-C4-1-R and TIN2014-55206-R of the Spanish government and by the grant program for the recruitment of doctors for the Spanish system of science and technology (PAID-10-14) of the Universitat Politècnica de València.

References

1. Aamodt, A., Plaza, E.: Case-based reasoning: foundational issues, methodological variations, and system approaches. AI Commun. **7**(1), 39–59 (1994)
2. Bravo, J., Cook, D., Riva, G.: Ambient intelligence for health environments. J. Biomed. Inform. **64**, 207–210 (2016)
3. Costa, A., Heras, S., Palanca, J., Novais, P., Julián, V.: A persuasive cognitive assistant system. In: Lindgren, H., De Paz, J.F., Novais, P., Fernández-Caballero, A., Yoe, H., Ramírez, A.J., Villarrubia, G. (eds.) ISAmI 2016. AISC, vol. 476, pp. 151–160. Springer, Cham (2016). doi:10.1007/978-3-319-40114-0_17
4. El-Sappagh, S., Elmogy, M.M.: Medical case based reasoning frameworks: current developments and future directions. Int. J. Decis. Support Syst. Technol. (IJDSST) **8**(3), 31–62 (2016)
5. Green, N.: Implementing argumentation schemes as logic programs. In: The 16th Workshop on Computational Models of Natural Argument. CEUR (2016)

6. Hakkarainen, P.: 'No good for shovelling snow and carrying firewood': social representations of computers and the internet by elderly finnish non-users. New Media Soc. **14**(7), 1198–1215 (2012)
7. Heras, S., Botti, V., Julián, V.: Argument-based agreements in agent societies. Neurocomputing **75**(1), 156–162 (2012)
8. Holzinger, A., Ziefle, M., Röcker, C.: Human-computer interaction and usability engineering for elderly (HCI4AGING): introduction to the special thematic session. In: Miesenberger, K., Klaus, J., Zagler, W., Karshmer, A. (eds.) ICCHP 2010. LNCS, vol. 6180, pp. 556–559. Springer, Heidelberg (2010). doi:10.1007/978-3-642-14100-3_83
9. Lindley, S., Wallace, J.: Placing in age. ACM Trans. Comput. Hum. Interact. **22**(4), 1–39 (2015)
10. Ricci, F., Rokach, L., Shapira, B.: Recommender systems: introduction and challenges. In: Ricci, F., Rokach, L., Shapira, B. (eds.) Recommender Systems Handbook, pp. 1–34. Springer, New York (2015)
11. Walton, D., Reed, C., Macagno, F.: Argumentation Schemes. Cambridge University Press, Cambridge (2008)

e-Coaching for Intensive Cardiac Rehabilitation

A Requirement Analysis

Aldert Nooitgedagt[1(✉)], Robbert Jan Beun[2], and Frank Dignum[2]

[1] Sportpoli, Utrecht, The Netherlands
a.nooitgedagt@sportpoli.nl

[2] Utrecht University, Utrecht, The Netherlands
{r.j.beun,f.p.m.dignum}@uu.nl

Abstract. In this paper, the rationale and requirements are presented for an e-coaching system in the domain of intensive cardiac rehabilitation. It is argued that there is a need for a personalized program with close monitoring of the patient based on medical needs and needed lifestyle changes in a setting with other participants such as family and a human coach. Two roles are distinguished for the e-coach: support of the organizational process of the program (e.g. collecting and analyzing data) and support of the patient's process of making lifestyle changes (e.g. triggering and motivational support). Motivational interviewing is introduced as a means to minimize the discrepancy between undesired behavior and future goals of the patient. It is concluded that digital products that offer e-coaching support for these types of programs should coexist with human coaching.

Keywords: Intensive cardiac rehabilitation · e-Coaching · Requirements · Motivational interviewing · Lifestyle change

1 Introduction

Every day a thousand people are hospitalized in The Netherlands due to heart disease. Fortunately, the large majority (90%) of these patients survive. Currently, there are one million people with a heart disease in The Netherlands [1, 2]. This indicates that a large number of people has to rehabilitate from some kind of heart problem and adjust their lifestyle.

Cardiac rehabilitation has proven to be an effective secondary prevention measure for cardiovascular disease (CVD) [3, 4]. However, lifestyle changes necessary to modify risk factor profiles seem to be difficult to maintain long-term. Up to 60% of patients relapse within 6 months. 1.5 years after discharge from the hospital most beneficial effects of cardiac rehabilitation on risk factor profiles have been lost. Fortunately, lifestyle modification programs based on self-regulation theories seem to have more lasting effects [5]. The EUROASPIRE IV study [6] concluded that a new approach to cardiovascular prevention is required which integrates cardiac rehabilitation and secondary prevention into modern preventive cardiology programs with appropriate adaptation to medical and cultural settings.

P.W. de Vries et al. (Eds.): PERSUASIVE 2017, LNCS 10171, pp. 31–42, 2017.
DOI: 10.1007/978-3-319-55134-0_3

In this paper, we argue that there is a need for a personalized rehabilitation process that is closely monitored and adjusted to each patient's needs. However, due to the massive amount of heart patients it is impossible to have such a personalized rehabilitation process supervised by doctors or other caregivers alone. Thus, we advocate an e-coach to support and monitor the patients during and after the rehabilitation period.

Here, we discuss the rationale and basic requirements for such an e-coach. In Sect. 2, we will analyze the relation between lifestyle change and cardiac rehabilitation and especially the elements that are of importance for coaching the patients. We will describe an intensive cardiac rehabilitation program (ICR) that tries to combine the rehabilitation with lifestyle changes. In Sect. 3, we will elaborate on the process of coaching in relation to behavioral models. In Sect. 4, we discuss how these elements are important requirements for the e-coach design and we present a first design for an e-coach for ICR. Finally, in Sect. 5, we draw some conclusions.

2 Cardiac Rehabilitation and Lifestyle Change

As imposed by the Dutch Health Care Inspectorate [7], we assume that in most cases lifestyle change is promoted in cardiac rehabilitation programs in The Netherlands. However, many factors influence participation in rehabilitation programs and also the uptake and maintenance of healthy behaviors [8, 9].

First of all, there are many *causes of heart disease* that may require a different approach for the rehabilitation process. A patient that has had a heart attack due to arteriosclerosis can often be treated by inserting a stent, recover quickly and likely perform all tasks normally. A patient that has cardiomyopathy might not recover completely, certainly cannot get an instant treatment through an operation, and may have to learn to cope with a low heart efficiency.

Secondly, the *overall health condition* of patients can differ widely. Some patients are fanatic sportsmen and have a good condition, while others can be elderly and not very active outside their homes. Of course, cardiac rehabilitation programs should be adjusted to these individual conditions.

Thirdly but not lastly, many heart problems, like heart attacks, seem to appear suddenly, but are the result of a *long process* caused by an unhealthy lifestyle, genetic factors or other 'hidden' factors. This may influence commitment to change lifestyle. Choosing to change lifestyle after a heart attack is probably more difficult, because the relation between cause and effect is long-term and indirect. Unfortunately most studies have not attempted to link behavioral change after cardiac rehabilitation to any particular period of time or event [10]. Also, older adults tend to confuse the cause of their symptoms of their unhealthy lifestyle to normal aging or other chronic disease processes and may therefore be less likely to participate in cardiac rehabilitation [3]. On the other hand, when a heart problem is treated with a surgical (eg. acute) intervention, a high degree of cardiac rehabilitation participation is seen [1]. The acuteness of the intervention may trigger a need for rehabilitation, but not necessarily a need to change lifestyle. These assumptions are coherent with the acute and chronic models of the Common-Sense Model of Self-Regulation (CSM) [11].

2.1 Personalization in Cardiac Rehabilitation Programs

The above issues lead to the conclusion that there is a strong need for a *personalized* rehabilitation process that is *closely monitored* and *adjusted* to the needs of the individual patient. In order to personalize a cardiac rehabilitation program, the following steps have to be taken.

First, one has to establish the current situation of the patient and the desired end situation in order to establish possible plans to get from one situation to the other. Although this sounds simple, there are some major issues. First of all, there may be an important discrepancy in the perception of the current situation between the cardiologist and the patient. To some extent, the cardiologist will be aware of the medical situation of the patient, but the patient may be unaware of his own medical condition. Even though the specialist can indicate the patient exactly what happened, what caused it, etc., the interpretation of this medical condition will still be radically different for the cardiologist and the patient. A cardiologist can compare the situation with many other cases and judge how severe the situation is (or was). The patient usually does not have this comparison and is inclined to judge his medical condition by how he 'feels' and how it affects his life, an area where the patient may be informed better than the cardiologist. Does the patient exercise or used to exercise? How socially active is the patient? Does the patient have a partner? And so on. Even if the patient tries to inform the cardiologist as well as possible, their subjective perceptions of the situation will influence the starting point and the rehabilitation process differently. Thus, a first step in the rehabilitation process will be to create common understanding and align the initial situation.

Similar steps should be taken to establish the goal of the rehabilitation. Again, this end situation has two sides: the first part is the medical condition of the patient's heart and the second is the lifestyle of the patient that should be geared towards preventing further heart problems. Especially lifestyle is difficult to manage. Maybe, the patient should become physically active for at least 30 min each day, start to eat healthy and avoid stress at work. But that might not be feasible in the current situation of the patient. A realistic goal should therefore be set, such that both the patient and the cardiologist feel that the goal is feasible and satisfies some minimal requirements on healthy living.

Finally, the patient and cardiologist have to determine a path to get from the initial situation to the desired situation. The cardiologist has knowledge about steps that will rehabilitate the heart condition, while the patient has to consider whether these steps are feasible and sustainable. It is important that the plan is frequently monitored and adjusted whenever necessary.

Setting goals is linked with better disease management in cardiac patients. Self-regulation theories propose that these goals should be hierarchically organized, such that if (healthy lifestyle) behaviors are not innately interesting, they can still be engaged in if they satisfy or reward another goal [5]. Leventhal's CSM model explicates this process, but it needs methods that allow this process to be translated for patient care [11]. The model starts with mental constructs that steer behavior and focuses on '*what*' the specific rules for solving problems are in the current context and timeframe. It will help to interpret fear, attitude, focus, trust and attention of the patient. In the following section, we will elaborate on some of these methods.

2.2 An Intensive Cardio Rehabilitation Program

We will now briefly discuss an existing successful cardio rehabilitation program propagated by Ornish [12–14] and analyze which elements can be used as the basis for our e-coach design. This program has its focus on lifestyle change and incorporates the issues with lifestyle change discussed above. This so-called intensive cardio rehabilitation (ICR) program is based on a combination of experiencing (acting) and awareness (lessons). It consists of 18 sessions of 4 h, which are currently delivered within 18 weeks as required by the Social Security Act in the United States. Each session, there are four components: exercise, stress management, nutrition and group support. They correspond to the different areas of a lifestyle that should be considered in order to live a healthy life. By having lectures as well as practice, the program creates awareness and moments of feedback. By performing all exercises regularly, they can become part of habits of everyday life. Group support is an incentive to stick to the program in between meetings, to create a health-oriented peer group and to give support while coping with new situations. Participants of the program adhere to the program very well after 1 and 5 years. They show good reduction of risk factors and some show significant reversal of angiographic changes.

The most important issue from a coaching perspective is that patients are participating in extensive meetings outside their normal environment and experience the difference of living a healthy lifestyle in those meetings. The meetings stretch over a longer period and are repeated 18 times, making a habit breaking behavior possible. Because of the setting in which the program runs (USA hospital based), group support is felt as positive and safe; the new healthy behavior is associated with a positive attitude and thus more likely to persist. Interesting to see is that for male patients more partners participate in the program in order to sustain the changes in the home environment than for female patients. Patients that had a participating partner had more positive results than the ones that did not have their partners participate. Thus, including family and social environment in the program seems an important component [14]. One can suspect that rather homogeneous groups would be the most effective. If differences are large, the groups might lead to discouragement (e.g. 'I cannot keep up with them'), detachment and isolation (e.g. 'I don't fit with those young people having children at home'), as patients may have completely different concerns in their home environment and have completely different health prospects.

The cultural context of the program may be problematic for the Dutch culture. The ICR program assumes that people can and like to meet 18 times at moments of the day that include dinner time. In a Dutch situation, people may try to avoid any sessions that include dinner time. Dinner is an important social and family experience. One could also question whether people would join a support group for cardio rehabilitation at all. Making autonomous decisions about behavior is very important in the Dutch culture. Becoming part of the ICR program can be felt as being forced, because the primary reason will be their heart problem and not their personal choice. The medical necessity of a rehabilitation program might be acknowledged, but patients probably want to be autonomous about the timing of lifestyle changes. This leads to the demand for persuasion technology that supports patients individually in their lifestyle change while maintaining the persuasive elements of the collective ICR program.

3 Coaching the Patient

As we have explained, a patient and a cardiologist have to find a mutual understanding of the initial situation, set realistic goals and create a comprehensive action plan including monitoring of the outcome. The intended change during cardiac rehabilitation is usually one of the general goals of cardiovascular risk management. E.g. low-density lipoprotein cholesterol < 100 mg/dL, body mass index < 25 kg/m^2, smoking cessation and at least 150 min of physical exercise weekly [15].

Patients should only choose what they are motivated for. Then they can start figuring out 'how' they want to reach their chosen goals. This could include creating to-do-lists tailored to the patient's personal goals, circumstances, wishes, believes, drive and history. Motivational Interviewing (MI) [16] is a method that doctors can use to motivate patients in healthcare and seems a natural choice to start the ICR process.

3.1 Motivational Interviewing

Motivational interviewing (MI) has proven to be a successful guiding style for caregivers to help their patients, while they are in the process of resolving ambivalent feelings towards lifestyle changes. MI is 'a collaborative, person-centered form of guiding to elicit and strengthen motivation for change' [17]. It is testable and has scientifically proven positive effects on motivation and adherence; not only in cardiovascular prevention [18]. The effect of MI does not depend on the characteristics of the caregiver, but on a combination of empathy and a structured way of working. MI departs from traditional Rogerian client-centered therapy [19] through its use of direction, in which caregivers attempt to influence patients to consider making changes, rather than to let them explore themselves.

The caregiver should use the ambivalence towards unhealthy behavior to start 'moving' the patient. Thus, the caregiver should ask questions in a way that the patient feels competent to resolve the discrepancy. The main instrument of this communication style is empathic active listening and not to inform or to advise. The goal is to uncover sources of ambivalence, to avoid resistance or discussion and to give the existing ambivalence a positive direction. The positive internal voice is given more room to grow, while the negative counterpart is muted. Any resistance from a patient should be interpreted by the caregiver as a signal to change strategy, because the caregiver should try to stay away from the persuasion trap in which he argues for change and the patient argues against it.

MI can be divided into five steps: Step 1: Ask permission to talk about (un)healthy behavior. Step 2: Ask the 'why' question: Ask why a patient himself would like to change any behavior, but first why his current behavior is useful to him. Step 3: Ask the motivation-score question: listen for reasons or urges to change and rate the motivation. Step 4: Ask the confidence-score question: this is similar to step 3, but this time relates to their confidence to be able to change. Step 5: Ask the 'when' question: Ask at what moment they see themselves start concretely.

After this first step two things are achieved. The patient is aware of the situation and has made explicit why a change is desirable. And the patient has resolved the discrepancy together with the caregiver, who now can function as an external reference point for the patient to motivate and include social pressure. This is fundamentally different from performing the same process without accountability to the caregiver, which would lead to a much lower commitment to the results. At this point, the caregiver will have sufficient reason to offer patient-specific scientific information to support any motivation to change and to enhance self-confidence.

3.2 Coaching for Behavior Change

Besides MI there are other methods that can be used in the process of coaching and self-determination. Although it is what people 'believe' in that drives them, humans can also moderate decisions by conscious reasoning. This point of view is applied in Cognitive Behavioral Theory (CBT). CBT is the most widely used evidence-based practice for treating mental disorders [20] and will in some form be used by coaches in regular group and individual meetings of ICR programs. At regular moments in an ICR program, patients can reflect on behavior outside normal circumstances and decide (consciously) upon a course of action.

Still, most of our daily behavior consists of schema's and habits. A habit will develop consistently, if a specific behavior is performed repeatedly in an unvarying context. Once formed, habits may be difficult to inhibit, even when they conflict with conscious (healthy) intentions, because the habits bypass the intentions [21]. Thus, this type of behavior is not regulated by conscious decision making and cannot easily be changed by CBT. What is needed is some additional coaching method that prevents the patient from getting into situations that trigger the unhealthy behavior or have some trigger set up, that leads to healthy behavior in those situations. Forcing a patient to exercise more every day might be done by creating a cue, for example parking the car a few blocks away. This cue makes it harder to use the car and easier to use the bike. One can also get partners involved. For example, have them buy healthy food, such that unhealthy alternatives are not present any more.

When trying to change a behavior in a particular situation, the trigger for the healthy behavior will have to be part of this situation. Having a human coach present (or a buddy) to trigger behavior, is an option, but expensive and not always possible. A virtual coach, however, can be omnipresent and can remind the patient of the right course of action at the right time and situation.

4 e-Coaching

A recent systematic review and meta-analysis of web-based interventions showed that these interventions can be effective in improving cardiovascular risk factor profiles of middle-aged and older people, but effects are modest. Sustainability is of particular importance, because long-term effects are required for primary and secondary prevention to truly contribute to the prevention of cardiovascular disease. Web-based interventions combined with human support are considered to be more promising than electronic-only interventions [22].

Although pursuing a health promotion intervention based explicitly on a habit formation model is still rare [21], we argue that e-coaching combined with specific human support at regular intervals and based on a habit formation model will be the most effective way to affect lifestyle change in an ICR.

4.1 e-Coaching for a Rehabilitation Program

We will now discuss what the above means for e-coaches that support a cardio rehabilitation program. Especially at what moments they can support the patient and what the requirements are for the interaction such that they work effectively. How could the process of a behavioral change be enhanced by using an electronic support? Which tasks can computers do better than or just as good as humans? How can behavioral theories help us to structure an e-coach in a logical and practical form? From the previous discussion, we see a role for an e-coach in:

1. the organizational process of the cardio rehabilitation program: establish common grounds between the patient and caregiver about the initial situation; build trust during the process; establish a desired end situation; balance the program intensity and content according to the goals and taxability of the heart; create regular habit breaking situations to change the lifestyle of the patient; involve the social environment of the patient; include a close monitoring system to quickly react to non-adherence of the program and to detect progress or deterioration of the patient.
2. the patient's process of making lifestyle changes: determine the motivation of a patient ('why'), assist with setting goals ('what'), implement a planning process and support execution of new habit formation ('how')

An e-coach is well suited for offering structured information based on patient-specific data and can therefore add value to coaching. Also, an e-coach can support with asking questions. Based on the answers given, it can put together new questions. This results in a so-called 'digital consultation' (see e.g. [23]). A precondition is that the expected answers fall within predictable response categories. Subsequently, a logical follow-up question can be constructed. It seems that step 1 and 5 of MI are suitable for such a digital consultation. Steps 2 through 4 will depend on live-sessions with the caregiver. A digital consultation is not a questionnaire or survey, but rather a structured interview. An e-coach could also add a meaningful and time saving contribution by administering a questionnaire to gather data concurrent with the live sessions [24, 25]. Due to the increasing sophistication of natural language processing methods for automated coding, computers are learning to understand and encode human language better. Currently, this is used in research to encode MI interviews. In the future, it may serve 'digital consultation', when it would permit less rigid question-response category preconditions or when it could select follow-up questions according to the perceived mood [26, 27].

In general, an e-coach contributes to the experience of guidance and is patient-friendly as it allows for answering questions at a self-chosen moment. Furthermore, documenting answers and small steps of self-evaluation makes the patient aware of even the smallest successes achieved, which is essential to manage self-motivation.

The second part of an e-coach contribution could be monitoring and guiding the process that concerns planning and execution of 'how' to develop healthy habits. To develop a habit, a conscious desire needs to transform from an intention to a decision, to a behavior, to a context-dependent repetition of this behavior and finally into a habit. The Fogg Behavior Model states that three elements must converge at the same moment for a behavior to occur: Motivation, Ability, and Trigger 'B = MAT' [28]. Coherent with the preceding arguments in this article, we could argue to add a humanistic factor to the equation, 'B = MATH', that would acknowledge the influence of human mind setting as described above.

Following the behavioral point of view, an e-coach can assist the process of cardio rehabilitation by *setting cues* and *triggers* and *implementation intentions* (rigid if/then patterns) to initiate desired behavior. Triggers can be based on different contexts, e.g. location, time, a preceding event or emotion. The most effective cues for implementation intentions seem to be distinct events in daily life which are unlikely to be missed. It is also better to associate former cues with a new healthy response than to a non-response [21]. For example, an e-coach could trigger a response based on GPS location and remind patients to use the stairs instead of the elevator when they arrive at work. Humans tend to 'forget to remember' so-called prospective memory tasks. Electronic support systems are better at 'remembering' than people. So, to enhance intended medication compliance in an ICR, the e-coach could trigger an alarm to remember to take medication, e.g. medicine boxes with alarms. The e-coach can also help setting up a schedule or planning or help structure and plan 'implementation intentions'. For example, every day before I have dinner, I will go for a walk for 30 min. Or, every time I go to the fridge for a snack I will take 500 ml of cold water and wait for three minutes to see if I still want the snack. Another example where an e-coach could be beneficial is when it would assist patients with chunking of desired behavior by asking sensible questions. Behavior that is perceived as hard will be perceived as more doable when chunked. Additionally, if 'chunked' behavior is linked in a sequence by planning, doing one chunk of behavior can trigger the next chunk. For example, people who have initiated 'going for a run' by putting on their running shoes and leaving the house are more likely to continue with their run than those who have not initiated the process with the previous 'chunk' of behavior.

An e-coach can also *facilitate repetition of behavior* e.g. compliance with therapy, because it can give guidance for a prolonged time. Often, the effectiveness of conventional behavior change interventions is constrained because, when the active intervention period ends, so does attention for the desired target behavior, like going for a walk, eating more vegetable, attending yoga classes, etc. With conventional intervention, short-term behavioral gains tend to get lost in the long-term [21].

An e-coach can *monitor behavior* or outcomes and attach consequences to them, like praising and reprimands to assist with automation of behavior. Satisfaction is likely to be important in maintaining novel behavior, while dissatisfaction typically disengages people. The gaming industry uses sophisticated cues and triggers to keep people from disengaging.

Finally, an e-coach can contribute its strengths of *digital data processing* to the empathic and therapeutic management by healthcare professionals. Healthcare professionals tend to get overwhelmed by the vast amount of data that is produced in health care nowadays, including data produced by the world of sensors and wearables that people carry with them in their smartphones, watches, domotics or other applications. An e-coach can assist with careful selection and clear presentation of useful data, e.g. measurements of heartrate, weight, blood pressure, glucose levels or activity level. Tracking of how good or how bad patients are doing in between (healthcare) visits can offer healthcare information that would not be available otherwise.

4.2 Design of an e-Coach for ICR

Given the previous discussions, the contents of an ICR program and the possibilities for an electronic coach to support the patient and caregiver, we arrive at the following requirements that should be covered by an e-coach.

1. Collecting data about the initial situation of the patient at the start of the ICR program. This data consists of three parts:
   a. Medical data that can be imported from the electronic patient file concerning diagnosis, severity and all information that can influence the contents or course of the ICR;
   b. Data measured at the patient in a particular period that are linked to the secondary prevention goals in cardio vascular risk management (see Table 1);
   c. Data concerning the social situation of the patient that could influence adherence to the ICR program, for example, whether a partner smokes or not.
2. Support during consultations with the caregiver at the start of the ICR program.
   a. Show how some combination of habits and physical conditions bring about heart conditions and how this changes when the habits are changed.
   b. Support the determination of the 'why' of the intended change of habits and explicitly show contributing factors (e.g. work stress, unhealthy eating, no activity and smoking).
   c. Support the caregiver and patient to determine achievable goals and optimize what is medically possible (e.g. 30 min of activity could be swimming, running, playing soccer with children, …with a heartrate of 120–150 bpm).
3. Monitoring and triggering of the patient during the ICR program.
   a. Monitoring can be done by sensors that automatically collect data that can show progress of secondary prevention goals in cardio vascular risk management. To enhance adherence, it seems important to include sensors that patients already use, e.g. apps, smartwatches and phones. Data can also be collected through forms from the patient and possibly relatives and/or friends. Furthermore, results can be obtained at designated points in time (once a day or week). The results of the monitored data should be visible for the patient at any time. Under certain conditions data can be sent to the caregiver or to other designated relatives or friends for feedback on a patient's behavior.
   b. The e-coach can give triggers for activities when it is time for these actions according to the schedule or based on other contexts, e.g. location, time, a preceding event or emotion. It should be possible to automatically schedule activities in the agenda such that new appointments can be taken care of.

4. If the ICR program includes meetings with patient groups, a social media group or intranet can be created for these patients, such that they can keep contact outside these meetings for feedback, for accountability to their goals and to celebrate victories and results together.

Table 1. Secondary prevention goals for CHD

Factor	NHG goal	ACC goal
Systolic Blood Pressure (SBP)	≤140 mmHg*	<140 mmHg
Diastolic Blood Pressure (DBP)	<90 mmHg*	<90 mmHg
Low-Density Lipoprotein (LDL)	≤2.5 mmol/L	<100 mg/dL
Fasting Blood Glucose	<6 mmol/L	<100 mg/dL
Body Mass Index (BMI)	≤25 kg/m^2**	18.5–24.9 kg/m^2
Waist circumference	♂ < 94 cm; < 80 cm	♂ < 102 cm; < 88 cm
Physical activity (brisk)	5 times 30 min/wk	≥150 min/wk
Smoking, Food, Stress	Guideline compliance	Guideline compliance
Medication	Compliance***	Compliance

Abbreviations: ACC, American College of Cardiology; NHG, Nederlands Huisartsen Genootschap.
*≥80y SBP: 150–160 mmHg; With chronic kidney disease or proteinuria: <130/80 mmHg.
**>70y weigh (dis)advantage of medication against life expectancy; >70y BMI goal ≤ 30 kg/m^2.
***Medication is required when SBP > 180 mmHg or TC/High-density-lipoprotein ratio > 8 mmol/l.

From these concrete points mentioned above, it can be concluded that the first meetings of the ICR program between patient and caregiver are very important. The result of these meetings is a personalized recovery program of which the patient is convinced that it is achievable and medically useful (because he can see the basis from which it is constructed and thus also trust what the caregiver gives as advice) and for which the patient is motivated (because the patient has set the goals himself and can see that these are achievable and useful in his personal situation).

5 Conclusions and Future Work

To conclude, we envision an electronic support for intensive cardiac rehabilitation, that is driven by our belief that we can help patients change their unhealthy behavior by adding value to the physical process of coaching. A caregiver should discover together with the patient 'why' the patient wants to change and agree upon medical rehabilitation goals that are coherent with the patient's motivation. An e-coach can assist with processing data and offering clear data presentation. An e-coach may offer digital consultations and create triggers when people tend to forget. The possibility of close monitoring through a mobile device, forms and sensors gives the patient continuous feedback on progress, gives assurance and can give triggers to avoid bad habits. If data is shared in a digestible format with others, this can also build up a community that supports good behavior.

Digital products tend to be based on rules, but rules are never a substitute for thinking, so coaching and e-coaching should learn to coexist. Therefore, an e-coach should adapt to behavioral mechanisms and caregivers should learn to incorporate e-coaches in their workflow, where digital support adds value or saves time.

References

1. Koopman, C., Van Dis, I., Vaartjes, I., Visseren, F.L.J., Bots, M.L.: Hart- en vaatziekten in Nederland 2014, cijfers over kwaliteit van leven, ziekte en sterfte. Hartstichting, Den Haag (2013)
2. Zinnige Zorg - Screeningsfase: Systematische Analyse hart- en vaatstelsel. Zorginstituut Nederland, Diemen (2015)
3. Keib, C.N., Reynolds, N.R., Ahijevych, K.L.: Poor use of cardiac rehabilitation among older adults: a self-regulatory model for tailored interventions. Heart Lung. **39**(6), 504–511 (2010)
4. De Vries, H., Kemps, H.M.C., Van Engen Verheul, M.M., Kraaijenhagen, R.A., Peek, N.: Cardiac rehabilitation and survival in a large representative community cohort of Dutch patients. Eur. Heart J. **36**, 1519–1528 (2015)
5. Janssen, V., De Gucht, V., Van Exel, H., Maes, S.: A self-regulation lifestyle program for post-cardiac rehabilitation patients has long-term effects on exercise adherence. J. Behav. Med. **37**, 308–321 (2014)
6. Kotseva, K., et al.: EUROASPIRE IV: A European Society of Cardiology survey on the lifestyle, risk factor and therapeutic management of coronary patients from 24 European countries. Eur. J. Prev. Cardiol. **23**, 636–648 (2016)
7. Diemen-Steenvoorde van, J.A.M.: Instroom hartrevalidatie en naleving richtlijn op onderdeel leefstijlbegeleiding onvoldoende verbeterd. Inspectie voor de Gezondheidszorg, Utrecht (2013)
8. Kelly, S., Martin, S., Kuhn, I., Cowan, A., Brayne, C., Lafortune, L.: Barriers and facilitators to the uptake and maintenance of healthy behaviours by people at mid-life: a rapid systematic review. PLoS ONE **11**, e0145074 (2016)
9. Murray, J., Fenton, G., Honey, S., Bara, A., Hill, K., House, A.: A qualitative synthesis of factors influencing maintenance of lifestyle behaviour change in individuals with cardiovascular risk. BMC Cardiovasc. Disord. **13**, 48 (2013)
10. Murray, J., Honey, S., Hill, K., Craigs, C., House, A.: Individual influences on lifestyle change to reduce vascular risk: a qualitative literature review. Br. J. Gen. Pract. **62**, 403–410 (2012)
11. Leventhal, H., Phillips, L.A., Burns, E.: The Common-Sense Model of self-regulation (CSM): a dynamic framework for understanding illness self-management. J. Behav. Med. **39**(6), 935–946 (2016)
12. Ornish, D., Brown, S.E., Billings, J.H., Scherwitz, L.W., Armstrong, W.T., Ports, T.A., McLanahan, S.M., Kirkeeide, R.L., Gould, K.L., Brand, R.J.: Can lifestyle changes reverse coronary heart disease? Lifestyle Heart Trial. Lancet. **336**, 129–133 (1990)
13. Ornish, D., Scherwitz, L.W., Billings, J.H., Brown, S.E., Gould, K.L., Merritt, T.A., Sparler, S., Armstrong, W.T., Ports, T.A., Kirkeeide, R.L., Hogeboom, C., Brand, R.J.: Intensive lifestyle changes for reversal of coronary heart disease. JAMA **280**, 2001–2007 (1998)
14. Billings, J.H.: Maintenance of behavior change in cardiorespiratory risk reduction: a clinical perspective from the Ornish Program for reversing coronary heart disease. Heal. Psychol. **19**, 70–75 (2000)

15. Ton, V.-K., Martin, S.S., Blumenthal, R.S., Blaha, M.J.: Comparing the New European cardiovascular disease prevention guideline with prior American heart association guidelines. Clin. Cardiol. **36**, E1–E6 (2013)
16. Rollnick, S., Miller, W.R., Butler, C.: Motivational Interviewing in Health Care: Helping Patients Change Behavior. Guilford Press, New York (2008)
17. Miller, W.R., Rollnick, S.: Ten things that motivational interviewing is not. Behav. Cogn. Psychother. **37**, 129–140 (2009)
18. Thompson, D.R., Chair, S.Y., Chan, S.W., Astin, F., Davidson, P.M., Ski, C.F.: Motivational interviewing: a useful approach to improving cardiovascular health? J. Clin. Nurs. **20**, 1236–1244 (2011)
19. Rogers, C.R.: Client-Centered Therapy: Its Current Practice, Implications and Theory. Constable, London (1951)
20. Gaudiano, B.A.: Cognitive-behavioural therapies: achievements and challenges. Evid. Based Ment. Health **11**, 5–7 (2008)
21. Gardner, B.: A review and analysis of the use of "habit" in understanding, predicting and influencing health-related behaviour. Health Psychol. Rev. **9**, 277–295 (2015)
22. Beishuizen, C.R., Stephan, B.C., van Gool, W.A., Brayne, C., Peters, R.J., Andrieu, S., Kivipelto, M., Soininen, H., Busschers, W.B., Moll van Charante, E.P., Richard, E.: Web-based interventions targeting cardiovascular risk factors in middle-aged and older people: a systematic review and meta-analysis. J. Med. Internet Res. **18**, e55 (2016)
23. Beun, R.J., Brinkman, W.P., Fitrianie, S., Griffioen-Both, F., Horsch, C., Lancee, J., Spruit, S.: Improving adherence in automated e-coaching. In: Meschtscherjakov, A., De Ruyter, B., Fuchsberger, V., Murer, M., Tscheligi, M. (eds.) Persuasive Technology, pp. 276–287. Springer International Publishing, Salzburg (2016)
24. Friederichs, S.A.H., Oenema, A., Bolman, C., Lechner, L.: Long term effects of self-determination theory and motivational interviewing in a web-based physical activity intervention: randomized controlled trial. Int. J. Behav. Nutr. Phys. Act. **12**, 101 (2015)
25. Geissler, H., Hasenbein, M., Kontouri, S., Wegener, R.: e-Coaching: conceptual and empirical findings of a virtual coaching programme. Int. J. Evid. Based Coach. Mentor. **12**, 165–186 (2014)
26. Tanana, M., Hallgren, K.A., Imel, Z.E., Atkins, D.C., Srikumar, V.: A comparison of natural language processing methods for automated coding of motivational interviewing. J. Subst. Abuse Treat. **65**, 43–50 (2016)
27. Can, D., Marín, R.A., Georgiou, P.G., Imel, Z.E., Atkins, D.C., Narayanan, S.S.: "It sounds like…": a natural language processing approach to detecting counselor reflections in motivational interviewing. J. Couns. Psychol. **63**, 343–350 (2016)
28. Fogg, B.J.: A behavior model for persuasive design. In: Chatterjee, S., Dev, P. (eds.) Persuasive 2009 Proceedings of the 4th International Conference on Persuasive Technology. ACM, Claremont (2009)

A Scoped Review of the Potential for Supportive Virtual Coaches as Adjuncts to Self-guided Web-Based Interventions

Mark R. Scholten(✉), Saskia M. Kelders, and Julia E.W.C. van Gemert-Pijnen

Department of Psychology, Health and Technology, Center for eHealth and Wellbeing Research, University of Twente, Enschede, The Netherlands
m.r.scholten@utwente.nl

Abstract. This study aimed to explore supportive capabilities of VAs with the potential benefit in mind that users of self-guided eHealth interventions could be better supported. Spontaneous empathy and the explicitly expressed intention of *non-responsive VAs* to deliver user support is likely capable to engage and motivate users. *Responsive VAs* have even larger potential. However, they are more costly to realize and have a higher risk of failure. Effective user frustration detection and mitigation by Responsive VAs has been empirically demonstrated, but so far within artificial contexts. Altogether it makes sense to further explore the option to add VAs as adjuncts to self-guided eHealth interventions a potential remedy to low adherence.

Keywords: Virtual agent · Embodied conversational agent · Virtual human · Persuasive technology · ehealth

1 Introduction

Research [1] has suggested that Virtual Agents (VAs), taking on the role of coach or company on, have the potential to assist users of eLearning and eHealth solutions by engaging them. As self-guided eHealth interventions often face low adherence scores it is worthwhile to explore the motivational features and capabilities of VAs. Future self-guided eHealth interventions could potentially profit from VAs as adjuncts to engage and support users which could potentially offer a remedy to low adherence.

2 Methods

This study reviews and interprets the available literature on the potential of VAs by means of a scoped review. The rationale for choosing a Scoped Review is that the subject is broad, diverse and largely unexplored which warrants a scoped review methodology in which it is sought to present an overview of a such potentially large and diverse body of literature pertaining to a broad topic [2].

P.W. de Vries et al. (Eds.): PERSUASIVE 2017, LNCS 10171, pp. 43–54, 2017.
DOI: 10.1007/978-3-319-55134-0_4

2.1 Study Selection: Opportunities of Virtual Coaches to Deliver Support Within Web-Based Interventions for Health or Learning

The search aimed to create a generic idea of the capabilities of VAs (VAs) for supportive purposes. The Scopus and Web of Science databases were searched with a combination of the concepts 'VAs, 'web-based intervention', and 'support'. For each of the concepts, multiple key words were used. As VAs are often used within a e-learning context, it was decided to include studies on Intelligent Tutoring Systems (ITS) as well.

Inclusion criteria were:

- Papers had to address VAs interacting with users

Exclusion criteria were:

- Papers that solely focused on the effects of VAs in Virtual Reality.
- Papers that described computer simulations with agents/during which interaction between human users and VAs were absent

The systematic search resulted in a limited number of studies (8). Moreover, these studies addressed a wide range of topics; from physical attributes [3], architecture [4], route planning [5], non-verbal behavior [6], virtual museum guide [7], empathy [8], to theoretical models [9] and articulation rates [10]. None of the studies provided a high-level picture of the capabilities of VAs with regards to support delivery. Therefore it was decided to expand the number of articles by means of hand search. We started the hand search by checking references within the 8 articles and searching on terms found within the 8 articles in Google Scholar.

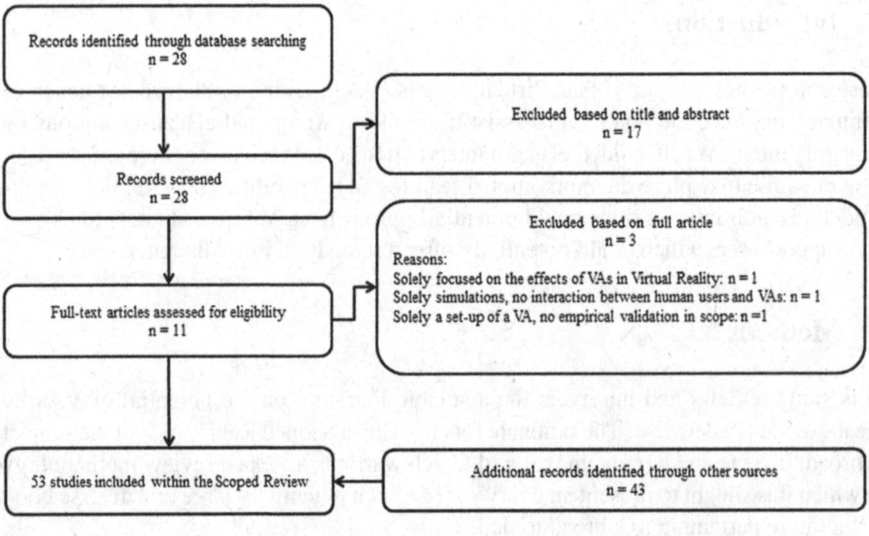

Fig. 1. Flow diagram of the study selection of the scoped review

The hand search had the following aims:

(a) Finding synthesizing information on VAs within a health or pedagogical context with a focus on the delivery of support and motivating users. We started with the information found in [8] and additionally searched for meta studies on VAs.
(b) Finding additional (founding) articles on the CASA effect as mentioned within [3, 8].
(c) Finding addition information on relationship building [10] and measures of relationship building as shortly described in [6, 10].
(d) Finding additional information on theoretical models related to VAs as touched upon in [9].

The search procedure resulted in **53** included articles (see Fig. 1 below)

2.2 Data Extraction

Studies were analyzed and the various VA aspects were categorized within themes. The themes were chosen as a means to provide insight in the various aspects of VAs that relate to user motivation. Secondly, the themes served to aggregate information that was distributed over different articles found. For an overview of the articles included.

2.3 Results

Table 1 shows the themes that were found in the included studies. Below Table 1 these themes are further described.

Theme 1: Computers as Social Actors (CASA). A large body of studies on VAs refer to the CASA effect [13, 15] as a cornerstone for studying human-computer interactions and especially human-VA interactions. The CASA effect demonstrates that humans treat media – in some respect- in the same way as they treat other humans. Various manifestations of this effect have been described such as:

- Computers that display flattery texts towards their users are preferred by their users compared to computers that do not display such texts
- Computers that textually praise other computers are better liked than computers that praise themselves, and computers that 'criticize' other computers are disliked compared to computers that criticize themselves
- Users who are partnered with an computer on basis of a color (e.g. the blue team) will have a more positive opinion on the computer and cooperate more with it than users who have to partner with a computer of the opposite, differently colored team

As an explanation of the CASA effect, it has been proposed that humans have a strong innate tendency to make social connections with other humans and other living creatures such as pets. This human tendency becomes real when objects such as personal computers demonstrate activities that could be socially interpreted by their users [15]. Although pc's can act socially, human users are logically aware of their non-social and non-living status. This seems a paradox: why would a human user socially respond to a pc while at the same time realizing that a pc does not warrant it? Nass and Moon [12]

Table 1. Table 1 Themes and articles for supportive VAs.

Theme	Explanation	Sources
1. Computers As Social Actors (CASA)	Humans treat media in the same way as they treat other humans	Systematic search: [8] Hand Search: [11–15]
2. Open dialogue between user and computer	VAs have the ability to have an open verbal dialogue with users	Systematic search: [7] Hand Search: [14–16]
3. Visible conversation partner	Interaction with a 'talking face' leads to more trust and believability.	Systematic search: [3, 5, 8, 9] Hand Search: [19, 23]
4. Human-Computer relationship	Interactions with an agent can lead to a relationship, which is important to keep users engaged over time	Systematic search: [10] Hand Search: [1, 24–30]
5. Measures of the Human-Computer relationship	Human-VA relationship quality can be measured	Systematic search: [6] Hand Search: [1, 22, 32, 37]
6. Responsive verbal and non-verbal communication	Computers should have the ability to notice and respond to verbally and non-verbally expressed emotions from their user, in order to create a more natural interaction	Systematic search: [8] Hand Search: [27, 31–38]
7. Impact of VAs on User motivation	There is evidence that VAs can motivate users, which is highly dependent on VA implementation, context, task etc	Systematic search: [4] Hand Search: [21, 38, 42–46]
8. Methodological issues within VA research	Most experiments into VAs face similar methodological issues which have to be taken into account when interpreting the research	Hand Search: [45–49]

refer to 'mindless' (automatic, largely unaware) human behavior that the machine can trigger. This mindless behavior will be displayed as long as it remains socially acceptable. This phenomenon is also associated with the notion of 'suspension of disbelief', meaning that up to a certain point humans are willing to apply social rules to non-human yet communicative objects, irrespective of their non-living status.

Theme 2: Open Dialogue between user and computer. A following theme is the ability of computers and VAs to have an open verbal (textual or speech) dialogue with users. Within regular, day to day Human-Computer Interaction events, a user who interacts with their IT system will typically activate pre-defined menu options such as the 'save as' option within Microsoft Word. Subsequently, the computer will respond to the request by presenting a pop-up window which will enable the user to type in the file name of the document. In such a closed dialogue scenario, the interactions between user and software traditionally have a task-specific character (e.g. serve to reach a specific goal such as saving a document), have a short duration and are typically initiated by the user (and not by the computer). In contrast, VAs enable more open-ended and more relationship-oriented interactions. Interactions between VAs and users can span multiple question and answer pairs and can therefore be interpreted as a dialogue. The ELIZA study [18] described an early version of a textual psychotherapists that gave 'canned' responses to user questions as a result of quickly processing the input text provided and create a response out of it without realizing what the user had said (e.g. a question like: "Eliza, I feel miserable today" and an answer: "How often do you experience feelings of being miserable?"). Later studies create richer dialogue contexts to explore the capabilities of computers interacting with humans. Examples are first a study that has shown that a robot taking the role of museum guide who uses e.g. empathy and humor in his conversation style led to a more positive attitude towards the robot than the same robot without this enhanced conversation style [7]. A second study showed that a VA with high dialog capabilities reached more accurate answers when interviewing a subject than an agent with less dialog capabilities [16]. A third study [17] aimed to explore where open-dialogue options between users and VAs would lead to. The authors report that when learners are given opportunities to guide an open conversation, they especially ask off-topic questions. For example, learners often want to know about the agents' operating systems, design, purpose, and capabilities. Such conversations seem to serve the 'testing' of agents' abilities during which learners are attempting to discover the boundaries, limits, and capabilities of agents through 'game-like' inquiry.

Theme 3: Visible conversational partner. The following theme is the visibility of the conversational computer depicted as a (either static or animated) human face. According to Lisetti [24] the human face has a special status in human to human communication as it has often been identified as the most important channel for conducting trust and believability. As Lisetti states, the face as a communication channel has a higher status than bodily regions such as posture and gesture [20]. Multiple studies have supported this notion by demonstrating that users preferred to interact with a 'talking face' instead of a text only interface [28], an anthropomorphic agent together with a human voice has led to greater agent credibility [19], visible agents have led to greater positive motivational outcomes [27] and task performance [29].

Theoretical support for a visible, human-like personal computer is provided by the Social Agency Theory [25] and Social Modelling/Social Learning Theory Jordine et al. [3, 9]. Nonetheless the visibility subject is somewhat controversial. Strong claims against the human face are provided by Norman [14] by his statement that a human face triggers false mental models and thus creates wrong user expectations. Other critique is provided by Rajan et al. [26] who demonstrated that it is first and foremost the voice (and not the visibility of the VA) that is responsible for positive learning effects. Mayer [25] criticizes the benefits of a visible, but static human face on screen. What is important for learning according to Mayer is the level of animation of the VA, which makes the agent engaging to the user.

Theme 4: Human-VA Relationship. A fourth theme is the concept that regular human-computer interaction events result in a relationship. Routine interactions between a user and their computer should be regarded as contributions to this human-computer relationship, as is argued by Bickmore et al. [1]. Although this relationship may be implicit, it has an impact on the user. The relationship plays a role even in case no relationship skills (e.g. empathy, humor) have been designed and built into the machine.

The question arises whether a VA with a relationship-focused design could behave and be perceived as a competent social actor. This quality of the VA as a conversational partner is impacted by:

- Interaction duration. As described by Krämer et al. [35] getting people engaged with VAs is easy, but keeping then engaged over time is much more challenging. Bickmore et al. [1] (on physical activity) and Creed et al. [31] (on fruit consumption) conducted emotional virtual coach studies that spanned more than 28 days. They both found that deploying the emotional VA did not result in user behavior changes, but that users in general preferred to interact with the emotional virtual coaches.
- Natural vs forced interaction. Gulz [33] suggests that most VAs studies force the human-computer relationship too much. Users have no other option than to interact with the VAs they are confronted with.
- User personality. Von der Pütten et al. [36] make clear that it depends on the personality of the user how the human-computer relationship will develop. They demonstrated that 5 user personality factors were better predictors for the evaluation outcome of VAs than the actual behavior of the VA.

Theme 5: Measures of the Human-VA Relationship. The literature found mentions two regular measures with regards to the Human-VA Relationship.

- Measure 1: Working Alliance

Working Alliance is a construct that originates from the psychotherapy literature and has been described as "the trust and belief that the helper and patient have in each other as team-member in achieving a desired outcome" [37]. Bickmore et al. [30] applied the working alliance inventory in their 30-day longitudinal study with a VA acting as an exercise coach. Participants who interacted with a VA with relational behavior enabled (empathy, social chat, form of address, etc.) scored the VA significantly higher on the Working Alliance Inventory compared to participants who interacted with the same VA with the relational behaviors disabled.

- Measure 2: Rapport

A second important human-computer relationship measure is rapport. Rapport has been described as "the establishment of a positive relationship among interaction partners by rapidly detecting and responding to each other's nonverbal behavior" [32]. Measurement of rapport has been conducted by Gratch et al. [32] in their evaluative VA study. Their results showed that the experience of rapport was of a comparable level compared to a face-to-face (i.e. human interlocutor) condition.

Theme 6: Responsive verbal and non-verbal communication. Within human to human communication, the exchange of non-verbal information plays a key role. Social psychologists assert that more than 65% of the information exchanged during a person-to-person conversation is conveyed through the non-verbal band [39, 45]. The non-verbal channel is said to be especially important to communicate socio-emotional information. Socio-emotional content [40] is vital for building trust and productive human relationships that go beyond the purely factual and task-oriented communication. D'Mello et al. [40] describe the mutual impact of user and (synthetic) computer emotions as an affective loop which is pictured as follows:

- The user first expresses their emotion through verbal and physical interaction with the machine, e.g. through detectable gestures, usage of the keyboard or spoken language
- Then, the system responds by generating affective responses, through words, speech, animation and theoretically also colors and haptics
- This response affects the user in such a way that they become more involved in their further interaction with the computer

Concerning the importance of the affective loop, there are two stances:

- Stance 1: Responsiveness of VAs (affective loop) is a critical condition for prolonged user interaction. Doirado et al. [42] confirm the importance of the affective loop mechanism and state that a VA that lacks the capacity to understand the user and the capability to adapt its behavior (a non-responsive VA) will break the user's suspension of disbelief.
- Stance 2: Autonomy of VAs (no affective loop) is a sufficient condition for prolonged user interaction. Rosenberg-Kima et al. [27] deployed an autonomous (i.e. non-responsive) VA that introduced itself and provided a twenty-minute narrative about four female engineers, followed by five benefits of engineering careers. The VA was animated and its voice and lip movements were synchronized. The VA acted autonomously; interaction between participants and VA was purely restricted to the user clicking on the button for text topic. The results showed that the self-efficacy of the users and of their interest in the subject presented was significantly higher within the VA + voice condition compared to the voice-only condition. In support of these results, Baylor et al. [19] state that people are willing to interact with anthropomorphic agents even when their functionality is limited. As she indicates the mere visual presence and appearance will in some contexts be the determining factor and not so much its supportive, conversational or animation capabilities.

Theme 7: Impact of VAs on user motivation. Meta-studies and reviews [33, 45, 49, 50, 53] have reported on claims and evidence for positive VAs effects on learning, engagement and motivation. Schroeder et al. reviewed 43 studies and conclude that pedagogical agents have a small but significant effect on learning as ultimate outcome. Within their study, Schroeder et al. [45] did not make a distinction between responsive and non-responsive VAs. Specific research with regard to motivating users has also been conducted by deploying responsive VAs with the task to notice user frustration and empathically respond to it. Autonomous delivery of warmth and empathy by VAs towards users has shown positive effects, and studies show that this effect may be larger at the time the user experiences frustration [38, 50].

All together the evidence for VAs that are capable of motivating users is mixed and inconclusive. VAs, whether they are non-responsive or responsive, provide a positive user experience as a result of their entertainment capabilities. Responsive VAs when specifically designed to detect user frustration and to empathically respond to it, have also empirically demonstrated positive effects on user attitudes. However, these positive effects have not yet been found in ecologically valid context but only within constrained contexts such as games with clear win and lose rules and as a result of system-generated moments of user frustration.

Theme 8: Methodological issues within VA research. The inconclusiveness regarding VA evidence as mentioned within the previous theme is claimed to be caused by methodological issues [50, 53]. Methodological issues make it difficult to compare study results and to draw generic conclusions. One of those issues is the difference in set-ups amongst VA studies. To name a few:

- Different modalities used for output: (synthesized or natural) speech or text
- Different levels of responsive emotional behavior; from textual responses projected alongside a static VA to fine-grained VA facial expressions intended to mirror the user's facial expressions
- Different roles: tutor, peer, interviewer, coach
- Different implementations/different computer code applied as Artificial Intelligence to steer the VA with code based on different behavioral theories

Many of these issues can be resolved by using a common, open research platform for VAs, such as the Virtual Human platform as provided by USCT [51]. Other issues can potentially be resolved by a common design framework for VAs as proposed by Veletsianos et al. with their EnALI framework [52].

Concerning the duration of the change programs several studies (e.g. [30, 31]) stress that the majority of virtual coaching studies concern short time spans of minutes or hours, which makes it difficult to study the development of the human-computer relationship and to realize effects on user behavior. Both Bickmore et al. and Creed et al. ([30, 31]) conducted emotional virtual coach studies that spanned more than 28 days. They both found that deploying the emotional VA did not result in user behavior changes, but that users in general preferred to interact with the emotional virtual coaches. Altogether Dehn and van Mulken [50] summarize the situation as follows: "… the simple question as to whether an animated interface improves human-computer interaction does not appear to be the appropriate question to ask. Rather, the question to ask is: what kind of animated agent used in what kind of domain influence what aspects of the user's attitudes or performance".

3 Conclusions and Discussion

This Scoped Review aimed to give insight into the potential of VAs to deliver effective related support to humans.

On a high level, the following two kinds of VAs were distinguished:

- *Non-responsive (autonomous) VAs*. These VAs are not endowed with senses to 'see' or 'hear' the verbal or non-verbal signals that the user expresses, and logically also lack the capacity to interpret these signals and respond to them. Instead, the VA is visually present to send out motivational messages intended to keep the spirits up, irrespective of how the user feels or what he does. Pro: these kinds of VAs have demonstrated that they can engage users. Con: forced presence of the VA runs the risk of annoying the user and can therefore become counter-productive. As a solution to keep the benefits and mitigate the drawbacks, users should be given control over the presence of the non-responsive VA.
- *Responsive VAs*. These VAs have the capability to capture and analyze the verbal and/or non-verbal signals sent by the user and emotionally respond to them. These VAs are set up with the intention to understand the user and to adapt their behavior accordingly. Pro: these VAs can tap into the rich sources of verbal and non-verbal information as spontaneously and freely provided by humans. This emotion-related information is key for human to human communication and it therefore makes sense to find ways to use this kind of information for productive HCI. Con: realizing a VA that does understand the user is a heavy task, requiring costly computational modeling of user BDI (Believe, Desire and Intentions) and affective loop facilities with a high chance of failure.

Altogether it makes sense to further explore the option to add VAs as adjuncts to self-guided eHealth interventions a potential remedy to low adherence. For preventing complex and costly experimental set-ups, it is advisable to start further experimentation with non-responsive VAs.

References

1. Bickmore, T., Gruber, A., Picard, R.: Establishing the computer-patient working alliance in automated health behavior change interventions. Patient Educ. Couns. **59**(1), 21–30 (2005). doi:10.1016/j.pec.2004.09.00
2. Pham, M.T., Rajic, A., Greig, J.D., Sargeant, J.M., Papadopoulos, A., McEwen, S.A.: A scoping review of scoping reviews: advancing the approach and enhancing the consistency. Res. Synth. Methods **5**(4), 371–385 (2014). doi:10.1002/jrsm.1123
3. Jordine, K., Wilson, D-M., Sakpal, R.: What is age's affect in collaborative learning environments? In: Proceedings of the International Conference on Universal Access in Human-Computer Interaction (2013)
4. Ieronutti, L., Chittaro, L.: Employing virtual humans for education and training in X3D/VRML worlds. Comput. Educ. **49**(1), 93–109 (2007). doi:10.1016/j.compedu.2005.06.007
5. Hofmann, H., Tobisch, V., Ehrlich, U., Berton, A.: Evaluation of speech-based HMI concepts for information exchange tasks: a driving simulator study. Comput. Speech Lang. **33**(1), 109–135 (2015). doi:10.1016/j.csl.2015.01.005
6. Novick, D., Gris, I.: Building rapport between human and ECA: a pilot study. In: Kurosu, M. (ed.) HCI 2014. LNCS, vol. 8511, pp. 472–480. Springer, Heidelberg (2014). doi: 10.1007/978-3-319-07230-2_45
7. Bickmore, T.W., Vardoulakis, L.M.P., Schulman, D.: Tinker: a relational agent museum guide. Auton. Agent. Multi-Agent Syst. **27**(2), 254–276 (2013). doi:10.1007/s10458-012-9216-7

8. Amini, R., Lisetti, C., Yasavur, U., Rishe, N.: On-demand virtual health counselor for delivering behavior-change health interventions. In: 2013 IEEE International Conference on Proceedings of the Healthcare Informatics (ICHI) (2013)
9. Apostol, S., Soica, O., Manasia, L., Stefan, C.: Virtual pedagogical agents in the context of virtual learning environments: framework and theoretical models. Elearn. Softw. Educ. (2), 531–536 (2013)
10. Schulman, D., Bickmore, T.: Modeling behavioral manifestations of coordination and rapport over multiple conversations. In: Proceedings of the International Conference on Intelligent Virtual Agents (2010)
11. Mori, M., MacDorman, K.F., Kageki, N.: The uncanny valley [from the field]. IEEE Robot. Autom. Mag. **19**(2), 98–100 (2012)
12. Nass, C., Moon, Y.: Machines and mindlessness: social responses to computers. J. Soc. Issues **56**(1), 81–103 (2000). doi:10.1111/0022-4537.00153
13. Nass, C.I., Brave, S.: Wired For Speech: How Voice Activates and Advances The Human-Computer Relationship. MIT press, Cambridge (2005). 0262140926
14. Norman, D.A.: Emotional Design: Why we Love (or Hate) Everyday Things. Basic Books, New York (2005). 0465051367
15. Reeves, B., Nass, C.: How People Treat Computers, Television, and New Media Like Real People and Places. CSLI Publications and Cambridge University Press Cambridge, Cambridge (1996)
16. Conrad, F.G., Schober, M.F., Jans, M., Orlowski, R.A., Nielsen, D., Levenstein, R.: Comprehension and engagement in survey interviews with virtual agents. Front. Psychol. **6** (2015). doi:10.3389/Fpsyg.2015.01578, Artn. 1578
17. Veletsianos, G., Russell, G.S.: What do learners and pedagogical agents discuss when given opportunities for open-ended dialogue? J. Educ. Comput. Res. **48**(3), 381–401 (2013). doi: 10.2190/Ec.48.3.E
18. Weizenbaum, J.: ELIZA—a computer program for the study of natural language communication between man and machine. Commun. ACM **9**(1), 36–45 (1966)
19. Baylor, A.L., Ryu, J., Shen, E.: The effects of pedagogical agent voice and animation on learning, motivation and perceived persona. In: Proceedings of the World conference on educational multimedia, hypermedia and telecommunications (2003)
20. Cowell, Andrew, J., Stanney, Kay, M.: Embodiment and interaction guidelines for designing credible, trustworthy embodied conversational agents. In: Rist, T., Aylett, Ruth, S., Ballin, D., Rickel, J. (eds.) IVA 2003. LNCS (LNAI), vol. 2792, pp. 301–309. Springer, Heidelberg (2003). doi:10.1007/978-3-540-39396-2_50
21. Kim, C.M., Baylor, A.L.: A virtual change agent: Motivating pre-service teachers to integrate technology in their future classrooms. Educ. Technol. Soc. **11**(2), 309–321 (2008)
22. Krämer, N.C., von der Pütten, A., Eimler, S.: Human-agent and human-robot interaction theory: similarities to and differences from human-human interaction. In: Zacarias, M., de Oliveira, J.V. (eds.) Human-Computer Interaction: The Agency Perspective. Studies in Computational Intelligence, vol. 396, pp. 215–240. Springer, Heidelberg (2012)
23. Lester, J.C., Converse, S.A., Kahler, S.E., Barlow, S.T., Stone, B.A., Bhogal, R.S.: The persona effect: affective impact of animated pedagogical agents. In: Proceedings of the ACM SIGCHI Conference on Human Factors in Computing Systems (1997)
24. Lisetti, C., Amini, R., Yasavur, U.: Now all together: overview of virtual health assistants emulating face-to-face health interview experience. KI-Künstliche Intelligenz **29**(2), 161–172 (2015)
25. Mayer, R.E.: Cognitive theory of multimedia learning. In: The Cambridge Handbook of Multimedia Learning, p. 43 (2014)

26. Rajan, S, Craig, S.D., Gholson, B., Person, N.K., Graesser, A.C., Tutoring Research Group.: AutoTutor: incorporating back-channel feedback and other human-like conversational behaviors into an intelligent tutoring system. Int. J. Speech Technol., **4**(2), 117–126 (2001)
27. Rosenberg-Kima, R.B., Baylor, A.L., Plant, E.A., Doerr, C.E.: The importance of interface agent visual presence: voice alone is less effective in impacting young women's attitudes toward engineering. In: kort, y, IJsselsteijn, W., Midden, C., Eggen, B., Fogg, B.J. (eds.) PERSUASIVE 2007. LNCS, vol. 4744, pp. 214–222. Springer, Heidelberg (2007). doi:10.1007/978-3-540-77006-0_27
28. Sproull, L., Subramani, M., Kiesler, S., Walker, J.H., Waters, K.: When the interface is a face. Hum. Comput. Interact. **11**(2), 97–124 (1996). doi:10.1207/s15327051hci1102_1
29. Zanbaka, C., Ulinski, A., Goolkasian, P., Hodges, L.F.: Effects of virtual human presence on task performance. In: Proceedings of the International Conference on Artificial Reality and Telexistence 2004 (2004)
30. Bickmore, T.W., Picard, R.W.: Establishing and maintaining long-term human-computer relationships. ACM Trans. Comput. Hum. Interac. (TOCHI) **12**(2), 293–327 (2005)
31. Creed, C., Beale, R., Cowan, B.: The impact of an embodied agent's emotional expressions over multiple interactions. Interact. Comput. **27**(2), 172–188 (2015). doi:10.1093/iwc/iwt064
32. Gratch, J., Wang, N., Gerten, J., Fast, E., Duffy, R.: Creating rapport with virtual agents. In: Pelachaud, C., Martin, J.C., André, E., Chollet, G., Karpouzis, K., Pelé, D. (eds.) IVA 2007. LNCS (LNAI), vol. 4722, pp. 125–138. Springer, Heidelberg (2007). doi:10.1007/978-3-540-74997-4_12
33. Gulz, A.: Benefits of virtual characters in computer based learning environments: claims and evidence. Int. J. Artif. Intell. Educ. **14**(4), 313–334 (2004)
34. Kang, S.H., Gratch, J., Wang, N., Watt, J.H.: Does the contingency of agents' nonverbal feedback affect users' social anxiety? In: Proceedings of the 7th International Joint Conference on Autonomous Agents and Multiagent Systems, vol. 1 (2008)
35. Kramer, N.C., Eimler, S., von der Putten, A., Payr, S.: Theory of companions: what can theoretical models contribute to applications and understanding of human-robot interaction? Appl. Artif. Intell. **25**(6), 474–502 (2011). doi:10.1080/08839514.2011.587153
36. Pütten, A.M., Krämer, N.C., Gratch, J.: How our personality shapes our interactions with virtual characters - implications for research and development. In: Allbeck, J., Badler, N., Bickmore, T., Pelachaud, C., Safonova, A. (eds.) IVA 2010. LNCS, vol. 6356, pp. 208–221. Springer, Heidelberg (2010). doi:10.1007/978-3-642-15892-6_23
37. Horvath, A.O., Greenberg, L.S.: Development and validation of the working alliance inventory. J. Couns. Psychol. **36**(2), 223–233 (1989). doi:10.1037/0022-0167.36.2.223
38. Baylor, A.L., Rosenberg-Kima, R.B., Plant, E.A.: Interface agents as social models: the impact of appearance on females' attitude toward engineering. In: Proceedings of the CHI 2006 Extended Abstracts on Human Factors in Computing Systems (2006)
39. Berry, D.C., Butler, L.T., de Rosis, F.: Evaluating a realistic agent in an advice-giving task. Int. J. Hum Comput Stud. **63**(3), 304–327 (2005). doi:10.1016/j.ijhcs.2005.03.006
40. D'Mello, S., Picard, R., Graesser, A.: Towards an affect-sensitive autotutor. IEEE Intell. Syst. **22**(4), 53–61 (2007)
41. D'Mello, S., Olney, A., Williams, C., Hays, P.: Gaze tutor: a gaze-reactive intelligent tutoring system. Int. J. Hum. Comput. Stud. **70**(5), 377–398 (2012). doi:10.1016/j.ijhcs.2012.01.004
42. Doirado, E., Martinho, C.: I mean it!: detecting user intentions to create believable behaviour for virtual agents in games. In: Proceedings of the 9th International Conference on Autonomous Agents and Multiagent Systems, vol. 1 (2010)

43. Lisetti, C., Amini, R., Yasavur, U., Rishe, N.: I can help you change! an empathic virtual agent delivers behavior change health interventions. ACM Trans. Manag. Inf. Syst. (TMIS) **4**(4), 19 (2013)
44. Picard, R.W., Picard, R.: Affective Computing. MIT Press, Cambridge (1997)
45. Schroeder, N.L., Adesope, O.O., Gilbert, R.B.: How effective are pedagogical agents for learning? a meta-analytic review. J. Educ. Comput. Res. **49**(1), 1–39 (2013)
46. Brave, S., Nass, C., Hutchinson, K.: Computers that are care: investigating the effects of orientation of emotion exhibited by an embodied computer agent. Int. J. Hum. Comput. Stud. **62**(2), 161–178 (2005). doi:10.1016/j.ijhcs.2004.11.002
47. Klein, J., Moon, Y., Picard, R.W.: This computer responds to user frustration: theory, design, and results. Interact. Comput. **14**(2), 119–140 (2002)
48. Moreno, R.: Software agents in multimedia: an experimental study of their contributions to students' learning. In: Proceedings of Human-Computer Interaction, pp. 275–277 (2001)
49. Beale, R., Creed, C.: Affective interaction: how emotional agents affect users. Int. J. Hum Comput Stud. **67**(9), 755–776 (2009). doi:10.1016/j.ijhcs.2009.05.001
50. Dehn, D.M., van Mulken, S.: The impact of animated interface agents: a review of empirical research. Int. J. Hum Comput Stud. **52**(1), 1–22 (2000). doi:10.1006/ijhc.1999.0325
51. Hartholt, A., Traum, D., Marsella, S.C., Shapiro, A., Stratou, G., Leuski, A., Morency, L.P., Gratch, J.: All together now. In: Proceedings of the International Workshop on Intelligent Virtual Agents (2013)
52. Veletsianos, G., Miller, C., Doering, A.: Enali: a research and design framework for virtual characters and pedagogical agents. J. Educ. Comput. Res. **41**(2), 171–194 (2009)
53. Veletsianos, G., Russell, G.S.: Pedagogical agents. In: Spector, J.M., Merrill, M.D., Elen, J., Bishop, M.J. (eds.) Handbook of Research on Educational Communications and Technology, pp. 759–769. Springer, New York (2014)

Augmenting Group Medical Visits with Conversational Agents for Stress Management Behavior Change

Ameneh Shamekhi[1(✉)], Timothy Bickmore[1], Anna Lestoquoy[2], and Paula Gardiner[2]

[1] College of Computer and Information Science, Northeastern University, Boston, MA, USA
{ameneh,bickmore}@ccs.neu.edu

[2] Department of Family Medicine, Boston Medical Center, Boston, MA, USA
{Anna.Lestoquoy,Paula.Gardiner}@bmc.org

Abstract. Individuals with chronic stress can improve their overall wellness by making lifestyle changes. We describe the design and evaluation of a computer-animated conversational agent that is used in conjunction with Integrative Medical Group Visits (IMGV) to help individuals with chronic pain and depression manage their stress. The agent coaches patients in between group visit sessions, reviewing how to manage stress using non-pharmaceutical techniques (e.g., mindfulness, meditation, yoga self-massage). We conducted a longitudinal clinical trial with 154 participants in which patients in the intervention group interacted with the conversational agent at home on a touch screen tablet to reinforce what they learned in group medical visits. Results from the trial indicate that the conversational agent, in conjunction with medical group visits, leads to significantly more positive stress management behaviors in patients compared to outcomes that arise in usual care, after 9 and 21 weeks.

Keywords: Conversational agent · Health behavior change · Meditation · Stress management · Complementary and alternative medicine

1 Introduction

Group medical visits, also known as "shared medical appointments", are an increasingly popular way of delivering medical care and behavioral interventions for chronic health conditions in a more economical way than one-on-one medical appointments [1]. In these visits, patients receive individualized care, but also participate in group health education and self-management instruction, as well as social time with other patients. Typically, 8–12 patients meet with one or two clinicians for 2 h on average. These visits are not only cost effective, but they allow patients to share stories, tips, and accomplishments with each other and to provide mutual social support to promote empowerment [2, 3]. Despite their advantages, patient adherence to recommendations for self-care tasks such as stress management techniques (mindfulness and meditation) are rarely followed exactly between group visits, especially over the long term. Automated systems may be able to bridge the time between group visits by reinforcing information delivered during visits, guiding patients through self-care practices and procedures,

P.W. de Vries et al. (Eds.): PERSUASIVE 2017, LNCS 10171, pp. 55–67, 2017.
DOI: 10.1007/978-3-319-55134-0_5

motivating adherence to the overall integrated intervention, and providing a source of social support.

"Chronic stress" is the chronic overactivity or ineffective managing of adaptive processes in people, and can predispose individuals to be more susceptible to acquiring new disease or exacerbate their existing diseases [4]. Standard treatments for chronic stress and associated symptoms include: medications, exercise, counseling, teaching coping skills. Increasingly, mindfulness-based approaches to stress management—including meditation and yoga—are being recommended to individuals who suffer from chronic stress.

Beginning in 2010, Boston Medical Center began an experimental program called Integrative Medicine Group Visits (IMGVs) to provide non-prescription self-management techniques and health services to patients who had been diagnosed with chronic pain. Individuals with chronic pain and depression typically have chronic stress as one of the primary symptoms they have to manage. IMGV contains 8–10 weekly clinical appointments where patients meet with doctors and health professionals, and are taught how to care for themselves. This program combines Mindfulness-Based Stress Reduction (MBSR) with patient education and integrative medicine therapies. Patients attend facilitated discussions about numerous health topics relevant to their care such as nutrition and physical activity in 2.5-hour visits. The visits also involve instruction in self-care management topics, spanning mindfulness-based stress reduction (meditation), and complementary non-medical pain therapies (exercise, self-massage, yoga [5]).

In this paper, we describe our work in designing, developing, and evaluating a conversational agent that assists patients with stress and chronic pain in conjunction with the weekly group visits. A conversational agent is a computer animated character that interact with users with simulated face-to-face conversation, including nonverbal behavior such as hand gestures and facial displays. Patients interact with the conversational agent at home during the intervention, and use it to review the material covered in each visit in more detail, as well in order to practice self-care skills such as meditation and yoga.

First, we will review the related work, and then describe the design of the conversational agent to help patients to better manage their stress. Next, we will present the results of a randomized clinical trial, before concluding.

2 Related Work

Numerous technologies have been developed to promote healthy behaviors such as physical activity, healthy eating and stress management. Here we briefly review related work in technologies for stress management, including prior research on conversational agents.

2.1 Technology Promoting Stress Management

A wide range of technologies, from internet-based programs to mobile applications and virtual reality, have been designed to promote stress management and related behaviors [6]. For example Gaggioli et al. developed a mobile platform to help users by tracking their stress, via heart rate and self-reports, and providing relaxing multimedia [7].

Several games have also been designed to promote stress management by providing stress-related biofeedback in an entertaining way [8]. A few interventions have also been developed for promoting mindfulness or relaxation, using techniques from complementary and alternative medicine. Gromala developed "The Virtual Walk" to teach users mindfulness-based stress reduction using VR [9]. RelaWorld is a nueroadaptive virtual reality meditation system that provides a relaxation experience by adapting the VR elements to the user's brain activity level [10]. Finally, Rector describes a system to provide yoga coaching to blind users [8].

2.2 Conversational Agents for Health Behavior Change

A growing body of research investigates the role of embodied conversational agents as virtual health coaches to engage users in persuasive and motivational interactions. Several agent-based systems have been developed in recent years to assist patients in managing their health problems in general, and to enable chronic disease self-care in particular [11]. Researchers have shown the positive effects of interaction with conversational agents on health behavior change. Agents developed in the past provide one-on-one counseling for patients and consumers for motivating exercise promotion [12], weight loss [13], breastfeeding [14], insomnia therapy [15], and preconception care [9], with generally positive results.

Bickmore et al. demonstrated that utilizing relational skills such as empathy and social dialogue in health behavior change interventions can significantly improve the users' experience and desire to continue over multiple interactions [16]. However, Schulman found that social dialogue actually made a health counseling conversational agent less persuasive [17].

Beun et al. developed a sleep-care e-coach mobile application that enhanced users' adherence for sleep behavior change by providing just in time, personalized feedback. They reported that increasing motivation plays a key role in coaching systems [2]. Shulman, et al., examined the effect of using motivational interviewing techniques in a conversation on users' self-efficacy [18].

Several conversational agents have also been developed specifically to help individuals manage chronic health conditions. Monkaresi, et al., developed the "IDL coach", an embodied conversational agent that helps individuals with diabetes manage prescribed exercise, nutrition, blood glucose monitoring, and medication adherence, although no evaluation studies are reported [19]. Bickmore, et al., developed an agent to help individuals with schizophrenia to manage their condition, to increase physical activity, and continue to take prescribed antipsychotic medication, and promising results from a quasi-experimental pilot evaluation have been reported [20]. ICT's SimCoach is an embodied conversational agent that was designed to address depression and/or post-traumatic stress disorder (PTDS), although results from a large randomized controlled trial with 333 patients failed to find any clinically significant benefits of using SimCoach [21].

2.3 Conversational Agents for Stress Management

Conversational agents have also been developed specifically to help users manage their stress. Hudlika developed a virtual coach that provides mindfulness meditation training and the coaching support necessary to begin a regular meditation practice [22]. Shamekhi describes a conversational agent that played the role of a meditation coach, using information from a respiration sensor to tailor its feedback to users [23].

Stress management education also has been shown to be more effective when a conversational agent is used. Results of a study by Jin indicates that the presence of a conversational agent causes significantly higher engagement and enjoyment in an stress management education system, and this engagement mediates learning outcomes [24].

All of these prior efforts seek to provide automated support for a single technique for stress management; none provide an integrated intervention that incorporates several best practices such as meditation, yoga, self-massage, and nutrition. In our intervention design, we seek to provide such an integrated intervention, with a conversational agent as a persistent, unifying persona across all of these activities.

3 Design of the Agent-Based Intervention for Stress Management

The conversational agent-based intervention was designed with monthly input from a patient advisory group (PAG) of patients, suffering from chronic pain and depression, who, at the time, were already participating IMGV at Boston Medical Center. The PAG provided input on how best to deliver and facilitate the IMGV, as well as the interaction features, intervention media, and dialogue content for the conversational agent; the PAG continued to provide guidance during the pilot studies and clinical trial whenever new challenges emerged.

3.1 The Conversational Agent "Gabby"

The conversational agent used in our work named "Gabby" was designed based on feedback from the PAG, and a pilot study [25], and focus group interviews. The agent is animated and rendered in the Unity game engine using custom animation software, running on a dedicated-use 8" Dell touch screen tablet computer (Fig. 1). The agent's appearance was designed to appear to be a racially ambiguous female in her mid-forties. The agent speaks using synthetic speech from a commercial Text to Speech (TTS) system with synchronized nonverbal conversational behavior. The agent's nonverbal conversational behavior is generated using BEAT [26], and includes beat hand gestures and eyebrow movements for emphasis, as well as a range of iconic, emblematic, and deictic gestures, alterations in the agent's gaze for signaling turn-taking, and posture shifts to mark topic boundaries, synchronized with speech. Additional media elements, including images and video clips, are also integrated into the agent's environment for pedagogical purposes. Dialogues are scripted using a custom hierarchical transition network-based scripting language. User input is obtained via multiple choice selection of utterances on the touch screen [27].

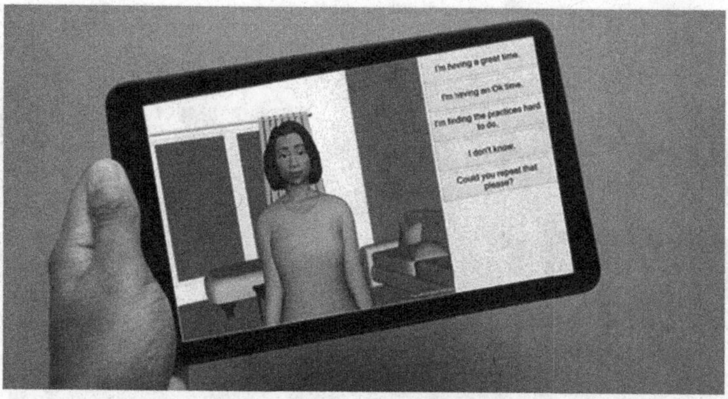

Fig. 1. Touch screen tablet computer with conversational agent

3.2 Intervention Design

The agent-based intervention was designed to guide users through nine weeks of new material along with the weekly group visits, followed by an arbitrary number of weeks of "maintenance", in which patients can continue to review material learned in the group visits. During this intervention, patients also have the option of completing practice activities with the agent (Fig. 2). The IMGV model utilizes principles of adult learning and engagement to allow for experience and knowledge sharing. This collective experience empowers and motivates patients to create an individualized treatment plan for their pain. This model aims to promote behavior change by providing a learning environment for the principles of principles of Mindfulness Based Stress Reduction, enriched with reinforcement. Each week, following a group visit, the agent reviews the new material just learned, walks patients through practice sessions, and allows them to review material covered in prior weeks. Home practice topics include meditation, self-massage, acupressure, and yoga. The agent also reviews educational information on nutrition, physical activity, pain, stress, sleep, and depression.

The first conversation with the agent starts with Gabby introducing herself and then summarizing the various services she can provide. Each subsequent conversation starts with a greeting, followed by two questions about how the participant is doing mentally (mood) and physically (how comfortable they are). They are then asked whether they have already reviewed the past topics or not, followed by a question about how their practices are going. Based on their responses, they are offered either to review past sessions or go through the new content in coordination with the material covered in the group that week. Gabby's talks are designed to be interactive and user-directable. The final coordinated session is allocated to behavioral goal setting, wherein patients can set new goals, review their goals or check their goals. The system is also designed to detect multiple logins per day. When a user returns to talk to Gabby in less than six hours, she skips the greeting and asks if the user wants to continue from where he/she left off.

In order to induce the feeling of a personal interaction, Gaby calls users by their given name, and asks questions such as, "How is your Wednesday going?". In addition,

Fig. 2. Screenshots of Gabby guiding a chair yoga (left) and a meditation session (right)

to avoid repetition in conversations, in several dialogue scripts, including greetings, farewells, and other dialogue that is frequently used, the system randomly selects agent utterances among several options for each interaction.

Mindfulness-Based Stress Reduction. Mindfulness Based Stress Reduction (MBSR) is a set of techniques that incorporate mindfulness to assist people with chronic pain to manage their self-care by practicing exercises such as mindfulness meditation, body awareness and mindful yoga [28]. The main curriculum consists of nine weekly group visits guided by a trained instructor, which the last session is more focused on mindfulness practice. MBSR has been proven to have positive effect on stress reduction [29], anxiety and depression [30].

In the intervention design we paid particular attention to the incorporating MBSR techniques into the system. The agent is designed to guide users through several meditation practices and a yoga session (Fig. 2). To increase the efficacy of these sessions guided by Gabby, we manipulated Gabby's voice and visual and acoustic background while she was leading the user through a meditation. Since mindfulness activities require participants to be focused and mindful, any distraction can disrupt the process. The instructor's voice is a potentially crucial element when people are trying to concentrate. Thus, we used speech synthesis markup language (SSML) to make Gabby's voice as natural and calming as possible. Calming music was also added as a background sound during these sessions to improve the experience. Based on feedback from the patient advisory group, we also decided to display a calming background picture while users are meditating. To generate visual variety, Gabby selects a random picture for each meditation session. from a set of 15 pictures. Gabby also guides a chair yoga session. All instructions were designed to be most effective for the target population with chronic pain.

4 Stress Reduction Intervention Clinical Trial

In order to evaluate the fully-developed system and examine how the conversational agent influences patient performance and satisfaction, a longitudinal between-subjects randomized clinical trial study was conducted. The patients in the control group received the usual care (non-IMGV) for chronic pain and depression. They attended regular meetings with their primary care physician, while those randomized to the intervention condition attended group medical visits for nine weeks, and were provided with Gabby on a tablet computer which they took home for 21 weeks (12 weeks after the group visits have ended), as well as with access to a website with review materials.

Group Visits Structure: Groups are led by a clinician facilitator and a co-facilitator, who use a facilitative leadership style as part of the group. Each session follows the same structure each week. In each visit new material about stress management, physical activity or nutrition are reviewed by the facilitators. Part of the visits are also allocated to the stress reduction exercises such as mindfulness meditation or yoga.

4.1 Participants

Study participants were recruited through flyers, placed in the family medicine and adult medicine departments of three study sites, and through Primary Care Providers (PCP), who referred patients for the study. Participants were required to be suffering from chronic pain and depression, 18 years of age or older, and English-speaking.

A total of 338 patients were screened across three sites, among whom 160 participants were enrolled. Six patients have withdrawn to date, resulting in 154 participants (75 in the intervention group, and 79 in the control group) enrolled. Participants are 22 to 84 years old (mean = 51 SD = 12.3), 86% female, 58% African-American, 29% white.

4.2 Measures

Besides sociodemographic we also assessed the stress management skill development by asking participants what they have done to relax or manage their stress in the past, at baseline, 9 weeks and 21 weeks. They were asked about the different types of stress management or relaxation techniques they used including both positive (e.g. deep breathing and walking), and negative behaviors (e.g. smoking and drinking alcohol). This question was presented as a 10-item checkbox that they could select as many items as they want. The participants' impression of the whole system and Gabby were assessed through a 17-item self-report measures. Participants were asked to fill a questionnaire on their perception of Gabby and her role in improving their health behaviors such as stress. The questionnaire contained four 7-point likert scale items on Gabby and stress, nine 7-point likert scale items on Gabby herself, a few open-ended question about users' experience with Gabby and a yes/no questions regarding whether they use Gabby's suggestion to manage their stress or not.

5 Results

The analysis of log files from tablet computers suggests that intervention participants interacted with Gabby an average of 105 min, with the average number of 8.8 logins, over the first 9 weeks they had the system.

Table 1. Stress management responses coding

Stress management behaviors	Positive behavior (Deep breath, exercise, listening to music, praying, spending time friends, shopping, walking)	Negative behaviour (Drinking alcohol, smoking cigarette, smoking marijuana)
Score	+1	−1

Table 2. Self-report rating of Gabby in helping users to manage their stress (N = 68), single sample Wilcoxon signed ranked test demonstrates ratings were significantly greater than neutral.

Question	Anchor 1	Anchor 7	Week 9	p-value	Week 21	p-value
How satisfied were you with talking to Gabby about reducing stress?	Not at all	Very satisfied	5.21(1.6)	<.001	5.55(1.5)	<.001
How helpful was Gabby in relieving your stress?	Not at all	Very satisfied	5.14(1.6)	<.001	5.25(1.7)	<.001
Do you think that you will use some of Gabby's suggestions to improve your health in the future?	Not at all	Very much	6.28(1.1)	<.001	6.08(1.4)	<.001
How confident do you feel that you can continue the changes you made based on Gabby's recommenda tions after this study is finished?	Not at all	Very much	5.76 (1.2)	<.001	5.52(1.3)	<.001

Table 3. Self-Report ratings of Gabby at 9 weeks and 21 weeks points (N = 68), single sample Wilcoxon signed ranked test demonstrates most of the ratings were significantly greater that neutral.

Question	Anchor 1	Anchor 7	Week 9	P value	Week 21	P value
How easy was it to talk to Gabby?	Very difficult	Very easy	5.66(1.6)	<.001	5.66(1.7)	<.001
How much do you trust Gabby?	Not at all	Very much	5.95(1.3)	<.001	5.66(1.5)	<.001
How well did Gabby answer any questions that you had?	Not at all	Very well	5.47(1.5)	<.001	5.54(1.5)	<.001
Did you feel like Gabby provided feedback and information that was specific to you?	Definitly no	Definitly yes	5.62(1.5)	<.001	5.42(1.8)	<.001
Did you feel like Gabby provided support and encourageme nt to reach your goals?	Definitly no	Definitly yes	5.89(1.3)	<.001	5.69(1.6)	<.001
How did you feel about the amount of time that you spent with Gabby?	Too little	Too much	3.86(1.6)	.475	3.66(1.6)	.075
Would you like to interact with Gabby again?	Not at all	Very much	6.15(1.5)	<.001	5.89(1.5)	<.001
How much would you have preferred talking to a doctor or nurse than to Gabby?	Definitely prefer doctor/ nurse	definitely prefer Gabby	4.00(1.9)	.975	3.85(2)	.377
How likely is it that you would recommend Gabby to someone you know?	Not at all	Very likely	6.17(1.3)	<.001	6.14(1.3)	<.001

Stress Management: We categorized the ten behaviors under two groups: positive behaviours, and high-risk negative behaviors. We coded the checked positive behavior as +1 and checked negative behavior as –1 (Table 1).

Adding up the coded scores, we came up with three total scores for each participant, at baseline, 9 week, and 21 week assessment.

In order to compare the improvements, we calculated the changes in stress management scores after 9 weeks and 21 weeks (21week_change = 21w_score – baseline_score, 9week_change = 9w_score – baseline_score). Results of a T-test for independent samples showed that participants who used Gabby and the OWL website along the group visits (M = 1.1 SD = 1.6) had a significantly higher improvement than control group (M = 0.09 SD = 1.58) in managing their stress after 9 weeks, $t(136) = 3.74$, $p < 0.0001$. There is also a significant difference in the score changes for intervention (M = 1.22, SD = 1.86) and control (M = 0.29 SD = 1.34) after 21 weeks $t(136) = 3.42$, $p = 0.001$ (Fig. 3).

Gabby Evaluation: 81% of participants declared that they used at least one of Gabby's suggestion to reduce their stress. Tables 2 and 3 show the average responses to each item by 68 participants.

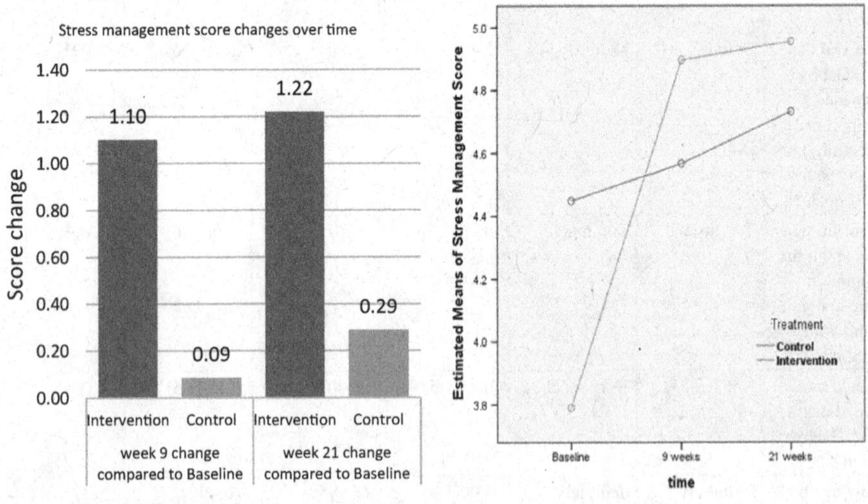

Fig. 3. Results of a t-test on independent samples to compare the control and intervention (left) and the changes of both groups over time (right)

Qualitative Results: Overall participants found Gabby as a very useful complement for the material covered during group visits. Many of them stated that Gabby provided the opportunity to review the class lessons and more detailed information at any time, and anywhere they wanted, without limitation. They felt that they had more time to review and understand the lessons with Gabby than they did during the group sessions. Here we report some comments from participants in regards with the stress management capabilities of Gabby. Most of the participants found the meditation, yoga and mindful sessions very "*useful*" and "*relaxing*". Many participants found Gabby very effective in motivating them to practice the mindfulness exercises. For example, P1 said that "*I was able to incorporate meditation in my life which is extremely important to me as I age. Before Gabby I felt like my heart was racing all the time with relationship issues. Just everything. It really calmed me down.*". P55 added "*Gabby got me motivated. The exercise the deep breathing, she would tell me to calm my nerves.*" Participants also found themselves "much calmer" and could "take life with less stress" than they used to, by practicing mindfulness with Gabby. Particularly P41 mentioned that "*Gabby improved it in terms of stress because of her calming voice and it was something out of his norm. Spoke nice and easy and allowed him to take the time*".

Participants were also asked about the difficulties of interacting with Gabby and what they would like to improve in the interaction. Consistent with the self-report data some participants reported that they did not have enough time to interact with Gabby. A few participants also said that they would like to be able to ask a "specific question" and "go beyond what was available". Some of them also reported facing difficulties using the program: "*hard to log onto because not as computer savvy*". We believe that this kind of issue might give rise to stress that could also negatively impact their satisfaction with Gabby.

6 Conclusion

We have demonstrated that a home-based conversational agent can be effective, when used in conjunction with medical group visits, in promoting stress management techniques over a 21-week period of time. The agent was well-accepted by patients, and appears to have helped reinforce the material taught during the group visits. We have demonstrated this with a cohort of patients with chronic pain and depression from a safety net hospital, where most patients are low income, disadvantaged minorities. Previous studies have found low levels of computer literacy in these populations [31], indicating that conversational agents are a good delivery medium for disadvantaged populations. Our study indicates that although individuals who have little experience with technology had some difficulties using the tablets, they find the conversational agent an acceptable and effective medium for receiving healthcare counseling and information.

Limitations of our study include the relatively small sample from health centers in the Boston area. We also are unable to disentangle the effects of the face to face weekly group visits from those of the home-based conversational agent, and plan future studies to tease these apart.

6.1 Future Work

Our future work includes conducting a longitudinal study with a control group attending the group visits without any technology and the intervention group using technology such as conversational agents, and compare the changes in outcomes to investigate the specific role of technology in such an intervention. We are also exploring the use of a variety of sensors to assist Gabby in helping patients through self-care practices, such as respiration sensors to help with meditation [23], and Kinect to provide more tailored yoga coaching. Finally, we are exploring ways in which the agent can more interactively demonstrate skills, such as yoga positions.

Acknowledgement. Research reported in this paper was partially funded through a Patient-Centered Outcomes Research Institute (PCORI) Award (AD-1304-6218).

References

1. Brennan, J., Hwang, D., Phelps, K.: Group visits and chronic disease management in adults: a review. Am. J. Lifestyle Med. **5**, 69–84 (2011)
2. Jaber, R., Braksmajer, A., Trilling, J.S.: Group visits: a qualitative review of current research. J. Am. Board Fam. Med. **19**, 276–290 (2006)
3. Trotter, K.J.: The promise of group medical visits. Nurse Pract. **38**, 48–53 (2013)
4. Chiesa, A., Serretti, A.: Mindfulness-based stress reduction for stress management in healthy people: a review and meta-analysis. J. Altern. Complement. Med. **15**, 593–600 (2009)
5. Gardiner, P., Dresner, D., Barnett, K.G., Sadikova, E., Saper, R.: Medical group visits: a feasibility study to manage patients with chronic pain in an underserved urban clinic. Glob. Adv. Health Med. **3**, 20–26 (2014)
6. Newman, M.G., Szkodny, L.E., Llera, S.J., Przeworski, A.: A review of technology-assisted self-help and minimal contact therapies for anxiety and depression: is human contact necessary for therapeutic efficacy? Clin. Psychol. Rev. **31**, 89–103 (2011)
7. Gaggioli, A., Cipresso, P., Serino, S., Campanaro, D.M., Pallavicini, F., Wiederhold, B.K., Riva, G.: Positive technology: a free mobile platform for the self-management of psychological stress. Stud. Health Tech. Inform. **199**, 25–29 (2014)
8. Al Osman, H., Dong, H., El Saddik, A.: Ubiquitous biofeedback serious game for stress management. IEEE Access. **4**, 1274–1286 (2016)
9. Gromala, D., Tong, X., Choo, A., Karamnejad, M., Shaw, C.D.: The virtual meditative walk: virtual reality therapy for chronic pain management. In: Proceedings of the 33rd Annual ACM Conference on Human Factors in Computing Systems, pp. 521–524. ACM (2015)
10. Kosunen, I., Salminen, M., Järvelä, S., Ruonala, A., Ravaja, N., Jacucci, G.: RelaWorld: neuroadaptive and immersive virtual reality meditation system. In: Presented at the 21st International Conference on Intelligent User Interfaces. ACM (2016)
11. Bickmore, T., Giorgino, T.: Health dialog systems for patients and consumers. J. Biomed. Inform. **39**, 556–571 (2006)
12. Bickmore, T.W., Silliman, R.A., Nelson, K., Cheng, D.M., Winter, M., Henault, L., Paasche-Orlow, M.K.: A randomized controlled trial of an automated exercise coach for older adults. J. Am. Geriatr. Soc. **61**, 1676–1683 (2013)
13. Watson, A., Bickmore, T., Cange, A., Kulshreshtha, A., Kvedar, J.: An internet-based virtual coach to promote physical activity adherence in overweight adults: randomized controlled trial. J. Med. Internet Res. **14**, e1 (2012)
14. Edwards, R.A., Bickmore, T., Jenkins, L., Foley, M., Manjourides, J.: Use of an interactive computer agent to support breastfeeding. Matern. Child Health J. **17**, 1961–1968 (2013)
15. Horsch, C., Brinkman, W.-P., van Eijk, R., Neerincx, M.A.: Towards the usage of persuasive strategies in a virtual sleep coach. In: Proceedings of UKHCI 2012 Workshop on People, Computers and Psychotherapy (2012)
16. Bickmore, T., Gruber, A., Picard, R.: Establishing the computer–patient working alliance in automated health behavior change interventions. Patient Educ. Couns. **59**, 21–30 (2005)

17. Schulman, D., Bickmore, T.: Persuading users through counseling dialogue with a conversational agent. In: Proceedings of the 4th International Conference on Persuasive Technology, pp. 25. ACM (2009)
18. Bickmore, T., Schulman, D.: An Intelligent Conversational Agent for Promoting Long-Term Health Behavior Change Using Motivational Interviewing. In: AAAI Spring Symposium: AI and Health Communication (2011)
19. Monkaresi, H., Calvo, R., Pardo, A., Chow, K., Mullan, B., Lam, M., Twigg, S.M., Cook, D.I.: Intelligent diabetes lifestyle coach. In: OzCHI, Adelaide, Australia (2013)
20. Bickmore, T., Puskar, K., Schlenk, E., Pfeifer, L., Sereika, S.: Maintaining reality: relational agents for antipsychotic medication adherence. Interact. Comput. **22**, 276–288 (2010)
21. Meeker, D.: SimCoach evaluation: a virtual human intervention to encourage service-member help-seeking for posttraumatic stress disorder and depression. Rand Corporation, Santa Monica (2015)
22. Hudlicka, E.: Virtual training and coaching of health behavior: example from mindfulness meditation training. Patient Educ. Couns. **92**, 160–166 (2013)
23. Shamekhi, A., Bickmore, T.: Breathe with me: a virtual meditation coach. In: Brinkman, W.-P., Broekens, J., Heylen, D. (eds.) Intelligent Virtual Agents, pp. 279–282. Springer International Publishing, Cham (2015)
24. Jin, S.-A.A.: The effects of incorporating a virtual agent in a computer-aided test designed for stress management education: the mediating role of enjoyment. Comput. Hum. Behav. **26**, 443–451 (2010)
25. Shamekhi, A., Bickmore, T., Lestoquoy, A., Negash, L., Gardiner, P.: Blissful agents: adjuncts to group medical visits for chronic pain and depression. In: Traum, D., Swartout, W., Khooshabeh, P., Kopp, S., Scherer, S., Leuski, A. (eds.) IVA 2016. LNCS (LNAI), vol. 10011, pp. 433–437. Springer, Heidelberg (2016). doi:10.1007/978-3-319-47665-0_49
26. Cassell, J., Vilhjálmsson, H.H., Bickmore, T.: BEAT: the Behavior Expression Animation Toolkit. In: Prendinger, H., Ishizuka, M. (eds.) Life-Like Characters, pp. 163–185. Springer, Heidelberg (2004)
27. Bickmore, T.W., Picard, R.W.: Establishing and maintaining long-term human-computer relationships. ACM Trans. Comput. Hum. Interact. **12**, 293–327 (2005)
28. Kabat-Zinn, J.: Full Catastrophe Living: Using the Wisdom of Your Body and Mind to Face Stress, Pain, and Illness. Bantam Books trade paperback, New York (2013)
29. Sharma, M., Rush, S.E.: Mindfulness-based stress reduction as a stress management intervention for healthy individuals: a systematic review. J. Evid. Based Complement. Altern. Med. **19**, 271–286 (2014)
30. Hofmann, S.G., Sawyer, A.T., Witt, A.A., Oh, D.: The effect of mindfulness-based therapy on anxiety and depression: a meta-analytic review. J. Consult. Clin. Psychol. **78**, 169–183 (2010)
31. Bickmore, T.W., Pfeifer, L.M., Byron, D., Forsythe, S., Henault, L.E., Jack, B.W., Silliman, R., Paasche-Orlow, M.K.: Usability of conversational agents by patients with inadequate health literacy: evidence from two clinical trials. J. Health Commun. **15**, 197–210 (2010)

Letters to Medical Devices: A Case Study on the Medical Device User Requirements of Female Adolescents and Young Adults with Type 1 Diabetes

Gillian M. McCarthy[1(✉)], Edgar R. Rodríguez Ramírez[1], and Brian J. Robinson[2]

[1] School of Design, Victoria University of Wellington, Wellington, New Zealand
{gillian.mccarthy,edgar.rodriguez-ramirez}@vuw.ac.nz

[2] Graduate School of Nursing, Midwifery and Health,
Victoria University of Wellington, Wellington, New Zealand
brian.robinson@vuw.ac.nz

Abstract. Adolescents and young adults with type 1 diabetes are required to use a variety of persuasive medical technologies to manage their health. However, adolescents' experiences with and preferences regarding these technologies, and the implications these have on self-management are not broadly recognised. In this case study six female adolescents and young adults wrote love letters or break-up letters to one of their medical devices. Four categories were constructed from a grounded theory analysis of the letters and follow-up interviews: acquiring and changing medical devices, requiring convenience and practicality for everyday contexts, collecting and using data, and corresponding with preferences and values. Young people are often excluded from research and development regarding medical devices, yet this method was successful in identifying experiences and preferences to inform the design of medical devices.

Keywords: Letter · Type 1 diabetes · Cultural probes · Adolescence · Medical technology

1 Introduction

The self-management regimen of young people with type 1 diabetes is intensive, including numerous blood glucose tests, monitoring and planning diet and exercise, and calculating and self-administering insulin [1]. Resultantly, engagement among adolescents with diabetes is low, with nearly two-thirds of adolescents failing to meet targets for glycaemic control, a concern for long-term health outcomes [2–4].

Adolescents' barriers to self-managing type 1 diabetes include fears of negative reactions from peers, pressure to fit in socially, developing independence, school issues and affective disorders [5–8]. However, there has been little research into how medical devices could be designed to facilitate self-management practices. Type one diabetes is managed largely through the use of medical technologies such as insulin pens, insulin pumps and blood glucose monitors. These devices can be considered persuasive technologies. The behaviours targeted with these devices span many of Fogg's behaviour

P.W. de Vries et al. (Eds.): PERSUASIVE 2017, LNCS 10171, pp. 69–79, 2017.
DOI: 10.1007/978-3-319-55134-0_6

types, with some individuals learning new behaviours, others striving to increase the frequency of self-care behaviours, and others maintaining behaviours [9].

These medical technologies occupy a space bridging clinical and consumer products and services. While devices must be clinically effective, they must also meet users' needs, so that they will be willing and able to use and purchase equipment. Designers of medical devices ought to consider users' contexts, motivations and preferences [10]. If adolescents' priorities are in conflict with the required treatment regimen, their engagement may be affected and they can become disengaged from effective self-management [11]. As such, it is pertinent to explore the effects of these persuasive medical technologies on self-management.

Previous research has described psychosocial barriers to insulin pumps and blood glucose meters, and explored adolescents' medical device compatibility and expectations [12–14]. However, there is very little research documented in this area from a design perspective. Designers need to better understand how to elicit adolescents' user requirements in a manner that informs the design of medical technologies.

Adolescence is characterised by biological, cognitive, social and emotional development within a changing context of family, peers, school and such [15]. This normative development results in adolescents valuing social norms and other people's perceptions of them, having difficulty anticipating the consequences of their behaviours, having feelings of invincibility, and taking risks [16–18]. Although adolescents have specific requirements of medical devices, their preferences can go unrecognised by researchers, with adolescents often using technology designed for children or for adults, resulting in decreased engagement and poorer health outcomes [19, 20]. Ethical restrictions around research practices with this group are viewed as a barrier [21], and commonly proxies such as parents, clinicians, healthy adolescents and device designers are used in the development and validation of new medical technologies [22]. Involving young people in the design of interactive technologies is an import requirement for patient-centred design [23]. Medical technology designed for adolescents' unfulfilled needs may also be attractive to both children and adults.

Previous studies have developed and employed user requirements to inform the design of medical and assistive technologies for a range of users with various health conditions [24–26]. The four-layer Needs and Aspirations of Design (NADI) model builds on this approach. The model classifies insights at four levels: solutions - the products people want, scenarios - the interactions people want, goals - why people want these interactions, and themes - the values that drive people's goals [27]. The deeper levels such as themes and goals are helpful for reframing design problems and opening up the solution space, while scenarios and solutions are helpful for incremental innovation and refining solutions [28].

The aim of this case study was to trial a letter writing method for eliciting female adolescents' wide-ranging user requirements for the self-management of type 1 diabetes. This method has previously not been documented or used to inform the design of persuasive medical technologies.

2 Methodology

Participants were six females aged 13 to 24 years who had been diagnosed with type 1 diabetes for between 6 months to 15 years. They were given a cultural probes pack to complete at home containing instructions, an activity card, stationery, and an envelope for returning the activity once completed. The activity card is pictured (Fig. 1). Adapted from a Smart Design group workshop method, participants were requested to write either a love letter or a break-up letter to diabetes technology [29]. The instructions read, "Ever had a medical device for diabetes that you love, or maybe one you hate? Your task is to choose the medical item that you feel the most strongly about, and write it a love letter or break-up letter. You could choose a medical device or any other piece of medical equipment (e.g., Accessory or phone app). You could choose a device you currently use, something you used to use, or something you would like to use one day. Tell the story of how you came to use it, explain why you love it, why you hate it, your expectations, what went wrong…"

Fig. 1. Activity card for letter-writing method.

Participants self-selected a device, as previous research points towards the existence of adolescent user requirements that can be generalised across devices and medical conditions [30]. This aligned with the study's aim to find broad adolescent user requirements, and rich description of experience that other cultural probes methods can struggle to achieve [31].

The letter-writing cultural probe was intended as a prompt for further discussion, and followed up through an interview to delve for further information and insights, and to confirm that the researcher's comprehension and interpretation of the cultural probe activity aligned with the participant's [32]. The letters and interviews were analysed together using a constructivist approach to grounded theory [33]. This included stages of open coding, focused coding, memo writing and theory construction.

3 Results

Seven letters were written by the six participants, with one participant writing both a love letter and a break-up letter. Four break-up letters were written to blood glucose meters and one to a syringe holder. One love letter was written to an insulin pump, and one to a glucose monitoring system. A 13-year-old participant had a parent present during the interview. During the interviews some participants talked about multiple devices, introducing more comparison within the information gathered. Three letters are included in full below with original spelling and grammar.

Dear, My very problematic CareSens™ N.
Though you are a very permanent fixture in my life - often meeting 4 to six times a day, I wish I could let you go. Even though you are the cheapest way of monitoring my blood glucose we are only together because the government decided you were the best option. I don't ask much from you - just correct readings and that you stop deciding to pack it in if you get too hot or cold. Your predecessor was reliable, sleek and always functional, but alas you do not give me the same curtosy. You make me second guess myself and my health which I don't appreciate. Having an ongoing medical condition is hard enough without faulty devices and constant questions marks. Sort it out and then we can talk. You're a bit outdated.
P.S. If only you were like the devices that didn't require pricks.

Dear insulin pump,
I love that you came into my life! You give me flexibility, confidence and happiness. Before you I did not realise how much I felt weighed down & suffocated by the weight of Type One Diabetes. I love you because you enabled me to experience in many areas of my life. I felt the freedom and confidence to travel, to eat when and what I want, to play, run and jump. To step forward in life with greater independence. Although some days are frustrating and you seem to endlessly test my patience, overall I love what you have brought into my life. Whilst having type one diabetes is no "walk in the park", (and I would love a cure) I appreciate your presence in my life. I love how you tell me how much insulin I have onboard, can match my food to bolus and are discreet. I like no longer being restricted by 24 h insulin, and can instantly drop or raise my basal rates. I am more confident to try new things, to be more active, and I love the support you offer me when I am unwell. Thanks pump, here's to a long and healthy future! (Unless a cure comes along!)
Love [Name] x

Rocket - syringe holder to inject insulin
You came into my life at such a low point and I counted on you for support. I relied on you to stay healthy and happy, where most days you did the complete opposite. I started using you when I was first diagnosed, to ease the pain when giving my injections. At first I was relieved, as I didn't have to physically inject myself, I could simply push a button. Little did I know the noise of that button would continue to cause me grief, even up until today. Every meal time I felt terrified at the thought of it, blasting my headphones to my favourite songs through my walkman just to avoid that noise. The pain wasn't any easier, as you hurt me over and over with the bounce back spring that gave me the darkest bruises I've had from injecting. I'm glad I only had you in my life for a short period of time. Even looking at you a few years back brought tears to my eyes with the horrible memories flooding back.

Four categories were constructed in a combined grounded theory analysis of the letters and interviews: acquiring and changing medical devices, requiring convenience and practicality for everyday contexts, collecting and using data, and corresponding with preferences and values. The analysis covers both participants' prior experiences with medical devices and desires for how medical devices could be improved.

3.1 Acquiring and Changing Medical Devices

Two participants discussed wanting a glucose monitoring system that they were aware of because they had friends currently using or trialling the device. However, the cost of the device, which is not currently subsidised[1], was the primary barrier to use. They described awareness but inability to access the device as "annoying". "I don't have the Freestyle Libre but I would really like to get one. It's so easy to scan and get your level without having to prick your finger. My friend has one and she can get her level before I even put the test strip in my meter." Many of the participants had been affected by a change of Government subsidies for blood glucose monitoring. Although the meters previously used by these participants were still available, the ongoing costs of consumables were no longer subsidised and now unaffordable for some. This forced change, to a meter some felt was less accurate and reliable, was frustrating for participants. "We used to have a different machine but it always worked, readings were good, always worked when hot and cold, wasn't temperamental. But then the Government was like we'll swap to this one because it's cheaper. You can still get the other one but it's insane, it costs like $50 for a thing of strips that would last you a week… That's a quarter of my rent. It's a constant grievance." Another form of forced change was described by one participant whose blood glucose meter was broken when a vehicle reversed over it. Her meter was replaced with a different model that she did not like as much.

[1] Pharmac, the New Zealand Government drug purchasing and funding agency who choose which medical devices and consumables are subsidised moved to a single supplier arrangement for blood glucose monitors and test strips in 2013.

In other instances participants stopped using a medical device as they had a choice to change devices. One participant described devices she used initially after diagnosis so that she did not have to see the syringes or insulin pen needles, and so the injection was carried out by pushing a button. While she found the device paired with the insulin pen very helpful, a similar device that paired with a syringe was one described in a break-up letter above. She objected to the pain caused by the device, and the noise it made, using loud music to attempt to cover the noise of the button. "I didn't really like anything about it. It was quite horrible, just big and clunky. I hated the noise, it had a bit of a spring back when you chucked the needle in… I couldn't bring myself to actually inject myself, just physically push the needle in, so I mean it was good, don't get me wrong… I just cringe from it still." This example demonstrates the significance of considering the multisensory experience of using medical devices.

While one participant wrote about her reluctance at the permanence of a medical device in her life, "Though you are a very permanent fixture in my life - often meeting 4 to six times a day, I wish I could let you go", another described her wish for a device to have a continued presence in her life if necessary "Here's to a long and healthy future! (Unless a cure comes along!)."

3.2 Requiring Convenience and Practicality for Everyday Contexts

The functions available on medical devices affected how they were used in everyday contexts. One participant described a blood glucose meter that did not make any beeping noises or have a light, which while discreet in some situations, made it hard to use at night without turning on the room lights, and caused frustration if sleeping in a shared room with her peers at sporting events. While she liked the aesthetics of the device, it didn't have the appropriate features for night-time use. "It was real flat and really fancy, but I think that they went too far and forgot the basic things." One lancing device was described positively because it was rapid to use. It required only a one button push to deploy, in comparison to others that require the lancet to be cocked first. Helpful features of an insulin pump included showing how much insulin remained in the pump, and matching an insulin bolus to food that was about to be eaten.

There were mixed feelings about insulin pens versus insulin pumps. Pens do not require constant invasive attachment but must be personally carried and take time to use. Insulin pumps do not require multiple needle insertions and are easy to use in many situations (e.g., in a moving vehicle), but require backup equipment to be carried. In describing a glucose monitoring system, one participant described not having to prick her fingers as often, and the speed of using the device as possible benefits.

Participants described the inconvenience of carrying medical devices and consumables with them. Insulin pumps and blood glucose meters were described as "chunky" and "uncomfortable". While there was a desire for these devices to be smaller or more streamlined, the screen size needed to be legible. One participant wore her insulin pump in her bra, while others using insulin pens used a small bag to store devices and consumables. One participant suggested that keeping the many small devices required for self-management contained within a small bag would be easier than loose within a larger bag. She thought this would make them easier to find and reassure her that she had everything required when leaving home, as in the past she has accidentally left devices at home when transferring them between bags.

One participant described how she does not always take her medical devices with her, particularly if she plans to return home during the day. She spoke about being reminded by a health professional and her mother that there may be emergency situations where she would need to access the devices. While she acknowledged the importance of this, and had coincidentally been "fine" in the past, describing one incident where she was unable to get to her home 50 km away from her current location, she found carrying all the devices with her inconvenient. "One time there was an earthquake and I was stuck in town and couldn't get to Kapiti, luckily I had stuff on me." This behaviour typifies adolescent propensity for risk taking, one of the many features of normative adolescent development that interacts with managing a long-term condition. It reflects adolescent prioritising of short-term benefits over possible consequences with a low-likelihood. Fear of a medical device breaking was also discussed, and one participant chose to use disposable insulin pens, rather than reusable ones while travelling in case of breakages, reasoning that she would still have a stockpile of pens if anything went wrong with one.

3.3 Collecting and Using Data

The reliability and accuracy of blood glucose meters was discussed frequently as a major frustration. Many participants described not trusting the accuracy of the blood glucose readings they were required to take to make decisions about management such as how much insulin to inject. One participant wrote, "The amount of inaccurate numbers you've had, makes you second guess every single number, and so that really does play on your mind and you end up being a little bit more conservative with your corrections or your actions so then you run the risk of still being too high or too low." Another wrote "It scares me with how wrong you can be." Participants reported that their confidence in the meter's accuracy eroded over time, and that mistrust of the reading led them to second-guess both the accuracy of the device, and of their own bodily feelings, making decision-making more difficult.

When faced with a reading that did not match the way she felt one participant would use a spare meter, then calculate insulin dosage based on which reading best represented how she currently felt. Duplicate measurements could differ by 2–3 mmol/L which can be clinically significant as the target blood glucose range is approximately 4–7 mmol/L before meals and 5–10 mmol/L after meals [34].

Another participant hypothesised that the inaccuracy of her blood glucose meter stemmed from not washing her hands prior to taking a blood glucose measurement. "The old Optium meters were fine. I used to use hand sanitiser all the time, but with this one you can't… I live on a farm when I go home. So if you're out all day on the bike or on the farm where's your soap and water? I think it's stupid, you know we don't live in these perfect worlds where you have a bathroom nearby all the time."

Some participants described a desire for data to be transformed into meaningful information. One wanted a glucose monitoring system that would regularly collect and visualise her blood glucose levels. She felt being able to see trends would help her and her parents better understand how her blood glucose levels fluctuate overnight. She had seen data visualised this way from continuous glucose monitor images on Facebook and wanted access to that technology. Another participant felt features of her blood glucose meter may be more helpful to people recently diagnosed with diabetes who did not have an insulin pump. "It would have a little arrow if you were too high, and a little down arrow if you were too low, and you could log your meals on it." This illustrates the importance of not only supplying data, but also transforming it into meaningful information.

Participants also discussed sharing blood glucose monitoring data with parents and health professionals. One participant thought a benefit of a glucose monitoring system that didn't require finger pricks would be allowing her parents to easily check her blood glucose concentrations if they were concerned during the night, relieving their distress. While another participant wanted to share a detailed log of her blood glucose data with a health professional to develop more insight into her own health, her blood glucose meter stopped providing data for a day because it wasn't within the required functioning temperature range. This participant attempted to warm the meter by putting it under her blanket and on her cat. While she was reasonably confident that she could tell without the blood glucose meter whether she was within the target range, she was also frustrated there was an entire day of her log data missing, and how the health professional would react to this missing data. "She'll probably tell me off but it's not my fault. I actually put a little disclaimer in my spreadsheet like 'um, did intend to do it this day, didn't actually work, sorry'."

3.4 Corresponding with Preferences and Values

As medical devices for type 1 diabetes are used in many contexts, their appearances and discreetness were described as important factors. One participant valued her blood glucose meter not making any noise and drawing attention in particular contexts, while another described how her insulin pump was far more discreet to use in public than injecting insulin using a pen. Multiple participants voiced a preference for their medical devices to appear visually more like consumer devices, with preferences for sleek and slim designs. A participant described the challenge of finding medical technology that suited her aesthetic preferences when moving into young adulthood "I feel it's hard the older you get to find things that don't look like they were made for children… it doesn't necessarily have to be discreet but as long as it just kind of blends in with everything else."

In addition to aesthetic preferences, it was important for medical devices to correspond with people's broader values. For instance, one participant described wanting to replace her disposable insulin pen with a reusable one to be more environmentally sustainable, and spoke about how issues around waste and recycling were often raised by her flatmate. While this participant preferred to have disposable insulin pens for travelling, her priorities and corresponding device preference had now changed.

When a medical technology is well matched to the user's needs, it can facilitate positive feelings and life experiences. Writing about her insulin pump, one participant wrote, "You give me flexibility, confidence and happiness… I love you because you enabled me to experience in many areas of my life." Accordingly, this participant saw her device as not only administering insulin, but allowing her the freedom to live her life unhampered by diabetes.

4 Discussion

This case study explored young people's experiences and preferences using medical devices to self-manage type 1 diabetes. While the methods used in this study identified user requirements, the number of participants was small and comprised only of females, necessitating the need to conduct further research with male and female participants to check the effectiveness of the method and whether the requirements can be generalised. There was a concern by the researchers that writing a break-up letter to a device that participants were required to use could be distressing. To alleviate this, participants were free to choose whether to write a love letter or a break-up letter to a device of their choice. During the follow-up interview some participants reported enjoying the activity and finding it cathartic.

The combination of writing love letters or break-up letters and a follow-up interview revealed information about adolescents' and young adults' experiences with medical devices and their expectations for these devices to function allowing them to effectively self-manage their long-term condition. The information gathered ranged from very specific (e.g., button noises) to broader feelings (e.g., confidence). The letter writing alone resulted in insights at all four levels of the NADI model. For blood glucose monitoring, a solution was a meter that did not require finger pricks, a scenario was the ability to function reliably at a range of temperatures, an important goal was to give users confidence to make diabetes-related decisions, and a broad theme was managing blood glucose effectively. As discussed previously, themes and goals, are helpful for reframing design problems and opening up the solution space, while scenarios and solutions are helpful for incremental innovation and refining solutions. We suggest that having participants write letters to a range of medical devices also contributed to the development of themes and goals, whereas for research focused more on incremental innovation, focusing on a single device may be preferable.

Designers and suppliers of medical devices could apply this combination of methods to gain insight into the user requirements of people with health conditions when used in relation to devices with which users have significant prior experience.

5 Conclusion

This paper trialled letter writing to explore medical device experiences and requirements. Resultant user requirements were arranged into four categories: acquiring and changing medical devices, requiring convenience and practicality for everyday contexts, collecting and using data, and corresponding with preferences and values. While the persuasive medical technologies described are intended to improve self-management of type 1 diabetes, using these research methods has uncovered their unintended effects and elicited users' experiences and expectations of these devices.

References

1. Strawhacker, M.: Multidisciplinary teaming to promote effective management of type 1 diabetes for adolescents. J. School Health **71**(6), 213–217 (2001)
2. Hilliard, M., et al.: Predictors of deteriorations in diabetes management and control in adolescents with type 1 diabetes. J. Adolesc. Health **52**(1), 28–34 (2013)
3. Diabetes Control and Complications Trial Research Group: Effect of intensive diabetes treatment on the development and progression of long-term complications in adolescents with insulin-dependent diabetes mellitus: diabetes control and complications trial. J. Pediatr. **125**(2), 177–188 (1994)
4. Cameron, F.: Teenagers with diabetes: management challenges. Aust. Fam. Physician **35**(6), 386–390 (2006)
5. Borus, J.S., Laffel, L.: Adherence challenges in the management of type 1 diabetes in adolescents: prevention and intervention. Curr. Opin. Pediatr. **22**(4), 405–411 (2010)
6. Pyatak, E.A., Florindez, D., Weigensberg, M.J.: Adherence decision making in the everyday lives of emerging adults with type 1 diabetes. Patient Prefer. Adherence **7**, 709–713 (2013)
7. Croom, A., et al.: Adolescent and parent perceptions of patient-centered communication while managing type 1 diabetes. J. Pediatr. Psychol. **36**(2), 206–215 (2010)
8. Wysocki, T., Greco, P.: Social support and diabetes management in childhood and adolescence: influence of parents and friends. Curr. Diab. Rep. **6**(2), 117–122 (2006)
9. Fogg, B.J., Hreha, J.: Behavior wizard: a method for matching target behaviors with solutions. In: Ploug, T., Hasle, P., Oinas-Kukkonen, H. (eds.) PERSUASIVE 2010. LNCS, vol. 6137, pp. 117–131. Springer, Heidelberg (2010). doi:10.1007/978-3-642-13226-1_13
10. Tatara, N.: Designing mobile patient-centric self-help terminals for people with diabetes. In: 11th International Conference on Human-Computer Interaction with Mobile Devices and Services, p. 97. ACM (2009)
11. Suris, J., et al.: The adolescent with a chronic condition. Part I: developmental issues. Arch. Dis. Child. **89**(10), 938–942 (2004)
12. Seereiner, S., et al.: Attitudes towards insulin pump therapy among adolescents and young people. Diab. Technol. Ther. **12**(1), 84–89 (2010)
13. Carroll, A.E., Downs, S.M., Marrero, D.G.: What adolescents with type I diabetes and their parents want from testing technology. CIN Comput. Inf. Nurs. **25**(1), 23–29 (2007)
14. McCarthy, G.M., Rodríguez Ramírez E.R., Robinson, B.J.: Dissonant Technologies: Health Professionals' Impressions Of Adolescents' Interactions with Medical Technologies For Managing Type 1 Diabetes. Wellbeing. 26–39. Birmingham City University, Birmingham (2016)
15. Warner, D.E., Hauser, S.T.: Unique considerations when treating adolescents with chronic illness. In: O'Donohue, W.T. (ed.) Behavioral Approaches to Chronic Disease in Adolescence, pp. 15–28. Springer, New York (2009)
16. Elkind, D.: All Grown Up & No Place to Go. Addison-Wesley, Reading (1984)
17. Irwin, C.E.: Adolescent health at the crossroads: where do we go from here? J. Adolesc. Health **33**(1), 51–56 (2003)
18. Millstein, S.G., Halpern-Felsher, B.L.: Perceptions of risk and vulnerability. J. Adolesc. Health **31**(1), 10–27 (2002)
19. Carter, B.: Tick box for child? The ethical positioning of children as vulnerable, researchers as barbarians and reviewers as overly cautious. Int. J. Nurs. Stud. **46**(6), 858–864 (2009)
20. Lang, A., et al.: The effect of design on the usability and real world effectiveness of medical devices: a case study with adolescent users. Appl. Ergonomics **44**(5), 799–810 (2013)

21. Money, A., et al.: The role of the user within the medical device design and development process: medical device manufacturers' perspectives. BMC Med. Inf. Decis. Making **11**(1), 15 (2011)
22. Lang, A., et al.: Not a minor problem: involving adolescents in medical device design research. Theor. Issues Ergonomics Sci. **15**(2), 181–192 (2012)
23. van der Velden, M., Culén, A.: Designing privacy with teenage patients: methodological challenges. In: Workshop on Designing for and with Teenagers, CHI 2013. ACM, Paris (2013)
24. Owen, T., Buchanan, G., Thimbleby, H.: Understanding user requirements in take-home diabetes management technologies. In: Proceedings of the 26th Annual BCS Interaction Specialist Group Conference on People and Computers, pp. 268–273. British Computer Society, Birmingham (2012)
25. Milewski, J., Parra, H.: Gathering requirements for a personal health management system. In: Proceedings of the 12th ACM International Conference Adjunct Papers on Ubiquitous Computing - Adjunct, pp. 415–416. ACM, New York (2010)
26. Lang, A., Martin, J., Crowe, J., Sharples, S.: Medical device design for adolescent adherence and developmental goals: a case study of a cystic fibrosis physiotherapy device. Patient Prefer. Adherence **8**, 301 (2014)
27. van der Bijl Brouwer, M., Dorst, K.: How deep is deep? A four-layer model of insights into human needs for design innovation. In: Design and Emotion. Ediciones Uniandes, Bogotá (2014)
28. van der Bijl-Brouwer, M.: The challenges of human-centred design in a public sector innovation context. In: International Conference: Future-Focused Thinking, DRS 2016. The Design Research Society, Brighton (2016)
29. Smart Design: Breakup Letter Creative Meeting. Presentation at the IIT Design Research Conference, IIT Institute of Design (2010)
30. Lang, A.R.: Medical Device Design for Adolescents [Thesis]. Nottingham University, Nottingham (2012)
31. Hassling, L., et al.: Use of cultural probes for representation of chronic disease experience: exploration of an innovative method for design of supportive technologies. Technol. Health Care **13**(2), 87–95 (2005)
32. Thoring, K., Luippold, C. Mueller, R.M.: Opening the cultural probes box: a critical reflection and analysis of the cultural probes method. In: 5th International Congress of International Association of Societies of Design Research, pp. 597–603, Tokyo (2013)
33. Charmaz, K.: Constructing Grounded Theory, 2nd edn. Sage Publications, London (2014)
34. Craig, M.E., et al.: National Evidence-Based Clinical Care Guidelines for Type 1 Diabetes in Children, Adolescents and Adults. Australian Government Department of Health and Ageing, Canberra (2011)

Personality, Personalization, and Persuasion

Deconstructing Pokémon Go – An Empirical Study on Player Personality Characteristics

Elke Mattheiss[1](✉), Christina Hochleitner[1], Marc Busch[1], Rita Orji[2], and Manfred Tscheligi[1,3]

[1] AIT Austrian Institute of Technology, Vienna, Austria
{Elke.Mattheiss,Christina.Hochleitner,Marc.Busch,Manfred.Tscheligi}@ait.ac.at
[2] University of Waterloo, Waterloo, Canada
rita.orji@uwaterloo.ca
[3] University of Salzburg, Salzburg, Austria
Manfred.Tscheligi@sbg.ac.at

Abstract. Pokémon Go can be considered a successful persuasive game to promote physical activity. This study provides an in-depth analysis of Pokémon Go players' characteristics that relate to adoption and continued play of the game. Based on online studies at two different points in time (at the initial release and three months later) we analyzed differences in personality traits between players and non-players, as well as between players who continued playing the game and those who stopped playing. The results show that people who played the game score lower in "Conscientiousness" and higher in "Player Motivation" than non-players. Furthermore, people who continued playing the game three months later have a lower score in "Neuroticism" than those who stopped playing. Insights into player characteristics from our empirical analysis of Pokémon Go contribute to answering the key research question regarding to which personality characteristics the persuasive game experience should be personalized.

Keywords: Persuasive games · Pokémon Go · Personality · Physical activity · Personalization

1 Introduction

Persuasive games aim to change human behavior and underlying attitudes towards a desirable direction [1], such as being more physically active. Considering the game's catch phrase "Get up and Go"[1] as well as numerous news stories about the game's success and how it increases the physical activity of its players, Pokémon Go can probably be considered as the most widespread and commercially successful persuasive game so far. Several of its gameplay elements motivate players to walk around outdoors in order to advance in the game, similar to other location-based, persuasive games developed in the research area (e.g. NEAT-o-Games [2]).

[1] www.pokemongo.com.

P.W. de Vries et al. (Eds.): PERSUASIVE 2017, LNCS 10171, pp. 83–94, 2017.
DOI: 10.1007/978-3-319-55134-0_7

However, due to its novelty, there is still a lack of scientific research on the game features and player characteristics that contribute to the success of the game. Investigating Pokémon Go players and their characteristics will shed light into why the game is successful and inform the design of other (persuasive) games and persuasive technology in general. On a more specific note, insight into Pokémon Go player characteristics will contribute to advancing research in the area of personalized persuasive games, which aims to increase the effectiveness/persuasiveness of games by tailoring game experiences to specific player groups [3]. At the beginning of many personalization endeavor stands the overall research question: *For which personality characteristics should we personalize the persuasive game experience to increase the effectiveness?* Considering the variety of characteristics of players that persuasive games can be personalized to, the first crucial step towards successful personalization is to pin down which characteristics are relevant for the adoption and continuation of playing a persuasive game.

To investigate Pokémon Go players and their characteristics that have contributed to their adoption of the game and hence reveal opportunities for tailoring game experience, we conducted two online survey studies. The first study involved 335 participants and was conducted within the first two months of the Pokémon Go release. The second study was conducted about three months after the beginning of the first study with 72 participants. Comparing players and non-players, as well as players who continued playing three months later and those who stopped, we found differences in Big Five personality traits and player motivations. This is – to the best of our knowledge – the first study to deconstruct the distinct personality characteristics of people who play Pokémon Go.

2 Related Work

Persuasive games, as a particular research area of persuasive technologies [4] have been successfully applied in various areas such as health [5], sustainability [6] or social benefits [7]. Although frequently applied, there is still a lack of knowledge on how to best design persuasive games [8] that will be adopted by users and at the same time effectively motivate behavior change. This has led to investigations into factors that contribute to the success of persuasive games [9] or that motivate people to play (persuasive) games [10].

2.1 Personality and Persuasive Games

Advances in recent years suggest that games in general and persuasive games in particular that take individual differences into account are more effective than "one-size-fits-all" approaches [11, 12]. Research on persuadability, i.e. the tendency to comply with a specific strategy, has targeted these differences [13] and user characteristics, such as personality traits [14]. As a consequence, personalized solutions could be built based on the relationship between player characteristics and preferences for specific games or game elements.

The Big Five Personality Traits. One area of player characteristics investigated with regard to persuasive interventions is personality. Most previous work employed the widely accepted and adopted Big-Five model [15] to identify relationships between personality and preferences for persuasive elements. The personality traits described in the Big-Five model are the following:

1. **Extraversion** is a tendency to be outgoing, seek out new opportunities, and ambitions. People who score high on this trait are easily excited, confident, energetic, and show high level of emotional expressiveness and talkativeness.
2. **Conscientiousness** is a tendency to be self-disciplined, well-organized and goal-oriented. Those high in this trait tend to follow norms and rules and prioritize tasks.
3. **Neuroticism** which is the opposite of Emotional Stability is a tendency to be nervous, fearful, sensitive, and emotionally unstable. Those who are high in this trait tend to be dynamic, have difficulty managing stress and are likely to be stressed and frustrated by minor situations.
4. **Agreeableness** is a tendency to be considerate, cooperative, and tolerant. Those high in this trait are very compassionate, caring, helpful, conflict avoiding, and may prefer cooperation to competition.
5. **Openness** is a tendency to be imaginative, creative and insightful. People who are high in this trait tend to be adventurous and open to new experiences.

Research on persuasive technologies in combination with the Big Five personality traits has indicated that certain traits are positively correlated with certain persuasive strategies. For example, in [14] relationships between six different persuasive strategies and the traits of the Big Five were found. Research in the area of health promotion indicates a positive correlation between the Big Five trait "Openness" and the strategy "Competition", as well as "Openness" and "Authority" and "Conscientiousness" and "Competition" [16]. In the particular field of persuasive games, research has shown several relationships between "Extraversion", "Openness to Experience" and "Neuroticism" and gameful elements such as points, levels, leaderboards and avatars [17].

Player Motivation Model. Besides the general personality models, dedicated player motivation models have also been used to characterize players. Based on a large-scale study of online players' behaviors, Yee discovered three main types of player motivation [10, 19]. His work was based on a frequently applied player type model created by Bartle [18]. Despite its high prevalence Bartle's player type model has several shortcomings, such as not being based on empirical data. To counteract these shortcomings, Yee conducted several studies on the questions used in [18] to identify the following three main components of player motivation [10, 19]:

1. **Social** motivation factors are seen as all activities contributing to social interactions related to the game: socializing, relationship, and teamwork.
2. **Immersion** is related to the feeling of being immersed in the game and all contributing factors: discovery, role-playing, customization, and escapism.
3. **Achievement** motivation factors include items related to achieving goals as part of the game: advancement, mechanics, and competition.

The model was validated with western and non-western cultures and results indicate a correlation between self-reported data and in-game behavior [10]. Therefore, it deemed appropriate for the self-reporting of personal player motivations when playing Pokémon Go.

2.2 Location-Based Persuasive Games

A common characteristic of persuasive games is that they aim to change the user behavior in the real world [20]. According to [21] all games are able to include persuasive elements, even if it is not the main objective of the game itself. Thus, several location-based games can be considered persuasive games, as their mechanics and strategies encourage the player to be more physically active. For example, Geocaching promotes physical activity through a real-world treasure hunt [22], while iDetective encourages movement by investigating a detective's case and finding a certain location [23]. Pokémon Go, with its recent success can be listed as another location-based persuasive game, whose main objective is not the movement itself, but nevertheless encourages players to be more physically active. Pokémon Go, although an atypical persuasive game, successfully persuades players without coercion [24].

Research, especially about long-term and real-world application of persuasive games, is scarce [22]. This is crucial, since the success of location-based persuasive games to a substantial extend relies on a large-scale and long-term adoption.

3 Methods

The current state-of-the-art stresses the importance of considering personality factors in the design of persuasive games. Valuable contributions have already been made by previous research into investigating *which personality characteristics that persuasive game experiences should be personalized to increase their effectiveness* [cf. 14,16,17]. Knowing which personality characteristics influence the playing behavior opens up possibilities to tailor game experiences to specific groups of players, either fostering adoption of the persuasive game in new target groups or promoting continuous play of active players.

However, previous research is mostly based on one-time measures and was often carried out in the context of games that were specifically created for the purpose of the study and hence, not usually played by a wider audience in situ and under real-life conditions. In the present research we investigate the game *Pokémon Go by Niantic*, played by millions of users across the globe for several months. We are able to compare players with non-players, as well as players who continued playing the game after three months and those who stopped playing over time. Therefore, we contribute to the ongoing research with new insights about relevant personality factors for personalizing persuasive games by investigating the following specific research questions:

(1) *Do Pokémon Go players and non-players differ in terms of personality characteristics?* Answering this question would shed light on the personalization factors that are necessary for players' adoption of persuasive games.

(2) *Do people who continued playing Pokémon Go after three months and those who stopped playing differ in terms of personality characteristics?* Answering this question would shed light on the personality characteristics relevant for player engagement and continuous play of persuasive games.

We conducted two online surveys to answer the above stated research questions. The first survey ran from end of July to beginning of September 2016, therefore starting within the first two months of the Pokémon Go release. The second survey was conducted with part of the participants of the first survey, about three months later, in November 2016. Both surveys were implemented online using LimeSurvey.

3.1 Participants

Survey One. The first survey was distributed in English and German to potential participants from North America and Europe respectively, via social media, mailing lists, and Amazon Mechanical Turk (AMT). As compensation for completing the survey, participants from AMT received a small amount of money (0.25 Canadian Dollars) and the other participants took part in a draw for 15 vouchers worth 30 Euros each. Survey one was completed by 380 participants, from which we excluded 45 participants, who completed the survey in below 3 min, entered impossible values or did not show variance in their answers. The remaining 335 participants (154 males, 180 females, 1 other; 136 English speaking, 199 German speaking) were between 13 and 69 years old (mean age = 29.53, SD = 9.05). 226 of the participants were active players of Pokémon Go and 109 were not playing Pokémon Go (see Table 1).

Table 1. Number of complete data sets in the first and second survey for the English and German speaking sample, depending on whether or not they played Pokémon Go.

	Survey one		Survey two		
	Playing	Not playing	Still playing	Stopped playing	Other[a]
English speaking	101	35	13	16	5
German speaking	125	74	17	26	8
Total	226	109	30	42	13

[a]Other encapsulates participants who started playing after the first survey and who still not play the game, both not relevant for the present study.

Survey Two. The second survey was directly sent to 239 participants (127 English speaking, 112 German speaking) from the first survey, who agreed in the first survey to participate in a follow-up study. In the first survey, no further details about the follow-up study were provided to avoid a bias in participants who agreed. All participants received a small amount of money for completing the survey (participants from AMT 0.20 US Dollar, all other participants a voucher worth 5 Euros). After sorting out 18 data sets (because of duplicated entries of participants or lack of variation in answers), 85 participants (50 females, 35 males; 43 German speaking, 29 English speaking) between 14 and 51 years (mean age = 29.47, SD = 8.45) remained for further analysis.

30 of the participants were still active players of the game, 42 stopped, 3 started after the first online survey, and 10 were still not playing the game (see Table 1).

3.2 Study Design

Survey One. Both, players of Pokémon Go and non-players could participate and we used validated scales in the survey. The Big Five personality traits were assessed with 15 items, i.e. three for each of the Big Five factors "Extraversion", "Conscientiousness", "Neuroticism", "Agreeableness", "Openness to Experience" [25]. Participants were required to indicate their level of agreement to statements (e.g. "I am a person that worries a lot" for "Neuroticism") on a 7-point Likert rating scale. The three player motivation subscales "Social", "Immersion" and "Achievement" were assessed with the 12-item scale by [10], asking participants to rate on a 7-point Likert scale how important specific game elements (e.g. grouping with other players) are to them. The German Big Five questionnaire [25] and the English player motivation scale [10] were translated to English and German, respectively, by an English native speaker also fluent in German.

We also assessed personal innovativeness in Information Technology (IT), which describes the willingness of a person to try out new IT, as a potential moderating variable in our study, with three items according to [26]. Besides, further information, which is not in the scope of the present paper[2], we also assessed how many hours the participants played (non-digital and digital) games per week as potential moderating variable, as well as the players' demographic information. The question sets for players and non-players differed in some questions but not on those relevant for this paper. Thus, active players had to answer a higher number of questions than non-players. The survey took on average 7 min to be completed (depending on whether participants where active players of Pokémon Go or not).

Survey Two. The second survey was conducted to assess which participants continued playing the game after three months and which stopped (and several variables not relevant for the present paper[3]). Depending on whether the participants played the game or not, completing the survey took on average 3 min.

3.3 Data Analysis

Data analysis was conducted with SPSS version 20.0.0. We conducted two multivariate analyses of variance (MANOVAs) to analyze personality differences between players and non-players, as well as players who continued playing the game three months later and those who stopped playing. We subsequently calculated two multivariate analyses of covariance (MANCOVAs) to control for two potential covariates: personal innovativeness in IT and hours of playing digital and non-digital games per week.

---

[2] Reasons to play Pokémon Go, ratings of specific game elements, estimated effect, concerns and psychic as well as physical wellbeing.

[3] Ratings of specific game elements of Pokémon Go, estimated effect, psychic and physical wellbeing, reasons for stopping to play and time of stopping.

4 Results

In order to identify personality factors relevant for playing Pokémon Go, and subsequently define meaningful personalization possibilities for persuasive games, we investigated differences between players and non-players (survey one), as well as between players who still played three months later and those who stopped playing (survey two) in terms of the **Big Five personality traits** and **player motivation**.

Survey One. The results (see Table 2) show that people who play Pokémon Go have a significantly lower score in the Big Five trait "Conscientiousness", as well as a higher score in the three player motivations "Social", "Immersion" and "Achievement" with small to large effect sizes compared to non-players. In contrast, no significant differences between players and non-players could be found for "Extraversion", "Neuroticism", "Agreeableness" and "Openness to Experience".

Survey Two. Players who already stopped playing the game three months later scored higher in "Neuroticism" with a medium effect size of part.η^2 = .064 than people who were still playing the game (see Table 2). Besides that, we found no significant differences between people who continued playing the game and those that have stopped playing.

Controlling for Covariates. To gain a further understanding about the differences in personality traits, we investigated whether controlling for the two potential moderating variables personal innovativeness in IT and the hours of playing (digital and non-digital) games per week, would change the above reported differences.

The MANCOVA with **personal innovativeness in IT** as covariate shows that the difference in player motivations between players and non-players remain significant for "Immersion" ($F_1 = 21.71$, $p < 0.001$, part.$\eta^2 = .061$) and "Achievement" ($F_1 = 37.64$, $p < 0.001$, part.$\eta^2 = .102$) with a medium effect. However, the difference between players and non-players in the subscale "Social" ($F_1 = 2.48$, $p = 0.116$, part.$\eta^2 = .007$) disappears when controlling for personal innovativeness. The difference between players and non-players in terms of their "Conscientiousness" score remains significant ($F_1 = 7.59$, $p = 0.006$, part.$\eta^2 = .022$), with a slightly larger effect when controlling for personal innovativeness. The (non-significant) differences between players and non-players in "Extraversion", "Neuroticism", "Agreeableness" and "Openness to Experience" remain unchanged.

Regarding the second survey, the difference in "Neuroticism" between people who continued playing the game three months later and those who stopped playing remained unchanged after controlling for personal innovativeness ($F_1 = 4.87$, $p = 0.031$, part.$\eta^2 = .066$). The remaining results of the second survey stayed the same after eliminating personal innovativeness as a mediator.

The MANCOVA results with the **hours per week of playing (digital and non-digital) games** as covariate did not change the original results presented in Table 2, both for survey one and survey two.

Table 2. Differences between different groups in the Big Five (BF) personality traits [12] and the player motivation (PM) scales [11] (M = mean; SD = Standard deviation; part.η^2 > .01 is considered as small, > .06 medium, .14 large effect size [27]).

	Survey One		**Survey Two**	
	Playing (n = 226)	*Not Playing (n = 109)*	*Still Playing (n = 30)*	*Stopped Playing (n = 42)*
Extraversion (BF)	M = 4.55 SD = 1.43	M = 4.54 SD = 1.54	M = 4.59 SD = 1.40	M = 4.64 SD = 1.44
	F_1 = 0.00, p = .966, part.η^2 = .000		F_1 = 0.03, p = .872, part.η^2 = .000	
Conscientiousness (BF)	M = 4.97 SD = 1.12	M = 5.24 SD = 0.92	M = 5.08 SD = 0.87	M = 4.99 SD = 1.14
	F_1 = 4.84, p = .028, part.η^2 = .014		F_1 = 0.12, p = .727, part.η^2 = .002	
Neuroticism (BF)	M = 3.79 SD = 1.43	M = 3.76 SD = 1.41	M = 3.54 SD = 1.14	M = 4.25 SD = 1.50
	F_1 = 0.03, p = .875, part.η^2 = .000		F_1 = 4.77, p = .032, part.η^2 = .064	
Agreeableness (BF)	M = 5.22 SD = 0.98	M = 5.16 SD = 1.07	M = 5.16 SD = 0.99	M = 5.13 SD = 1.01
	F_1 = 0.23, p = .634, part.η^2 = .001		F_1 = 0.01, p = .907, part.η^2 = .000	
Openness to Experience (BF)	M = 5.19 SD = 1.17	M = 5.17 SD = 1.26	M = 5.11 SD = 1.31	M = 5.51 SD = 1.28
	F_1 = 0.04, p = .841, part.η^2 = .000		F_1 = 1.66, p = .202, part.η^2 = .023	
Social (PM)	M = 4.03 SD = 1.55	M = 3.64 SD = 1.70	M = 4.20 SD = 1.55	M = 3.98 SD = 1.61
	F_1 = 4.45, p = .036, part.η^2 = .013		F_1 = 0.35, p = .557, part.η^2 = .005	
Immersion (PM)	M = 4.98 SD = 1.31	M = 4.00 SD = 1.69	M = 4.81 SD = 1.42	M = 5.08 SD = 1.31
	F_1 = 34.02, p < .001, part.η^2 = .093		F_1 = 0.69, p = .409, part.η^2 = .010	
Achievement (PM)	M = 4.85 SD = 1.13	M = 3.69 SD = 1.51	M = 4.81 SD = 1.35	M = 4.70 SD = 0.94
	F_1 = 61.07, p < .001, part.η^2 = .155		F_1 = 0.15, p = .696, part.η^2 = .002	

5 Discussion and Implications for Personalization

By identifying differences between players and non-players of Pokémon Go, as well as people who continued playing the game three months later and those who stopped, we can *deduce relevant personality characteristics for personalizing persuasive games* to foster adoption by new target groups and continuous play of active players.

Personality Trait Conscientiousness. The results of our study show related to the Big Five personality traits [15] that players of Pokémon Go describe themselves as significantly less conscientious than non-players. This indicates that Pokémon Go attracts people who characterize themselves as significantly more spontaneous and less organized, compared to the non-players. A possible explanation lies in the finding of [14] that conscientious people are seeking commitment, which is contradicted by the casual nature and the ad hoc game play elements of Pokémon Go, not demanding self-discipline and organization. Thus, these particular game elements (e.g. capturing of randomly spawning Pokémon) make it potentially even more attractive to spontaneous people. Knowledge about these distinct characteristics of players and non-players can be used to create persuasive games to increase their **adoption by new target groups**, in this case, more conscientious people. Therefore, our results suggest that personalization of a location-based persuasive game for this particular target group can be achieved by focusing on conscientiousness-oriented game elements. Therefore, game elements that do not involve spontaneity and require long-term planning or commitment at a time are expected to attract more conscientious people.

Player Motivation. Further opportunities for personalization can be found in our results of player motivation analysis [10], which show that players and non-players differ in player motivation scales. The smallest difference exists in the scale "Social", which completely disappears when controlling for the covariate personal innovativeness in IT. Thus, the difference between players and non-players in "Social" is a result of the difference in personal innovativeness between the groups, which correlates with "Social". Therefore, we can assume that social activities in games are equally important to players as they are to non-players of Pokémon Go and should be emphasized in persuasive games. The differences between players and non-players in "Immersion" and "Achievement" remain significant, even after controlling for personal innovativeness. This suggests that for non-players feeling immersed in the game play and game story, as well as achieving goals in the game are less important than for players. Central Pokémon Go game elements, such as exploring the world for Pokémon (immersion) and evolving Pokémon (achievement) are related to these motivational factors, which seem to be less attractive for non-players. These results indicate that a focus on social aspects of the game seems most promising to **reach non-players**, as there is no difference between players and non-players in rating social game elements (e.g. being part of a guild or chatting with other players). Although often played with friends, Pokémon Go has so far only limited social features affecting the social affordances of the game. This could be one of the reasons why it is not attractive to non-players. Interestingly, players who continued playing and those who stopped do not differ in terms of player motivation. Therefore, player motivation factors seem only relevant for adoption of the game.

Personality Trait Neuroticism. Regarding fostering **continuous play**, our results show a difference between players who continued playing and those who stopped playing in "Neuroticism". Thus, players who stopped playing the game are significantly more neurotic. This indicates that calm, emotionally stable and less dynamic individuals are more likely to continue playing the game in general. This significant finding is in line with [17], who suggest that "Neuroticism" is one of the traits that most significantly impact the design of a gameful intervention, as it correlates with preferences for several gameful elements, such as points and levels. Although we cannot compare their results with ours because the two studies focused on completely different game studies, this still emphasizes the importance of "Neuroticism" in the design of persuasive interventions. Therefore, to support continuous play in more neurotic people, carefully designed game elements to maintain steady interest (on-going activities), without stressing players and with low potential to frustrate players, need to be involved for long-lasting success of the game.

6 Conclusion and Future Work

In the present paper we investigate personality differences between players and non-players of a physical activity promoting game, Pokémon Go, as well as between people who continued playing the game a few months following the game's release and those who stopped. A deep understanding of player characteristics contributes to the selection of relevant personality characteristics for the personalization of persuasive games. We found that, to increase the adoption of persuasive location-based games, designers should emphasize on conscientiousness- and socially-oriented game elements. Neuroticism-oriented game elements should be emphasized to foster player engagement and continuous game play.

Our results have direct implications for the design of personalized, persuasive games.

However, a limitation of the study is the focus on one concrete location-based, persuasive game on which our claims are based. A generalization of our results still needs to be investigated. Future work should focus on the related design space of games and empirically investigate different game versions for people with different personality traits. Also, the identified differences should be verified for different games and cultural environments.

Acknowledgements. This research has partly been funded by the Austrian Research Promotion Agency, contract no. 844845 (GEMPLAY) and 849081 (Bike'N'Play).

References

1. Bogost, I.: Persuasive Games: The Expressive Power of Videogames. The MIT Press, Cambridge (2007)
2. Fujiki, Y., Kazakos, K., Puri, C., Buddharaju, P., Pavlidis, I., Levine, J.: NEAT-o-Games: blending physical activity and fun in the daily routine. Comput. Entertain. **6**(2), 21:1–21:22 (2008). ACM, New York
3. Kaptein, M.C.: Personalized persuasion in ambient intelligence. J. Ambient Intell. Environ. **4**(3), 279–280 (2012). IOS Press, Amsterdam
4. Fogg, B.J.: Persuasive Technology: Using Computers to Change What We Think and Do. Morgan Kaufmann Publishers, Inc., San Francisco (2003)

5. Thompson, D., Bhatt, R., Vazquez, I., Cullen, K.W., Baranowski, J., Baranowski, T., Liu, Y.: Creating action plans in a serious video game increases and maintains child fruit-vegetable intake: a randomized controlled trial. Int. J. Behav. Nutr. Phys. Act. **12**(39), 1–10 (2015). BioMed Central
6. Santos, B., Romão, T., Dias, A.E., Centieiro, P.: eVision: a mobile game to improve environmental awareness. In: Reidsma, D., Katayose, H., Nijholt, A. (eds.) ACE 2013. LNCS, vol. 8253, pp. 380–391. Springer, Heidelberg (2013). doi:10.1007/978-3-319-03161-3_28
7. Steinemann, S.T., Mekler, E. D., Opwis, K.: Increasing donating behavior through a game for change. In: Proceedings of the 2015 Annual Symposium on Computer-Human Interaction in Play - CHI PLAY 2015, pp. 319–329. ACM, New York (2015)
8. Kors, M.: Towards design strategies for the persuasive gameplay experience. In Proceedings of the 2015 Annual Symposium on Computer-Human Interaction in Play - CHI PLAY 2015, pp. 407–410. ACM, New York (2015)
9. De la Hera Conde-Pumpido, T.: A conceptual model for the study of persuasive games. In: DiGRA 2013 DeFragging Game Studies, Atlanta (2013)
10. Yee, N., Ducheneaut, N., Nelson, L.: Online gaming motivations scale: development and validation. In: Proceedings of the SIGCHI Conference on Human Factors in Computing Systems – CHI 2012, pp. 2803–2806. ACM, New York (2012)
11. Nov, O., Arazy, O.: Personality-targeted design: theory, experimental procedure, and preliminary results. In: Proceedings of the 2013 Conference on Computer Supported Cooperative Work - CSCW 2013, pp. 977–984. ACM, New York (2013)
12. Orji, R., Mandryk, R.L., Vassileva, J., Gerling, K.M.: Tailoring persuasive health games to gamer type. In: Proceedings of the SIGCHI Conference on Human Factors in Computing Systems - CHI 2013, pp. 2467–2476. ACM Press, New York (2013)
13. Kaptein, M., Lacroix, J., Saini, P.: Individual differences in persuadability in the health promotion domain. In: Ploug, T., Hasle, P., Oinas-Kukkonen, H. (eds.) PERSUASIVE 2010. LNCS, vol. 6137, pp. 94–105. Springer, Heidelberg (2010). doi:10.1007/978-3-642-13226-1_11
14. Alkış, N., Temizel, T.T.: The impact of individual differences on influence strategies. Pers. Individ. Differ. **87**, 147–152 (2015). Elsevier, Amsterdam
15. Goldberg, L.R.: The structure of phenotypic personality traits. Am. Psychol. **48**, 26–34 (1993)
16. Halko, S., Kientz, J.A.: Personality and persuasive technology: an exploratory study on health-promoting mobile applications. In: Ploug, T., Hasle, P., Oinas-Kukkonen, H. (eds.) PERSUASIVE 2010. LNCS, vol. 6137, pp. 150–161. Springer, Heidelberg (2010). doi: 10.1007/978-3-642-13226-1_16
17. Jia, Y., Xu, B., Karanam, Y., Voida, S.: Personality-targeted gamification: a survey study on personality traits and motivational affordances. In: Proceedings of the 34th Annual ACM Conference on Human Factors in Computing Systems - CHI 2016, pp. 2001–2013. ACM, New York (2016)
18. Bartle, R.A.: Players Who Suit MUDs. Mud, 1 (1999)
19. Yee, N.: Motivations for play in online games. Cyberpsychol. Behav. Impact Internet Multimedia Virtual Reality Behav. Soc. **9**(6), 772–775 (2006). Mary Ann Liebert, Inc., New Rochelle
20. Visch, V., Vegt, N., Anderiesen, H., van der Kooij, K.: Persuasive game design: a model and its definitions (2013). http://gamification-research.org/wp-content/uploads/2013/03/Visch_etal.pdf
21. Kors, M., van der Spek, E., Schouten, B.: A foundation for the persuasive gameplay experience. In: 10th Foundations of Digital Games Conference Proceedings (2015)
22. Gram-Hansen, L.B.: Geocaching in a persuasive perspective. In: Proceedings of the 4th International Conference on Persuasive Technology - Persuasive 2009, pp. 34:1–34:8. ACM, New York (2009)

23. Yoshii, A., Funabashi, Y., Kimura, H., Nakajima, T.: iDetective: a location based game to persuade users unconsciously. In: IEEE 17th International Conference on Embedded and Real-Time Computing Systems and Applications, pp. 115–120. IEEE Computer Society, Washington, D.C. (2011)
24. Hamari, J., Koivisto, J., Pakkanen, T.: Do persuasive technologies persuade? - A review of empirical studies. In: Spagnolli, A., Chittaro, L., Gamberini, L. (eds.) PERSUASIVE 2014. LNCS, vol. 8462, pp. 118–136. Springer, Heidelberg (2014). doi:10.1007/978-3-319-07127-5_11
25. Gerlitz J.-Y., Schupp, J.: Zur Erhebung der Big-Five-basierten Persönlichkeitsmerkmale im SOEP. In: DIW Research Notes 4. DIW Berlin (2005)
26. Agarwal, R., Prasad, J.: A conceptual and operational definition of personal innovativeness in the domain of information technology. Inf. Syst. Res. **9**(2), 204–215 (1998). INFORMS, Linthicum
27. Cohen, J.: Statistical Power Analysis for the Behavioral Sciences, 2nd edn. Lawrence Erlbaum Associates, Hillsdale (1988)

Personalized Assistant for Health-Conscious Grocery Shoppers

Chokdee Siawsolit(✉), Sarun Seepun, Jennifer Choi, An Do, and Yu Kao

Claremont Graduate University, Claremont, CA, USA
{chokdee.siawsolit,sarun.seepun,jennifer.choi,an.do,
yu.kao}@cgu.edu

Abstract. As the healthy eating trend gains momentum and the number of product choices explode, grocery shoppers face a distinctive challenge in making effective decisions. This article summarizes our efforts to improve consumer's ability to choose healthier food product choices. Through the framework of Health-Belief Model, we constructed a survey for shoppers and; based on the findings, designed a solution that adapts to each user's unique cocktail of preferences.

Proof-of-concept test was conducted on 12 female graduate students to verify the mock solution's persuasiveness. It was found that a noteworthy portion of subjects improved their selections after intervention. Finally, we propose a number of possible development avenues that may serve as our future directions.

Keywords: Persuasive technology · Health belief model · Grocery · Shopping assistant · Product choice · Health conscious · Personalized suggestion

1 Problem Statement

Today, grocery shoppers are presented with more product choices than ever. Whole Foods, Sprouts, Von's and Trader Joe's have all jumped on the healthy eating trend by offering wide varieties of health-conscious ingredients and ready-to-eat food products. While global sales of healthy food products are expected to reach $1 trillion in 2017 [1], the exploding number of choices –an average of 50,000 per store– have begun to negatively affect the industry. Consumer Reports suggests that nearly 80% of 2,800 respondents felt there were too many choices, and 36% reported they were "overwhelmed by the information they had to process to make a buying decision" [2].

In one of the largest studies ever conducted by Nielsen –a $1B net profit market research firm– in 2015 involving 30,000 consumers, data suggests that shoppers are ready to pay more for products claiming to boost health [3]. While eight out of ten Americans reported higher willingness to pay (WTP) for healthier foods, many did not know which products best suit their personal needs. This lack of connection between the product characteristics cited by manufacturers and the shoppers' perceived benefit may be the missing link to bridge the gap in further adoption of healthy food shopping; unleashing higher potentials in value generation to manufacturers, retailers, and consumers [4].

P.W. de Vries et al. (Eds.): PERSUASIVE 2017, LNCS 10171, pp. 95–106, 2017.
DOI: 10.1007/978-3-319-55134-0_8

2 Literature Review

Much work has been done in attempts to provide a solution in the form of third-party virtual shopping assistants. Ahn et al. in 2015 suggested an augmented-reality grocery shopping application designed to introduce dietary behavior interventions. They proposed a real-time recommendation system where users can set preferences such as food allergies. Improvements were measured in time duration to complete the grocery shopping as well as distance of paths taken within the store [5].

Even as far back as 2003, Shekar et al. had already proposed a client-proxy-server smartphone architecture to make healthy grocery shopping easier for its users by setting up shopping lists, mapping product locations in store and automating checkouts. Aptly named iGrocer, their system would allow users to select from medical conditions such as liver disorder or being pregnant etc, and offer recommendations accordingly [6].

To promote further adoption, we take a step back to view the challenge as a behavior change one and look for help through literature in persuasion psychology. The health belief model (HBM); first developed to understand widespread failure of people to accept disease prevention, stands out to offer an insightful framework embracing: perceived susceptibility, perceived severity, perceived benefits, perceived barriers and self-efficacy of a target health behavior [7].

The HBM stipulates that health-promoting behavior depends on the value placed by an individual on a particular goal as well as the individual's estimate of likelihood that a given action will achieve that goal. Previous studies have not indicated factors that may be preventing healthy grocery behavior change. Therefore, we shape our view on the subject matter according to the constructs of HBM (Fig. 1).

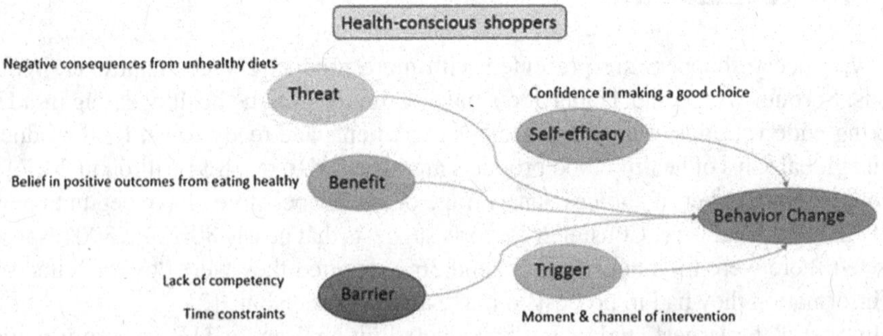

Fig. 1. Likelihood of adopting healthier food purchase decisions

Perceived susceptibility and perceived severity together forms the perceived threat to an individual. In our case, this threat refers to the negative consequences of maintaining unhealthy diet. An individual will perceive that a health-promoting behavior offers benefits if he or she believes in the positive outcomes derived from eating healthy.

Additionally, there may be factors potentially acting as barriers to performing the target behavior. Escoto, a research director under United Health Group, found in her 2012 study that "the most frequently reported barrier to healthful eating is lack of time, with young adults citing challenges in balancing work, school, and leisure schedules". Here we take the lack of time conclusions from Escoto et al. as well as lack of proficiency findings from Consumer Reports as our starting points [8].

The HBM also suggests significance of self-efficacy. This is described as the individual's perception on his or her ability to perform the target behavior. We view this as the confidence in making a well-informed and effective food product choice. In the end, we shall attempt to design a timely cue to action that provides a trigger to perform target behavior by means of a suitable channel. With our perspectives set, we then needed to verify our understanding of the problem through the eyes of actual grocery shoppers.

3 Survey Design

We proceed to design a survey to validate the existence of postulated barriers and potentially shed light on what we could offer as solutions. Outlined here are several objective questions to provide further insights into what constitutes the current barriers to consumers adopting healthier product choices.

First, we defined the motivation to make better eating decisions as a function of negative consequences of unhealthy diets (perceived threat) in conjunction with potential positive effects from performing the targeted behavior (benefits). Here, we did not explore in depth this intrinsic function unique to each shopper. The output of interest was simply how strongly consumers correlate healthy food choices to their well-being.

Next, we gauged basic knowledge concerning food product characteristics such as recommended dietary allowances (RDA) for sugar and sodium [9]. To verify the existence of problems suggested by Consumer Reports and Neilsen, we asked whether or not food shoppers actively check product labeling for nutrition facts and how well they understand the information presented (barrier).

Then, we assessed shopper's perspectives on their own ability to make healthy buying decisions. While this self-report ability may not be accurate, it provides an indicator for the shopper's level of confidence in performing target behavior (self-efficacy). For further development purposes, we also asked which nutritional characteristics most concern them, and which channel of intervention may best suit their lifestyles (trigger).

Finally, we wanted to know if some form of solution has the potential to translate into tangible values to product manufacturers. For those that considered nutrition facts in their purchase decisions, we asked if they were willing to try a different product offering 'better nutritional match' to their preferences. We also inquired whether customers would be more likely to visit the grocery store that offers a shopping assistant solution.

4 Survey Findings

Of the 25 surveys completed, roughly two-thirds were filled by females. We observed that women were more likely to be the designated grocery shopper of the household. The peaks in age group represent college students and adult shoppers at a local store.

When asked how well the participants correlate their food choices to their well-beings, up to 80% ranked themselves 4 or above on a scale of 1 to 5. It is conceivable to consider that if they cared about their health, then they have a precursor *motivation* to make healthier food product choices. Most shoppers exhibited some level of consciousness over the amount of sugar and sodium they consume. 9 out of 25 participants were able to correctly identify the recommended daily sugar intake (Table 1, left), but a significant amount believed they over consume sugar. In the case of sodium only 4 respondents produced correct answers even though the majority thought they were conscious (Table 1, right). Clearly, a feasible solution must be able to help users track the amount of sugar and sodium being consumed, as well as educating them on RDAs.

Table 1. Response distributions for selected questions pertaining to awareness

What do you think is recommended daily sugar intake?					
25g	50g	60g	85g	No clue	Other
36%	0%	4%	0%	44%	16%
How conscious are you of sugar content you consume?					
Very	Somewhat		Almost never		Varies
44%	28%		20%		8%
How much sugar do you think you consume daily?					
25g	50g	60g	85g	No clue	Other
4%	28%	12%	4%	40%	12%

What do you think is recommended daily sodium intake?					
500mg	1000mg	1500mg	3500mg	No clue	Other
31%	17%	17%	0%	35%	0%
How conscious are you of sodium content you consume?					
Very	Somewhat		Almost never		Varies
35%	35%		21%		9%
How much sodium do you think you consume daily?					
500mg	1000mg	1500mg	3500mg	No clue	Other
25%	13%	17%	0%	41%	4%

Over half of the survey samples responded they check nutrition information on at least some products on a regular basis. While this illustrates a point that there already exists a strong intention to adopt healthier food choices, just 18% of those surveyed reported they fully comprehend the information being presented on the product labels (Fig. 2).

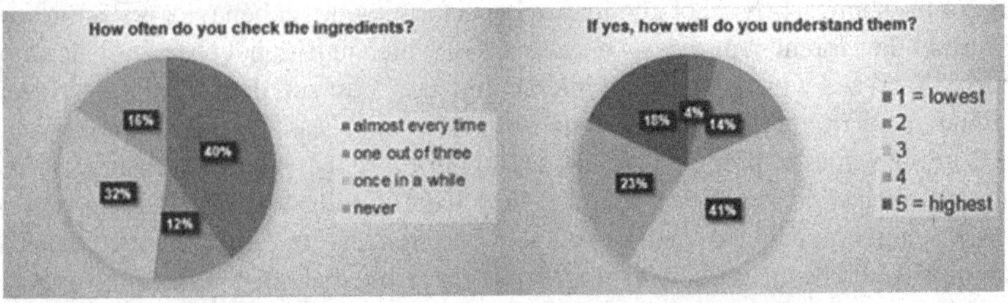

Fig. 2. Response distributions relating to current behavior and self-efficacy

This finding supports Consumer Reports data that many average shoppers lacked the necessary knowledge to make completely informed decisions. We confirmed that a subset of shoppers were willing to elaborate on the topic, and thus a central processing route may prove more effective over peripheral interventions [10].

Despite the fact that roughly 70% of respondents reported some level of consciousness over the amount of sugar and sodium they consume, only 4 out of 25 are fully confident that the products they pick up contain the best nutritional characteristics. This finding presents a business opportunity in assisting health-conscious customers to pick the right products that best suit their personal needs.

To further understand specific concerns of our target group, we asked shoppers to elaborate on nutritional characteristics that most concern them. We used a simple weighing algorithm on the top 5 rankings by assigning values of 5 for most concerned to 1 for fifth-ranked concerns. We learned that shoppers were less concerned whether or not a product was organic, while saturated fat received the highest weighted score. Thus, we needed to allow for flexible ways where each shopper can customize their preferences (Fig. 3).

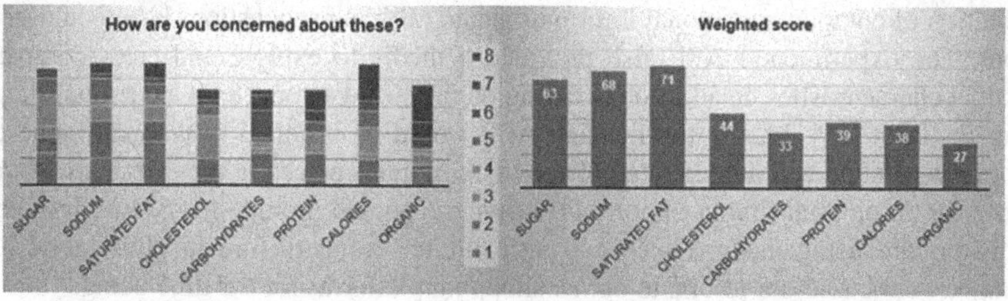

Fig. 3. Response distributions for concerns of specific product characteristics

In order to design an effective persuasion solution, we asked which recommendation channel would be most preferred. None of the respondents chose text messaging; which agrees with previous findings that texts were less preferred when offering suggestions [11]. Half of those who were willing to receive product recommendations preferred mobile application. For our purposes here a phone APP may be the most convenient, informative, and interactive channel.

Overall, we verified that nutritional characteristics do play a role in up to 71% of customers' buying decisions. Within those that consider nutrition facts, as much as 92% were willing to try a product with 'healthier' characteristics. While these self-reported figures may not fully convert into matching purchases, they represent a pre-cursor to perform behavior that may generate values for manufacturers offering healthy choices.

Lastly, we noted potential values to a critical player that bridges products to consumers. Over 80% of grocery shoppers would *selectively* visit retailers offering a shopping assistant solution. Seeing that there may be values extending from end users up to manufacturers, we proceed to designing our persuasive solution.

5 Solution Design

Given our findings from the survey, we formulated an experimental solution based on BJ Fogg's 8-step persuasive technology design process [12]. The target behavior is to make healthier food product choices during grocery shopping. Female graduate students who conduct their own groceries were chosen as subjects as we've identified their barriers to include time constraints and limited proficiency in nutritional grocery shopping. Most of these individuals have means of transportation, student budgets, and many have a part-time job while conducting research and enrolled full-time. We chose the intervention channel to be a smartphone APP as suggested by survey responses.

In the early design phase, we centered our attention on defining primary functionalities for the test APP. Following Fogg's behavior formula; behavior = motivation x ability x trigger, our mock solution focused on improving the user's ability as well as providing a well-timed trigger. Quantitative information was presented as to appeal to health-conscious users who were likely to elaborate on nutritional values. Rather than attempting to increase motivation, we aimed to shift health-conscious shoppers to the right along B = MAT curve referenced here [13].

To accomplish this goal, we employed a number of captological strategies including reduction of complex nutritional information and tailoring suggestions based on user's input. The experimental APP also provided a media to explore and cross-compare product characteristics, in addition to taking on the social actor role of a personal shopping assistant [14]. Ideally the finished APP would be installed on a smartphone or shopping cart and will automatically keep track of items placed in the cart. This would allow for an optimal-timing opportunity to intervene and provide triggers right when decisions are being made. Given currently limited infrastructure in physical stores though, we focused our efforts to online shopping platforms for the time being.

Quick visits to Apple's and Google's APP stores revealed numerous relevant examples. What we failed to find was a comprehensive solution that allowed users to bi-directionally define their preferences of nutritional characteristics, and offered personalized suggestions of available alternative products in grocery shopping settings. We also note the success of NuVal; a scoring system that represents overall nutritional values on a scale of 1 to 100 currently available in over 2,000 supermarkets [15]. Here we imitate the simplicity of a single numerical output, while adding complexity of personalization in determining which products better suit each customer's unique mix of health concerns and nutritional needs.

To verify the persuasive potentials of our solution, the team generated a mock-up prototype that captures the primary idea to be tested on subjects. For this proof-of-concept stage we constructed a simple linear algorithm that determines a rating from 1 to 5 stars based on the user's initial inputs. Future work will be needed to better develop and refine the rating algorithm; with expert perspectives taken from food scientists, nutritionists and perhaps medical doctors.

6 Proof-of-Concept Test

We used our web-based prototype to examine the effects of personalized suggestions on shopping behavior. The objectives included verification of the prototype's persuasiveness, understanding consumer spending behavior, as well as further developing the target for specific product groups.

New subjects were asked to specify their nutritional preferences (Fig. 4). The slider bar accounts for both directions as each shopper may have differing dietary needs. The price sensitivity bar does not play a role in the rating algorithm, but will help us later understand the correlation of willing-to-pay for healthier products.

The subjects were then tunneled through a mock-up shopping experience of six product groups including: hotdog, juice, bread, ice cream, vegetable, and fruit choices (Fig. 5). The product categories each contained unique characteristics other than nutrition to better capture effectiveness compatibility.

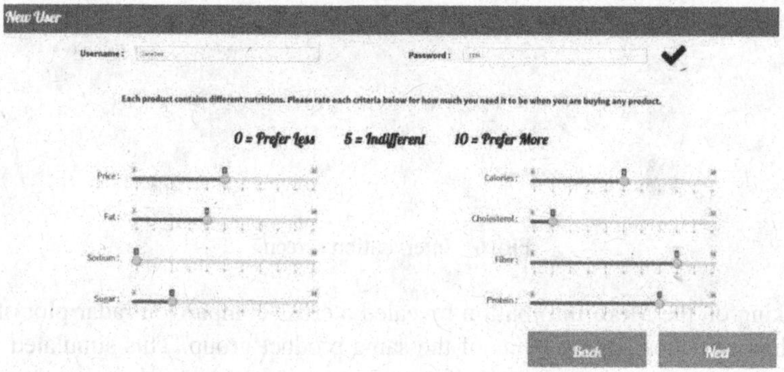

Fig. 4. Preference setup screen

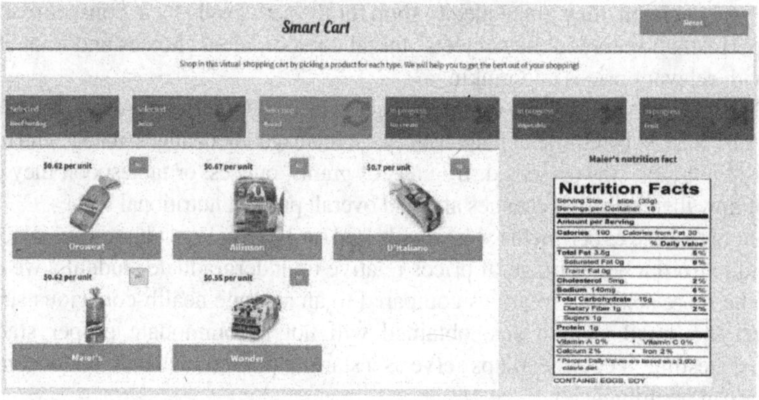

Fig. 5. Product selection screen

While all ice cream choices were chocolate, the fruit choices ranged from strawberries to bananas. Four of the groups contained branded products of the same type, while the latter two were unprocessed, different products within the same group. Unit price was displayed next to each selection, and clicking on the NF button revealed nutrition facts as seen on real-world product packaging.

After making an initial selection, a popup would display personalized ratings of each of the available food choices. Again the star ratings were personalized based on initial preferences and generated through our simple linear algorithm. The user were then able to decide whether they wish to change their current choice for a healthier product, or keep their selection as they please (Fig. 6).

Fig. 6. Intervention screen

Clicking on the More Info button revealed a cross-comparison radar plot of nutritional characteristics among items of the same product group. This simulated what a fastidious customer may go through if they were to compare products on the shelf in detail. Without juggling through multiple product screens, subjects could conveniently compare products of the same category based on selected parameters. Once subjects made a final decision, they proceeded to shop for the next products in consecutive order. The experimenter recorded the subjects' initial choices, ideal choices and final choices along with relevant prices and rankings.

At the end users were given a summary page of their shopping cart. A "receipt" summarizing individual subject's shopping history, percentage of changes made, and overall changes in nutrition was presented. By units of grams, ounces, or tablespoon they can be aware of how their selection changes affected overall product nutritional values.

We note that the experiments were conducted on 12 female graduate students. While they could afford a wider range of prices relative to undergraduate students, we expect them to be more sensitive to prices compared to an average health-conscious shopper. Although the small sample size obtained will not accommodate proper statistical hypothesis testing, it could perhaps serve as a starting point on what and how consumer behavior can be influenced.

7 Results and Discussion

Data collected from experiments are reviewed here to determine whether or not our solution method contained merits. Subjects were not instructed to decide primarily based on nutrition information and price alone. Despite this, many were self-driven in choosing healthy products and roughly 1 in 5 initial choices were optimal according to the APP's basic algorithm. Since we're interested in how much behavior change can occur, we exclude choices that already received 5 star ratings from further analyses.

Given that the initial choices were not optimal, we observed 46% of these choices were changed upon the APP's recommendation. Of the changes that occurred, over two thirds resulted in accepting top suggestions regardless of unit prices. The rest of alternative choices landed on products offering a better compromise on price and healthy ratings, or containing favorable traits not captured by test parameters.

Overall, 34 of the all-inclusive 72 possible final product choices ended up with a 5 star rating after intervention. Representative subjects ranged from savvy dieters whose initial choices were mostly on-point, to bullet-proof thrifty or brand-loyal students that scored very low initially and still declined to accept healthier alternative suggestions. The target audience for any commercial purposes would ideally be those who are motivated but lack time or proficiency to make healthy choices.

We proceed to analyze the possible financial implications of our recommendation system; as well as to identify any correlation in price and behavior change. Note the following results concerning spending outcomes are based on small samples of college-budget students and should be viewed accordingly.

Most subjects ended up spending less in total ($16.64) when compared to their initial choices ($18.16). We observed this to be the case most often when a product in the suggestion page offers better star rating at equal or lower unit prices. This may very well capture college students' budget-conscious mindsets. If suggestions were offered only for higher-priced products, the average spending per subject increases slightly ($18.54). Overall, the average star rating improved significantly after intervention (3.72) in comparison to the subjects' initial choices (3.01).

While recognizing that factors including brand, flavor, or buying habit among others also influence purchases, we introduce a measurement criteria Dollar-per-Star (DPS) to identify potential relationships between price and willingness to adopt new product choice. Here we observed noteworthy distinctions between choices that were accepted or rejected when products with higher unit price and star ratings were recommended.

As shown in Fig. 7, if the cost to adopt healthier choice is low, student subjects were more likely to accept suggestions. This finding-coupled with Neilsen's report that an average consumer would be willing to spend more on healthier products- could imply that a financially-independent shopper may accept higher DPS than observed here.

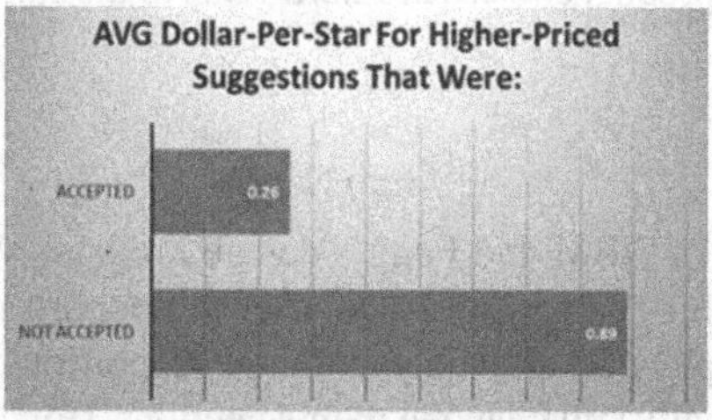

Fig. 7. DPS comparison for accepted and not accepted higher-priced suggestions

Finally, we summarize the effectiveness of our proof-of-concept prototype across the different product groups presented to subjects. There were instances when brand loyalty or taste preference override healthy suggestions, even when the changes would lead to lower prices. The HaagenDaaz brand emerged as a clear leader in this aspect while commanding a 2.5 fold premium in pricing and offering little nutritional improvements. Hot Dog brands also displayed stronger retention rates relative to other product groups (Fig. 8).

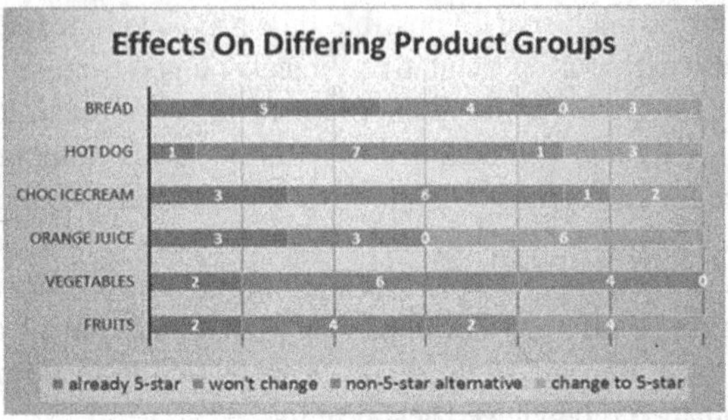

Fig. 8. Summary of persuasion effectiveness on each tested product group

The product group that proved most successful was orange juice. Half of the subjects were willing to try a different brand of orange juice, and a total of three quarter ended up with a 5 star choice. This may be attributed to the fact that orange juice brands are perceived as direct substitutes and offer little variations in terms of significant flavor differences. For fruit selections, test subjects were less resistive in opting for an alternate choice even when the taste and texture profiles could be drastically different from their initial choices. Our mock prototype performed worst in persuading college students to pick the healthiest vegetable choice. However, a few were willing to change to another alternative that offers better nutritional values.

Overall, we learned that there were strongly subjective characteristics such as dislikes toward certain vegetables or preferences for softness in bread. We also did not find conclusive correlation of initial price-sensitivity setting to the actual price-sensitivity behavior. Nonetheless, even the non-health conscious subjects that did little to adapt their choices were surprised to see their summary pages. Many subjects cited that this may lead them to contemplate on healthier product choices in future visits to the store.

8 Conclusion

Through the framework of HBM and Fogg's design process, we have achieved measurable levels of successful persuasion. Even after the opportunity to review nutrition facts as they would in a physical store, a significant proportion of test subjects who did not pick a top choice initially still changed their selections. Conceivably, product manufacturers may wish to offer healthier alternatives when the benefits can be uniquely communicated to each shopper. A flexible and adaptive personalized assistant APP may hold the same key as to why Netflix's or Amazon's suggestion algorithm had garnered critical acclaim.

As much as we have come to rely on customer review star ratings, health-conscious grocery shoppers are still left to balance their budget and unique nutritional needs on their own. To improve perceived benefits, a user-friendly star rating can offer both personalized and simplified quantitative nutritional information that appeals to the central processing route of health-conscious shoppers who are willing to elaborate on the choices.

Our recommendation strategy uniquely allows the user to remain involved in the decision-making process, while at the same time improving the user's ability to choose healthy products. Apart from looking for foods to avoid, shoppers can expose themselves to alternative products offering nutritional match to their preferences that are either set by the user themselves or through external expert help.

Proposed here are potential future research avenues for our efforts. A nutritional-based recommendation system may be useful to shoppers who are motivated to eat healthy but lack the time to compare products on their own. With flexible nutritional preference setup, savvy conscious-eaters can save time, and thrifty shoppers could curb spending while maintaining healthy diets. It may be worth exploring to distribute the solution as free-to-download application; perhaps by securing funding from governmental entities or wellness promotion NGO's.

Product manufacturers may also be interested in offering upgrade choices in attempts to introduce new health-oriented food products. Likewise, physical retailers may explore some form of in-store personal assistant technology as a differentiating factor to retain their customers. A health-based star rating solution stands ready to integrate upon proliferation of smart shopping carts with self-checkout capabilities; especially once item-level RFID sensors become adopted in the future following Amazon Go's launch.

Over the long run, healthier food choices may translate into lower medical expenses and reduced pressure on healthcare. One avenue of development could involve bulk licensing of a mobile APP to hospitals or dietitians as complementary services available to their patients. With better refined algorithms and sophisticated data security, a grocery assistant solution can theoretically be adapted to link with electronic health records or assigned by medical doctors to promote diets based on personalized health concerns.

References

1. Hudson, E.: Health and Wellness the Trillion Dollar Industry in 2017: Key Research Highlights. Euromonitor International, November 2012. blog.euromonitor.com
2. What to Do When There Are Too Many Product Choices on the Store Shelves? Consumer Reports Magazine, March 2014. www.consumerreports.org/cro/magazine/2014/03.htm
3. Nielsen Global Health and Wellness Report: The Neilsen Company, January 2015. https://www.nielsen.com/content/dam/nielsenglobal/eu/nielseninsights/pdfs/Nielsen%20Global%20Health%20and%20Wellness%20Report%20-%20January%202015.pdf
4. Moore, G.A.: Crossing the Chasm: Marketing and Selling High-Tech Products to Mainstream Customers. Harper Collins, New York (1991)
5. Ahn, J., Williamson, J., Gartrell, M., Han, R., Lv, Q., Mishra, S.: Supporting healthy grocery shopping via mobile augmented reality. ACM Trans. Multimedia Comput. Commun. Appl. **12**(1s) (2015). Article 16, 24 pages. doi:10.1145/2808207
6. Shekar, S., Nair, P., Helal, A.: iGrocer: a ubiquitous and pervasive smart grocery shopping system. In: Proceedings of the 2003 ACM Symposium on Applied Computing, pp. 645–652 (2003)
7. Rosenstock, I.: Historical origins of the health belief model. Health Educ. Behav. **2**(4), 328–335 (1974). doi:10.1177/109019817400200403

8. Escoto, K.H., Laska, M.N., Larson, N., et al.: Work hours and perceived time barriers to healthful eating among young adults. Am. J. Health Behav. **36**, 786–796 (2012)
9. Ross, A.C., Taylor, C.L., Yaktine, A.L., et al. (eds.): Dietary Reference Intakes for Calcium and Vitamin D. National Academies Press, Washington, D.C. (2011)
10. Petty, R.E., Cacioppo, J.T.: The elaboration likelihood model of persuasion. Adv. Exp. Soc. Psychol. **125** (1986). doi:10.1016/s0065-2601(08)60214-2
11. Li, H., Chatterjee, S.: Designing effective persuasive systems utilizing the power of entanglement: communication channel, strategy and affect. In: Ploug, T., Hasle, P., Oinas-Kukkonen, H. (eds.) PERSUASIVE 2010. LNCS, vol. 6137, pp. 274–285. Springer, Heidelberg (2010). doi:10.1007/978-3-642-13226-1_27
12. Fogg, B.J.: Creating persuasive technologies: an eight-step design process. In: Proceedings of the 4th International Conference on Persuasive Technology, pp. 1–6. ACM, Claremont (2009)
13. Fogg, B.J.: A Behavior Model for Persuasive Design. Persuasive Technology Lab, Stanford University. http://bjfogg.com/fbm_files/page4_1.pdf. Accessed 15 Nov 2016
14. Fogg, B.J.: Persuasive Technology: Using Computers to Change What We Think and Do. Morgan Kaufmann, San Francisco (2003)
15. Nutritional Scoring System: Nuval LLC, Massachusetts. www.nuval.com

Exploring the Links Between Persuasion, Personality and Mobility Types in Personalized Mobility Applications

Evangelia Anagnostopoulou[1(✉)], Babis Magoutas[1], Efthimios Bothos[1], Johann Schrammel[2], Rita Orji[3], and Gregoris Mentzas[1]

[1] ICCS of NTUA – National Technical University of Athens, Athens, Greece
{eanagn,elbabmag,mpthim,gmentzas}@mail.ntua.gr

[2] AIT – Austrian Institute of Technology, Vienna, Austria
Johann.Schrammel@ait.ac.at

[3] University of Waterloo, Waterloo, Canada
rita.orji@uwaterloo.ca

Abstract. Recent approaches on tackling the problem of sustainable transportation involve persuasive systems and applications. These systems focus on changing citizens' behavior towards adopting transportation habits that rely more on the use of public transportation, bicycles and walking and less on private cars. A main drawback of existing applications is the limited use or lack of personalization aspects that consider differences in users' susceptibility to persuasive strategies. In this paper, we explore two user traits that can be used for personalizing the persuasive strategies applied to end users: personality and mobility type. More specifically, we present the results of a study where we examined the perceived persuadability of eight persuasive strategies on users of five personality types and three mobility types.

Keywords: Personality · Persuasion · Mobility type · Personalization

1 Introduction

Persuasive systems can provide the means to respond to sustainability problems arising from human activities. This kind of persuasive systems are commonly referred as "persuasive sustainability systems" and their aim is to foster sustainable behaviors and raise individuals' awareness of their choices, behavior patterns and the consequences of their activities [1]. In this paper we focus on the problem of sustainable transportation. The impact of humans' transportation habits on the environment are grave: transportation accounts for 20% to 25% of the world energy consumption and carbon dioxide emissions [2], while these emissions are increasing faster than in other energy using sectors, especially in urban environments [3]. Increased urbanisation and mobility solutions highly dependent on private vehicles aggravate this situation.

Recent approaches on tackling the problem of sustainable transportation involve persuasive systems and applications. These systems focus on increasing travellers' awareness of the environmental impact of travel mode choices and changing citizens' behaviour towards adopting transportation habits that rely more on the use of public

P.W. de Vries et al. (Eds.): PERSUASIVE 2017, LNCS 10171, pp. 107–118, 2017.
DOI: 10.1007/978-3-319-55134-0_9

transportation, bicycles and walking and less on private cars [4]. Such choices lead to Green House Gas (GHG) emissions reduction, mitigating the effects on the environment while they are correlated with more healthy lifestyles, including increased exercise and less obesity [4]. Our recent review of related systems [5] has shown that persuasive systems in the context of personal mobility is an active area of research, with numerous approaches aiming to motivate more eco-friendly choices.

Typically, persuasive strategies are incorporated in applications that support the selection of routes and transportation means in everyday mobility needs [4, 5]. Common strategies include self-monitoring in the form of statistics that provide visualizations of users' past choices (e.g. graphs showing the emissions caused by the user's decisions), social comparisons where users are compared to peers and goal-setting along with rewards for achieving target behaviours (e.g. using more public transportation).

A main drawback of existing applications is the limited use or lack of personalization aspects [6] that consider differences in users' susceptibility to persuasive strategies. Personalized approaches can be more successful than "one size fits all" as they can adapt the selected persuasive strategies to specific users, rather than the general audience and can sustain users' interest over time while providing better results [7].

However, further exploration of persuasive strategies' personalization for behavioural change towards sustainable mobility is required. Most existing studies on this topic have focused on personalization of persuasive systems in the health domain [8, 9]. In this paper we explore two user traits that can be used for personalizing persuasive strategies: personality and mobility type. We present the results of a study where we examined the perceived persuadability of eight persuasive strategies on users of five personality types and three mobility types. We developed eleven models examining the relationships between persuasive strategies and personality/mobility types and created persuasive profiles, i.e. ranked lists of strategies that can be employed to motivate sustainable behavior for specific user personality/mobility types.

The remainder of the paper is organized as follows. Section 2 provides a short background on personality and mobility types and the related work on personalized persuasion in the domain of mobility. Section 3 describes our methodology and the process we followed to analyse the collected data. Section 4 presents the results of our analysis. In Sect. 5 we discuss our findings and we conclude in Sect. 6 with our final remarks and suggestions for future work.

2 Background and Related Work

In this section we present the main concepts we examine in this paper.

Personality Traits. Personality is a key driver behind people's interactions, behaviors, and emotions and it is related to preferences and interests [10]. One of the most widely used models in the area of psychology for studies encompassing personality and human behavior is the big five factor model, which measures a user's personality in terms of five dimensions, namely Openness, Conscientiousness, Extraversion, Agreeableness and Neuroticism [11] (also known with the acronym OCEAN). Each user's personality type is associated with certain characteristics as shown in Table 1.

Table 1. Personality Traits based on the Big Five model and associated characteristics.

Personality type	Characteristics
Openness	Appreciation for novelty or variety in experiences, diversity in interests
Conscientious-ness	Organized, consistent, cautious and dutiful, less creative
Extraversion	Appreciation for environments with higher levels of stimulation, high energy, more activity and social life
Agreeableness	Cooperative, Adaptable, submissive, tolerant, generous, modest and trusting
Neuroticism	High susceptibility to anger, frustration, insecurity, pessimism, anxiety and negative emotions

The measurement of users' personality is performed using specific instruments. The best known for contexts in which participant time is limited, is the Big Five Inventory proposed by Rammstedt et al. [12]. This instrument consists of ten 5-point Likert scale questions used to get a measure of the five personality dimensions.

In the domain of social psychology the association between personality traits and persuadability is still under exploration. In their preliminary study, Hovland et al. [13] found that people with high Neuroticism scores tend to be more susceptible to social influence and to be persuaded by social comparison. However, studies examining the relationship between Extraversion and persuasion have found contrasting results. In several studies a negative impact of Extraversion on persuasion has been reported [14]. On the other hand, other studies report that extraverts tend to be susceptible to influence given their needs to be socially desirable [15]. In addition, Hirsh et al. [16] examined whether message-person congruence effects can be obtained by framing persuasive messages in terms of Big-Five personality dimensions, using a sample of 324 survey respondents. The participants judged an advertisement emphasizing a particular motivational concern as more effective when it was congruent with their own personality traits. Hence, their results suggested that adapting persuasive messages to the personality traits of the recipient can be an effective way of increasing their impact, highlighting the value of personality-based communication strategies.

The Big-Five model was used in [17] to guide the design of persuasive systems for combating obesity trends in teenagers. Specifically, the authors used Big-Five traits to recommend games and to select persuasive messages to encourage users to play. In the health domain, Halko and Kientz [9] reported statistically significant relationships between personality traits and persuasive strategies. They found a positive correlation between Neuroticism trait and the cooperation strategy. On the contrary, negative correlations were found between the Conscientiousness trait and social persuasive strategies. This finding may indicate that conscientious people may be less susceptible to the use of social persuasive technologies in the health domain.

Mobility Types. The term mobility type is used in order to segment travel behaviors into potential "mode switchers" and support mobility management policies. A widely-adopted classification of mobility types is provided in [18], see also Table 2. Each class represents a unique combination of preferences, worldviews and attitudes, indicating that different groups need to be serviced in different ways to optimize the chance of influencing mode choice behavior.

Table 2. Mobility types as identified in [18].

Mobility type	Description
Devoted drivers	Prefer to use a car than any other mode of transport and they are not interested in reducing their car use
Image improvers	Like to drive, don't want their ability to drive to be restricted, but recognize that it would be good if they all reduced car use a little
Malcontented motorists	They want to cut down their car use but find that there are a lot of practical problems and issues with using alternative modes
Active aspirers	They feel that they drive more than they should and they would like to cut down
Practical travelers	They regard the car merely as a practical means of getting from A-B and largely use it only when necessary
Car contemplators	They do not have a car at the moment but would like one at some point in the not so distant future
Public transport dependents	Although they are not against cars in any way and think people should be allowed to use them freely, they don't like driving very much
Car-free choosers	They are not keen on driving and believe that cars and their impacts are something that need to be urgently addressed

For our analysis we grouped the mobility types of Table 2 into three classes: *Drivers*, i.e. people that actively use their car (Devoted Drivers and Image Improvers), *Potential non-Drivers*, i.e. people that want to avoid using their car (Malcontented Motorists, Active Aspirers, Practical Travelers) and *non-Drivers*, i.e. people who do not use a car (Car contemplators, Public Transport Dependents, Car-free Choosers).

Personalized Persuasion in Mobility Apps. Existing applications that focus on personalized persuasion in mobility apps try to personalize specific aspects of a single persuasive strategy and not the persuasive strategy per se. For example, in [19] an approach of personalizing challenges (competition strategy) is described, while in [20] an application that persuades users to make more sustainable choices through personalized suggestions and self-monitoring is implemented. Gabrielli et al. [21] describe a mobile application that motivates users to make sustainable transportation choices using goal-setting, self-monitoring and personalized notifications. Froehlich et al. [22], describe a transport application which adapts the graphics to provide visual feedback in order to reduce driving and to make personalized recommendations.

3 Study Design and Data

In this study we examine the perceived persuasiveness of eight persuasive strategies, namely *comparison, self-monitoring, suggestion, simulation, cooperation, praise, personalization and competition*, on users of different personality and mobility types. Out of the landscape of persuasive strategies we select those based on the work of Orji et al. [23] since these fit well to our envisaged application of motivating sustainable mobility choices. Table 3 provides an overview of how the selected strategies can be applied in the mobility domain.

Table 3. Selected persuasive strategies and how they can be used in the domain of mobility.

Strategy	Description
Comparison	Comparison of one's own mobility behavior to that of others
Self-monitoring	Tracking user behavior and providing feedback on the emissions caused by his/her choices
Suggestion	System generated suggestions that urge users to follow more environmental transportation modes
Simulation	Graphical representations presenting the impact of mode choices on the environment
Cooperation	Challenges such as a bicycle commuter challenge where users co-operate to meet certain targets (e.g., meeting a 100 km per month target per team)
Praise	Positive feedback when users achieve or exceed eco-efficiency goals
Personalization	Route planners that remember past preferences, frequently chosen means of transport and simplify the processes of finding the most sustainable travel choice
Competition	Online competitions to motivate mobility-related behavior change (e.g., taking the bicycle instead of the car, taking the train instead of the plane)

3.1 Procedure and Participants

In order to gather the data required for our analysis, we designed a simple questionnaire that captures personality, mobility type and user susceptibility to the persuasive strategies. The questionnaire consists of three parts. The first part contains demographics and ten questions about participants' personality as defined in [12] which provide a short instrument for measuring personality. The second part contains questions for identifying the mobility type of the participants. To collect these data, we use the scale defined in [18], along with the three classes of mobility types define in Sect. 2. The third part concerns user susceptibility to the selected persuasive strategies (i.e. persuadability). Following the approach described in [8], we present each persuasive strategy in a storyboard showing a character and her interactions with a mobile application that aims to support her to use more environmentally friendly transportation options. The use of storyboards alleviates the burden of implementing the application and then gathering feedback, provides the means to collect responses from diverse populations with a visual language that almost anybody can read and understand [24] and allows us to collect adequate volumes of data needed for building and validating our persuadability model. Figure 1 provides an example of one of the storyboards.

Fig. 1. Example storyboard representing the comparison persuasive strategy.

To measure the perceived persuasiveness of the strategies we adopt the measure described in [8], i.e. each storyboard is followed by a set of four questions as follows: (i) "The mobile app would influence me." (ii) "The mobile app would be convincing." (iii) "The mobile app would be personally relevant for me." and (iv) "The mobile app would make me use more environmentally friendly transportation means". The questions were measured in a 7-point Likert scale ranging from "1 = Strongly disagree" to "7 = Strongly agree". In order to ensure that participants understood the strategy depicted in each storyboard we added one comprehension question asking them to identify the illustrated strategy from a list of 4 different strategies ("What strategy does this storyboard represent?").

We collected data using Amazon Mechanical Turk (AMT) that provides access to a large user base. Past studies have shown that the quality of the results compares to that of laboratory experiments when the setup is carefully explained and controlled [25]. Note that we set geographic restrictions in order to involve users from the US and Europe only. We gathered a total of 320 responses and after filtering out incomplete responses and responses from participants who answered the comprehension question incorrectly, we retained 120 valid responses for our analysis, 50 from Drivers, 41 from Potential non-Drivers and 29 from non-Drivers.

3.2 Data Analysis

For the analysis of our data we follow the approach of [23] and use the Partial Least Squares Structural Equation Modelling (PLS-SEM) method [26, 27]. The PLS-SEM method allows estimating complex cause-effect relationship models with latent variables. It is based on a structural equation model which is composed of two sub-models: the measurement model and the structural model. The measurement model represents the relationships between the observed data and the latent variables, while the structural model represents the relationships between the latent variables. An iterative algorithm solves the structural equation model and estimates the latent variables by using the measurement and structural models in alternating steps.

We choose PLS-SEM for our analysis because it can test theoretically supported linear and additive causal models. SEM can be used to treat unobservable, hard-to-measure latent variables. PLS is a soft modelling approach to SEM with no assumptions about data distribution. Also, it is recommended when the sample size is small, for applications with little available theory, when predictive accuracy is paramount and when the correct model specification cannot be ensured.

Before feeding the collected data in the PLS-SEM and in order to examine the relationships between the user personality/mobility type and the persuadability of various strategies, we performed an EFA (Exploratory Factor Analysis, i.e. a statistical procedure that identifies the number of latent factors in a set of variables) to determine the appropriate number of factors in our data. First we verified the suitability of our data for factor analysis with the Kaiser-Meyer-Olkin (KMO) sampling adequacies [28]. The results (KMO value 0.94) showed that our data were suitable for factor analysis [29]. EFA was performed through Principal Component Analysis (PCA) using SPSS.

Regarding the PLS-SEM analysis, the first step concerns the creation of the method's measurement and structural models that were described above. In our case, we created the PLS-SEM models by using the SmartPLS 3.0 tool [26]. We used the PLS algorithm in the same tool to estimate the path coefficient (β) for each PLS-SEM model. Then, we used the bootstrap resampling technique to calculate standard error (SE) for each structural path. Finally, we calculated t-statistics and their corresponding p values were used for testing significant differences between path estimates.

To examine the perceived persuasiveness of the eight persuasive strategies on users of different personality types, we created eight different PLS-SEM models, one for each persuasive strategy. Figure 2 depicts as an example the PLS-SEM model created to examine the correlations between personality types and the comparison strategy.

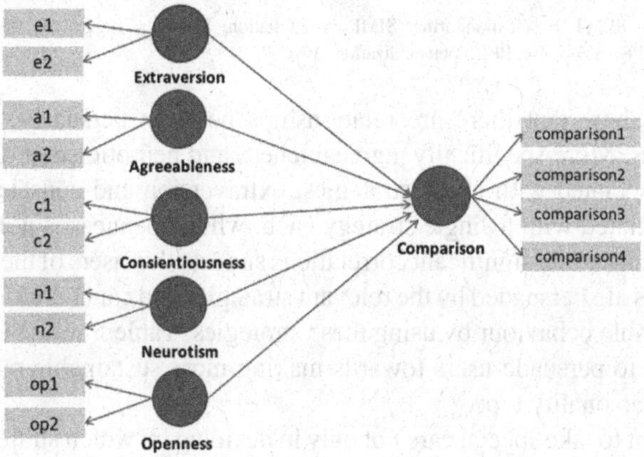

Fig. 2. PLS-SEM model examining the relationships between personality type and comparison

To examine the differences and similarity in the perception of the eight strategies by the three mobility types, we followed a similar approach. We created 3 different models, one for each mobility type, since in contrast to personality traits that characterise users in different degrees each, mobility types are mutually exclusive, i.e. each individual is characterised by exactly one of them. We therefore divided the participants into three groups based on their mobility type and for each group we created a model exploring the correlations between the relevant mobility type and the eight persuasive strategies.

4 Results

4.1 Results of Personality Type - Persuasive Strategies Relationship

The structural PLS-SEM models used in this case determine the perceived persuasiveness of the various strategies on users of different personality types. In order to measure how strong the various relationships between personality types and persuasive strategies are, we calculated the level of the path coefficient (β) and the significance of the path coefficient (p). Path coefficients measure the influence of a variable on another. The individual path coefficients (β) obtained from our models are summarized in Table 4. The table presents the results for $p < 0.05$.

Table 4. Relationships between personality type and persuasive strategies (* means $p < 0.1$).

Personality type	CMPR	SEMT	SIML	SUGG	CMPT	COOP	PRAS	PERS
Extraversion	–	–	–	–	–	–	–	−.17
Agreeableness	.19	.20*	.16*	.11	–	.19*	.23	.22
Neuroticism	.24	.23	.25	.27	.25	–	.20	.31
Openness	–	–	–	–	–	–	–	–
Conscientiousness	–	−.17*	–	–	–	–	–	–

CMPR = comparison, SEMT = self-monitoring, SIML = simulation, SUGG = suggestion, CMPT = competition, COOP = cooperation, PRAS = praise, PERS personalization

The results show that there are relationships between persuasive strategies and personality types. More specifically, agreeableness and neuroticism are positively and significantly associated with many strategies, extraversion and conscientiousness are negatively associated with a single strategy each, while openness is not related to any strategy. The positive and significant correlations suggest that users of the corresponding personality types are persuaded by the relevant strategies and can therefore be motivated to adopt sustainable behaviour by using these strategies. Table 5 shows the best and the worst strategies to persuade users towards making more sustainable mobility choices based on their personality type.

It is important to take special care not only in deciding on which strategies to employ to motivate behaviour performance for each personality type, but also which strategies to avoid in order not to deter users from performing the target behaviour. The results from our model will be used to decide the most appropriate strategy for each user in order to recommend sustainable routes in the most convincing way.

Table 5. Best/worst strategy for each personality type.

Personality type	Best strategies	Worst strategies
Agreeableness	Praise, Personalization, Self-monitoring, Cooperation, Comparison, Simulation, Suggestion	–
Neuroticism	Personalization, Suggestion, Competition, Simulation, Comparison, Self-monitoring, Praise	–
Conscientiousness	–	Self-monitoring
Extraversion	–	Personalization

4.2 Results of Mobility Type - Persuasive Strategies Relationship

The individual path coefficients (β) obtained from our PLS-SEM models in the case of mobility type are summarized in Table 6. The table presents the results for $p < 0.05$. The mobility types of Table 6 have been defined in Sect. 2.

Table 6. Relationships between the mobility type and persuasive strategies.

Mobility type	CMPR	SEMT	SIML	SUGG	CMPT	COOP	PRAS	PERS
Drivers	0.66	0.58	0.51	0.50	0.62	0.54	0.62	0.60
Potential non-Drivers	0.58	0.54	0.46	0.58	0.56	0.52	0.50	0.46
Non-Drivers	0.60	0.68	0.52	0.60	–	–	0.60	–

Table 7. Best strategy for each mobility type.

Mobility type	Best strategies
Drivers	Comparison, Competition, Praise, Personalization, Self-monitoring, Cooperation, Simulation, Suggestion
Potential non-Drivers	Comparison, Suggestion, Competition, Self-monitoring, Cooperation, Praise, Simulation, Personalization
Non-Drivers	Self-monitoring, Comparison, Suggestion, Praise, Simulation

The results summarized in Table 6 show that all mobility types are positively and significantly associated with many strategies. Table 7 shows the best strategies to persuade users towards making more sustainable mobility choices based on their mobility type.

5 Discussion

In our exploratory study we found statistically significant relations between personality types and persuasive strategies for all personality types except from openness. Our study shows that people with high scores in the personality trait of openness were not persuaded from any strategy.

In addition, we found that the best strategy for people with high agreeableness is praise, i.e. they prefer to get a reward, in order to change their behaviour. Personalization and self-monitoring are also good strategies to persuade these people, i.e. they may prefer to see their actual past and current states to meet their goals. A correlation between this personality type and comparison was also identified in [7]. Moreover, our results indicate that the best strategy to persuade people with high neuroticism scores is personalization. Furthermore, suggestion and competition are convincing strategies for these people.

We didn't find any positive correlation of people with a high score in the personality traits of conscientiousness and extraversion with any persuasive strategy. We found only negative correlations with self-monitoring and personalization strategies, respectively. Thus, people who are conscientious or extravert may not prefer any strategy to persuade them. Our results are also confirmed by Halko [7], who found only negative correlations between extraversion and persuasive strategies.

Table 8. Best strategies based on the personality and mobility type.

Personality/Mobility type	Best strategies
Drivers with high agreeableness scores	Praise, Comparison, Self-monitoring, Cooperation, Personalization, Simulation, Suggestion
Potential non-Drivers with high agreeableness scores	Comparison, Self-monitoring, Praise, Cooperation, Suggestion, Personalization, Simulation
Non-Drivers with high agreeableness scores	Self-monitoring, Praise, Comparison, Suggestion, Simulation
Drivers with high neuroticism scores	Personalization, Comparison, Competition, Praise, Self-monitoring, Suggestion, Simulation
Potential non-Drivers with high neuroticism scores	Suggestion, Comparison, Competition, Self-monitoring, Personalization, Simulation, Praise
Non-Drivers with high neuroticism scores	Praise, Self-monitoring, Suggestion, Comparison, Simulation

Our study highlighted differences in the perceived persuadability of the eight persuasive strategies on users of the three different mobility types. As it is observed the three most convincing persuasive strategies for Drivers are comparison, competition and praise. Potential non-Drivers are more susceptible to comparison, suggestion and competition, while non-Drivers to self-monitoring, comparison and suggestion.

Thus, based on the above results, knowing the mobility and personality type of users we can persuade them using the strategies that are more convincing. If the individual has high scores in Extraversion, Openness or Conscientiousness, which are not significantly and/or positively associated with any persuasive strategy, the selection of the most convincing strategy can be done based on her mobility type. In case the individual has high agreeableness or neuroticism scores she can be persuaded by taking into account both her personality and mobility types. Table 8 presents best strategies in case personality and mobility types of the individual are combined on the basis of an average of the corresponding correlations.

6 Conclusions

In this paper, we examined the perceived persuasiveness of eight persuasive strategies on users of five personality types and three mobility types. We found positive statistically significant correlations between Agreeableness and Neuroticism and seven persuasive strategies each. In addition, we found relationships between the three mobility types and almost all the persuasive strategies. Finally, we combined the results from our models in order to identify user persuadability on the basis of both personality and mobility types. Based on the developed personality-based and/or mobility-type based persuadability models, we will build a personalised persuasion service that will select the most appropriate persuasive strategies based on the personality and/or the mobility type of each user in order to nudge him/her towards more sustainable transportation modes. Our next step is to implement a mobile application that recommend routes and tries to persuade users to make more sustainable mobility choices based on the envisaged personalised persuasion service.

Acknowledgements. Research reported in this paper has been partially funded by the European Commission project OPTIMUM (H2020 grant agreement no. 636160-2).

References

1. Fogg, B.J.: Persuasive Technology: Using Computers to Change What We Think and Do. Morgan Kaufmann, San Francisco (2003)
2. IPCC Fourth Assessment Report: Mitigation of Climate Change, Chap. 5. In: Transport and its Infrastructure. Inter-governmental Panel on Climate Change (2009)
3. Pucher, J., Dijkstra, L.: Promoting safe walking and cycling to improve public health: lessons from the Netherlands and Germany. Am. J. Public Health **93**, 1509–1516 (2003)
4. Holleis P., Luther, M., Broll, G., Cao, H., Koolwaaij, J., Peddemors, A., Ebben, P., Wibbels, M., Jacobs, K., Raaphorst, S.: TRIPZOOM: a system to motivate sustain able urban mobility. In: 1st International Conference on Smart Systems, Devices and Technologies (2012)
5. Anagnostopoulou, E., Bothos, E., Magoutas, B., Schrammel, J., Mentzas, G.: Persuasive Technologies for Sustainable Urban Mobility. arXiv preprint arXiv:1604.05957 (2016)
6. Berkovsky, S., Kaptein, M., Zancanaro, M.: Adaptivity and personalization in persuasive technologies. In: PPT@ PERSUASIVE, pp. 13–25, April 2016

7. Halko, S., Kientz, J.: Personality and persuasive technology: an exploratory study on health-promoting mobile applications. In: Persuasive Technology, pp. 150–161 (2010)
8. Orji, R., Regan, L.M., Julita, V.: Gender and Persuasive Technology: Examining the Persuasiveness of Persuasive Strategies by Gender Groups. Persuasive Technology (2014)
9. Orji, R., et al.: Tailoring persuasive health games to gamer type. In: Proceedings of the SIGCHI Conference on Human Factors in Computing Systems (2013)
10. Youyou, W., Kosinski, M., Stillwell, D.: Computer-based personality judgments are more accurate than those made by humans. Proc. Nat. Acad. Sci. **112**(4), 1036–1040 (2015)
11. John, O.P., Donahue, E.M., Kentle, R.L.: The Big Five Inventory-Versions 4a and 54. University of California, Berkeley, Institute of Personality and Social Research, Berkeley (1991)
12. Rammstedt, B., Goldberg, L.R., Borg, I.: The measurement equivalence of Big-Five factor markers for persons with different levels of education. J. Res. Pers. **44**(1), 53–61 (2010)
13. Hovland, C., Janis, I., Kelly, H.: Communication and Persuasion: Psychological Studies of Opinion Change. C.T. Yale University Press, Yale (1953)
14. Gerber, A.S.: Big five personality traits and responses to persuasive appeals: results from voter turnout experiments. Political Behav. **35**(4), 687–728 (2013)
15. Eysenck, H.: The Structure of Human Personality. Methuen, London (1953)
16. Hirsh, J.B., Kang, S.K., Bodenhausen, G.V.: Personalized persuasion: tailoring persuasive appeals to recipients' personality traits. Psychol. Sci. **23**(6), 578–581 (2012)
17. Arteaga, S.M., Kudeki, M., Woodworth, A.: Combating obesity trends in teenagers through persuasive mobile technology. ACM SIGACCESS Access. Comput. **94**, 17–25 (2009)
18. Anable, J., Wright, S.D.: Golden Questions and Social Marketing Guidance Report (2013)
19. Jylhä, A., Nurmi, P., Sirén, M., Hemminki, S., Jacucci, G.: MatkaHupi: a persuasive mobile application for sustainable mobility. In: UbiComp 2013, Zurich, Switzerland (2013)
20. Bothos, E., Apostolou, D., Mentzas, G.: Recommender systems for nudging commuters towards eco-friendly decisions. Intell. Decis. Technol. **9**(3), 295–306 (2015)
21. Gabrielli, S., Forbes, P., Jylhä, A., Wells, S., Sirén, M., Hemminki, S., Jacucci, G.: Design challenges in motivating change for sustainable urban mobility. Comput. Hum. Behav. **41**, 416–423 (2014)
22. Froehlich, J., Froehlich, T.D., Klasnja, P., Mankoff, J., Consolvo, S., Harrison, B.: UbiGreen: Investigating a mobile tool for tracking and supporting green transportation habits. In: Proceedings of the SIGCHI Conference on Human Factors in Computing Systems 2009, pp. 1043–1052. ACM (2009)
23. Orji, R., Vassileva, J., Mandryk, R.: Modeling the efficacy of persuasive strategies for different gamer types in serious games for health. User Model. User-Adapted Interact. **24**(5), 453–498 (2014)
24. Lelie, C.: The value of storyboards in the product design process. Pers. Ubiquitous Comput. **10**(2–3), 159–162 (2005)
25. Paolacci, G., Chandler, J., Ipeirotis, P.G.: Running experiments on amazon mechanical turk. Judgm. Decis. Mak. **5**, 411–419 (2010)
26. Ringle, C.M., Wende, S., Becker, J.: SmartPLS.de—next generation path modeling (2005). http://www.smartpls.de. Accessed Dec 2016
27. Hair Jr., J.F., Hult, G.T.M., Ringle, C., Sarstedt, M.: A Primer on Partial Least Squares Structural Equation Modeling (PLS-SEM). Sage Publications (2016)
28. Kaiser, H.F.: An index of factorial simplicity. Psychometrika **39**, 31–36 (1974)
29. Henseler, J., Ringle, C.M., Sinkovics, R.: The use of partial least squares path modeling in international marketing. Adv. Int. Mark. **20**(20), 277–319 (2009)

Adapting Healthy Eating Messages to Personality

Rosemary Josekutty Thomas(✉), Judith Masthoff, and Nir Oren

Department of Computing Science, University of Aberdeen, Aberdeen, UK
{r02rj15,j.masthoff,n.oren}@abdn.ac.uk

Abstract. This paper considers how persuasive messages – within the healthy eating domain – should be communicated to individuals with different personality types. Following a personality assessment, subjects imagined themselves in a scenario and evaluated the effectiveness of messages constructed using Cialdini's principles of persuasion. Our results suggest that messages exploiting the principle of authority are the most effective across a range of personality types. In addition, personality had a statistically significant impact on the persuasiveness of messages, with "conscientious" subjects more willing to be persuaded than others. Finally, we found that positively framed messages were more preferred than negatively ones. We also found some interaction effects between personality traits and Cialdini's principles and framing on persuasiveness.

Keywords: Personalisation · Framing · Nutrition · Persuasion · Personality

1 Introduction

Food contributes towards obesity and other serious health conditions [1,2]. Ideally, to combat growing levels of obesity, people should receive individual support by a therapist to improve their eating habits, but this is hard to resource. Hence, there has been much work on developing digital behaviour change interventions to encourage healthy eating, which often take the form of a virtual healthy eating coach (e.g. [3,4]). We are investigating how to design such a coach which personalises to user characteristics, such as personality.

A coach can use different persuasive message types. In this paper, we investigate messages based on Cialdini's principles of persuasion [5]. Additionally, we investigate different message framings. For example, "Most people believe that eating a healthy breakfast contributes to a longer lifespan" and "Most people believe that eating an unhealthy breakfast contributes to a shorter lifespan" are messages of the same type but uses positive and negative framing respectively. People's characteristics such as personality are expected to influence the relative persuasiveness of messages [6,7]. Therefore, this paper will also investigate the effect of personality on message persuasiveness, and consider the interaction with type and framing.

P.W. de Vries et al. (Eds.): PERSUASIVE 2017, LNCS 10171, pp. 119–132, 2017.
DOI: 10.1007/978-3-319-55134-0_10

Table 1. Cialdini's six principles of Persuasion [8]

Principles	Description
Reciprocation (REC)	"People repay in kind." People are more likely to do something for someone, to who they feel they owe a favour.
Commitments and Consistency (COM)	"People align with their clear commitments." People will do something if they are committed to it. Also, they will act consistently with previous behaviour.
Consensus (CON)	"People follow the lead of similar others." People will do the same as other people who are similar to them.
Liking (LIK)	"People like those who like them." If a request is made by someone we like, we are more likely to say yes.
Authority (AUT)	"People defer to experts." If a doctor advises you to take medication, you are likely to comply.
Scarcity (SCA)	"People want more of what they can have less of." People will take the opportunity to do something that they can't leave until later

Section 2 covers related work; Sect. 3 describes the validation of persuasive messages that use framing and Cialdini's principles; Sect. 4 presents a study investigating how message type, framing and personality impact on persuasiveness; and Sect. 5 draws conclusions and proposes future work.

2 Related Work

Researchers have examined numerous behaviour modification techniques (for example, Michie et al. [9] identified 93). Cognitive and emotional approaches to persuasion can reinforce each other to achieve changes in behaviour [10]. This technique allows both cognitive and emotional content to be used is messaging.

The six principles of persuasion (Reciprocation, Commitments and Consistency, Social Proof, Liking, Authority and Scarcity) formulated by Cialdini [5] have been used to inspire persuasive messages. The principles are illustrated in Table 1. They can be easily implemented as messages to be applied in a virtual coach that encourages healthy nutrition [11]. For instance, people can be encouraged to modify their behaviour to eat fruit [12].

Different people respond to such behaviour modification strategies differently, thereby affecting the effectiveness of these strategies. There is an association between users' predisposition to various persuasive strategies, and their conformity to requests from implementations of these strategies. Kaptein et al. [13] studied the real-life relevance of persuasion profiles for encouraging people to make changes to their health behaviours with relation to physical activity and consumption of fruits. They observed significant favourable effect of persuasive health messages for individuals who are high persuadables (interest and intention values). In addition, an absence of a similar favourable effect or even an unfavourable effect was found for individuals who are low persuadables. Also, they observed a relationship between the anticipation and essential effort for behaviour change; that is individuals preferred activities which they anticipated required little effort. The effectiveness of behaviour interventions can be increased when adapted to persuasion profiles of the intended recipients. In addition, the simultaneous use of persuasive messages with other intervention strategies attempts to reduce the perceived effort and raise expected rewards [13–16].

In addition to message types and user personality, message framing can also have an impact on the persuasiveness of messages. According to Regulatory-focus theory, framed messages are segregated into gain-nongain and nonloss-loss categories [17]. Gain-nongain persuasive messages trigger promotion whereas nonloss-loss messages trigger prevention associated behaviours [18]. Also, people tend to become protective if messages are highly loss-framed; that is cases when there is a high degree of risk associated with the message. The success of loss and gain-framed messages relies on the self-relevance of message content. Loss-framed messages are more useful when content is highly self-relevant while gain-framed messages are more useful when content has low self-relevance [19,20]. Another vital factor is the individual's motivation along with their distinctions between losses and gains towards behavioural modifications [17].

3 Message Validation in Relation to Cialdini's Principles

3.1 Study Design

Our study aimed to validate a set of messages each of which instantiated only one of Cialdini's principles of persuasion [5] (see Table 1). We created 36 messages which instantiated each of the six principles. For each principle, there were three message pairs; each pair consisted of one positively and one negatively framed variant of the same message content (examples can be seen in Table 2). In the study, we validated which messages could be reliably classified as adhering to one Cialdini principle. For our further studies, we needed both messages in the pair to be validated in order to investigate the impact of framing.

Participants. Participants were recruited through Amazon Mechanical Turk [21], a crowd-sourcing internet marketplace. They needed to be located in the US, possess a 90% acceptance rating (denotes that 90% of their work is acknowledged as good value), and pass a Cloze Test for English fluency [22]. They were paid $1 for completing the study. 29 people participated: 19 male (4 aged 18–25; 8: 26–40; 7: 41–65); 9 female (5: 26–40; 4: 41–65); and 1 undisclosed (aged 26–40).

Procedure. Participants were introduced to the six Cialdini principles and their definitions (see Table 1). Next, the 36 messages were presented in a random order.

Table 2. Selection of messages for further studies

Category	Messages	REC	COM	AUT	LIK	OTH	SCA	CON	Kappa
COM	You have decided to add fruit and nuts to your cereals to make a nutritious breakfast. You will feel good if you keep your promise to yourself	0	27	0	1	1	0	0	0.84
COM	You have decided to add fruit and nuts to your cereal to make a nutritious breakfast. You will feel bad if you break your promise to yourself	2	25	1	0	0	0	1	0.70
COM	You've committed to eat a healthy breakfast. We hope you'll be sticking to it	0	24	3	1	1	0	0	0.64
COM	You've committed to eat a healthy breakfast. We hope you will not break your commitment	0	27	0	0	1	0	1	0.84
CON	Most people believe that cereals with fruit and nuts make a nutritious breakfast	0	0	0	1	0	0	28	0.92
CON	Most people believe that cereals without fruit and nuts make an unhealthy breakfast	0	1	0	0	1	1	26	0.77
CON	The majority of people believe that eating a healthy breakfast contributes to a longer lifespan	0	1	2	0	1	0	25	0.70
CON	The majority of people believe that eating an unhealthy breakfast contributes to a shorter lifespan	0	0	1	0	1	0	27	0.84
LIK	Your mom will be very glad if you add fruit and nuts to your cereal in order to make your breakfast nutritious	2	0	2	24	1	0	0	0.63
LIK	Your mom will be disappointed if you don't add fruit and nuts to your cereal in order to make your breakfast nutritious	4	0	2	22	1	0	0	0.52
LIK	Your friends will admire you if you eat a healthy breakfast	3	0	0	22	1	0	3	0.51
LIK	Your friends will be displeased with you if you don't eat a healthy breakfast	1	1	0	21	3	0	3	0.45
AUT	Dieticians recommend adding fruit and nuts to your breakfast cereal to stay healthy	0	0	29	0	0	0	0	1.00
AUT	Dieticians advise that cutting down on fruit and nuts in your breakfast cereal will adversely affect your health	0	1	28	0	0	0	0	0.92
AUT	Studies conducted by health experts have shown that eating a healthy breakfast keeps you energised	0	0	28	0	0	0	1	0.92
AUT	Studies conducted by health experts have shown that eating an unhealthy breakfast leads to fatigue	0	0	26	0	0	0	3	0.78

For each message, participants choose one of the six principles while viewing its definition, or 'other' if they felt the message did not follow any of the principles. They were informed that there were no correct or incorrect answers.

Validation Measure. The Free-Marginal Kappa [23] was utilised as a standard for demonstrating how effectively our messages were classified into the six principles. The Kappa measures the compliance between raters as follows: 1 denotes complete agreement, $0.7-1$ exceptional agreement and $0.4-0.7$ reasonable agreement. A message's Kappa had to be greater than 0.4 for a reasonable classification.

3.2 Results

In this study, Kappa ≥ 0.4 was achieved by 29 out of 36 messages, of which 16 scored a Kappa ≥ 0.7. The messages with highest Kappa for four principles which will be used in follow-on studies are presented in Table 2, which also shows how many participants selected each principle for each message.

We decided to exclude *reciprocity* and *scarcity* from the follow-on studies. Only 2 reciprocity messages validated with Kappa ≥ 0.4, and these were positive and negative framings of different message contents, making them hard to use for comparison in follow up studies. On reflection, we also felt that reciprocity is hard to apply in a system, as it requires a plausible favour (a message that validated was "We have spent a lot of effort and money in organising this "How to Eat a Healthy Breakfast" workshop. We will be disheartened if you don't eat fresh foods for breakfast."). Whilst 4 *scarcity* messages validated with reasonable agreement (Kappa ≥ 0.4), none validated with Kappa ≥ 0.7. Additionally, *scarcity* is also hard to use by an interactive healthy eating coach in a persuasive message, as messages such as - This is your last chance to eat your cereal with fruit and nuts today and replenish your body with important nutrients, may not be plausible in real life[1].

4 Adaptive Message Selection

4.1 Study Design

Our next study investigates the relationship between message properties (Cialdini principle and framing) and people's personality on the one hand and message persuasiveness on the other. We used a mixture of a within and between study design. Each participant saw all messages, using the four Cialdini principles and the two framing types. Personality was used as a between subject variable.

Participants. Participants were recruited through Amazon Mechanical Turk [21]. As before, they had to be based in the US, have an acceptance rate of 90%, and have passed a Cloze test for English fluency. Participants were paid \$1.5.

[1] Scarcity may still be good to use for reminders though.

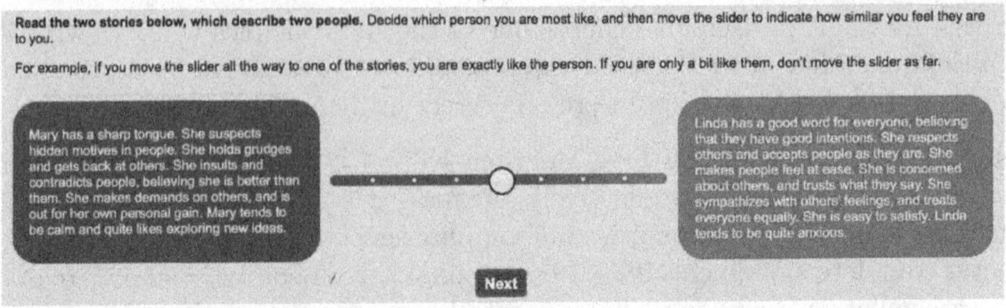

Fig. 1. Screenshot of the personality test

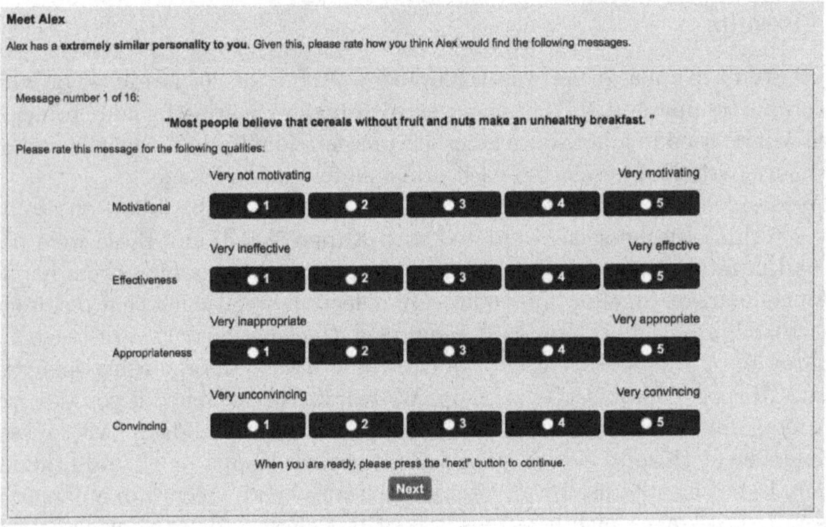

Fig. 2. Screenshot of rating scales of the experiment

152 participants took part: 89 male (22 aged 18–25; 48: 26–40; 19: 41–65); 62 female (8: 18–25; 35: 26–40; 18: 41–65; 1 over 65); and 1 undisclosed (aged 26–40).

Procedure. Participants provided their gender and age (which were optional). In addition, their attitude and behaviour towards food were examined by two questions. Next, we used a brief personality test applying personality sliders intended for the Five-Factor Model [24]. The participants were displayed stories created by Dennis et al. [25] that portray two extreme personalities (low and high) for each trait. The participants specified how similar they were to these personalities (see Fig. 1) by moving the slider. This produced a score for each trait out of 180. This personality measuring tool was previously validated as correlating with the 40-item mini markers test for the Five-Factor Model [26] as previously used by Smith et al. [8].

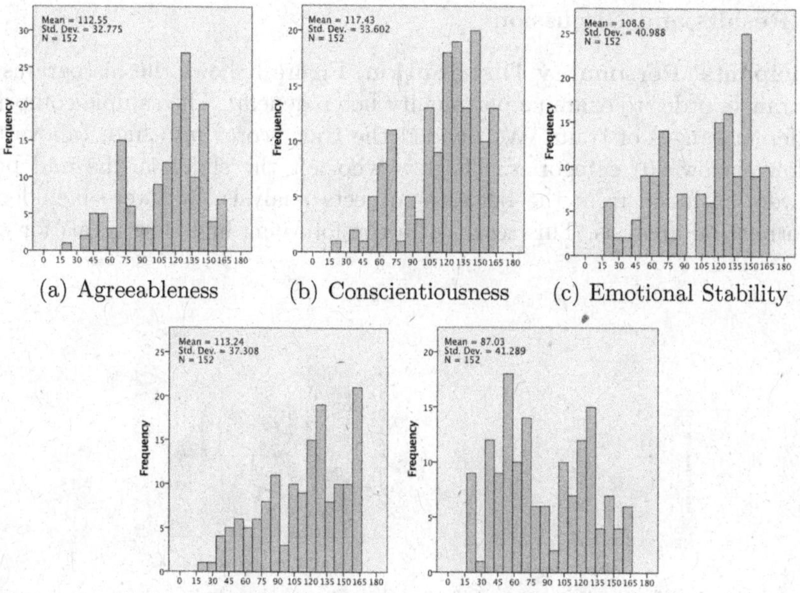

(a) Agreeableness (b) Conscientiousness (c) Emotional Stability

(d) Openness (e) Extraversion

Fig. 3. Distribution of personality traits

Next, participants imagined a fictional person 'Alex' who resembled their personality[2]. They rated the 16 messages in random order on how persuasive they would be for Alex using 4 criteria from [8]: motivational, effectiveness, appropriateness, and convincing (see Fig. 2) which were averaged for analysis. They were told that the messages would be used to encourage Alex to eat healthy breakfasts.

Hypotheses. We used the following hypotheses:

H1: Perceived persuasiveness ratings differ for different Cialdini's principles.
- H1a: Perceived persuasiveness ratings vary depending on personality.

H2: Perceived persuasiveness ratings differ for different framings.
- H2a: Perceived persuasiveness ratings vary depending on personality.

H3: There is an interaction effect.
- H3a: Between Cialdini's principles and framings on persuasiveness.
- H3b: Between personality, Cialdini's principles and framings on persuasiveness.

[2] The indirectness was used to avoid effects of participants' current eating habits.

4.2 Results and Discussion

Participants' Personality Distribution. Figure 3 shows the histograms for each trait in order to examine personality heterogeneity. The sample comprised of different ranges of traits. We divided the trait scores into high (above 90), and low (below 90) categories. The few who left the slider at the mid point (90) were excluded from the between-subjects analysis, but were included in the correlation analysis. This resulted in the following high/low totals for each

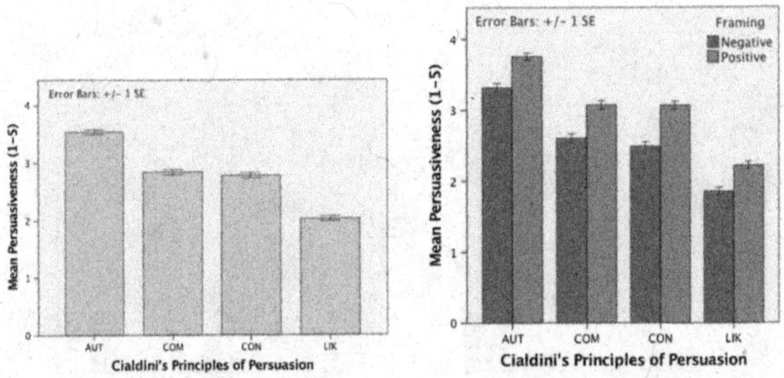

(a) Cialdini's Principles (b) Cialdini's Principles based on Framings

Fig. 4. Mean persuasiveness

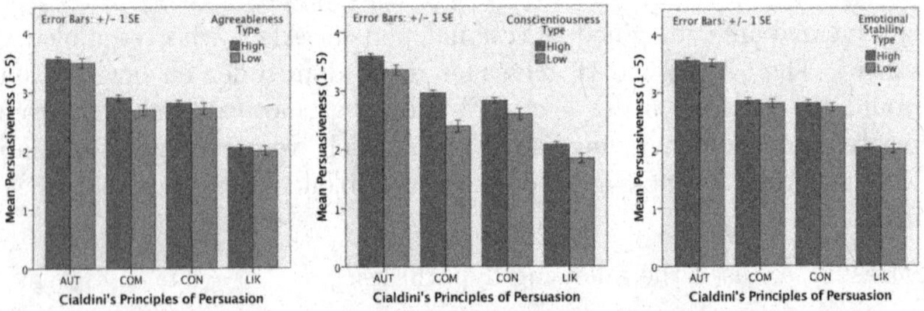

(a) Agreeableness (b) Conscientiousness (c) Emotional Stability

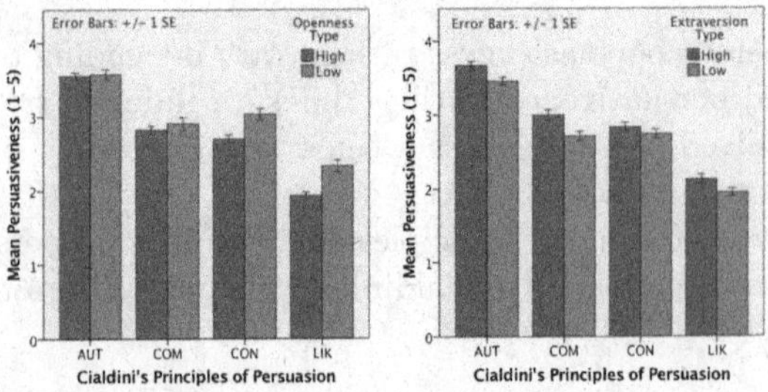

(d) Openness (e) Extraversion

Fig. 5. Mean persuasiveness of Cialdini's principles based on personality traits types.

personality trait: Extraversion: 68/82; Agreeableness: 113/38; Conscientiousness: 118/31; Emotional Stability: 99/51; and Openness: 107/39.

Effect of Cialdini's Principles on Perceived Persuasiveness. Figure 4(a) shows the mean persuasiveness of messages by principle used. AUT was the highest rated principle while LIK was the lowest. A one-way repeated measures ANOVA on principles showed a significant effect, $F(2.76, 416.10) = 158.75$, $p < 0.001$. AUT was significantly more persuasive than the other principles ($p < 0.001$) and LIK significantly less persuasive ($p < 0.001$). There was no significant difference between COM and CON. This supports hypothesis (H1) that perceived persuasiveness ratings differ for the different principles.

Effect of Framing on Perceived Persuasiveness. Figure 4(b) shows mean persuasiveness of messages by framing and Cialdini principle used. To test H2, we compared the persuasiveness of both framings of all the message pairs. On average, positively framed messages were rated higher ($M = 3.03$, $SD = 1.15$) than negatively framed ones ($M = 2.56$, $SD = 1.17$). This difference was significant, $t(1215) = 17.78$, two-tailed $p < 0.001$. This supports hypothesis, i.e., H2 that peoples' ratings of perceived persuasiveness differ for different framing.

Effect of Personality on Perceived Persuasiveness of Cialdini Principles. Figure 5 shows that there seems to be an effect of personality on perceived persuasiveness for Cialdini's principles. A one-way repeated measures ANOVA with principles as the within-subjects variable and personality traits as between-subjects variables showed that there is a significant effect of trait level for Conscientiousness ($F(1, 147) = 6.73$, $p = 0.01$), with higher persuasiveness for the high conscientious group than for the low conscientious group. There was no interaction effect with principles, though participants with high Conscientiousness seemed to rate COM messages higher. There were no significant main effects of trait level for the other traits. However, there was a trend for Openness ($p = 0.07$) and Extraversion ($p = 0.06$), with the low openness group and the high Extraversion group having higher persuasiveness ratings. There was a significant interaction effect of Cialdini's principles with the Openness trait level on persuasiveness ($F(2.79, 400.10)$, $p = 0.047$). Participants with low Openness rated both CON and LIK messages higher than participants with high Openness, as illustrated in Fig. 5(d). This provides some support for hypothesis H1a that perceived persuasiveness ratings for different principles vary depending on personality.

We also conducted a correlation analysis using the original numerical values for the traits (see Table 3). It shows many significant but small correlations between personality traits and Cialdini's principles. This further supports H1a.

For each trait, for the low and high values, a one-way repeated measures ANOVA showed a significant main effect of Cialdini's principles ($p < 0.001$). Pairwise comparisons showed that AUT was significantly more persuasive than the other principles ($p < 0.001$) and LIK significantly less ($p < 0.001$). There was no difference between COM and CON. This indicates that the results of the analyses of the whole dataset (related to H1) are still valid independent of personality.

Table 3. Personalities' correlation with the principles of persuasion

Personalities	AUT	COM	CON	LIK
Extraversion	0.140**	0.144**	0.050	0.112**
Agreeableness	0.094*	0.129**	0.104*	0.047
Conscientiousness	0.112**	0.152**	0.107**	0.058
Emotional stability	0.023	0.056	0.068	0.037
Openness	0.025	0.015	−0.095*	−0.194**

Effect of Personality on Perceived Persuasiveness of Framing. Figure 6 shows that there seems to be an effect of personality on perceived persuasiveness of framing. A one-way repeated measures ANOVA with framing as the within-subject variable and trait level as the between-subjects effects showed that there is a significant effect for Conscientiousness ($F(1, 147) = 6.73$, $p = 0.01$). Participants with high Conscientiousness rated both the framed messages much higher than those with low Conscientiousness. This is similar to what we found for Conscientiousness above, so highly conscientious people are more easily persuaded.

For both, low and high trait values, a t-test showed a significant effect of framing (two-tailed, $p < 0.001$). Participants tended to rate the persuasiveness of positively framed messages higher than negatively framed ones. This supports our earlier findings for H2.

It is interesting to note that high Emotional Stability participants rated negatively framed messages a little lower than low stability ones, but overall both groups rated positively framed messages higher than negative ones.

There were significant interaction effects between personality and framing for the trait level of Conscientiousness ($F(1, 147) = 5.30$, $p = 0.02$) and Emotional Stability ($F(1, 148) = 5.12$, $p = 0.03$), see Fig. 6. There was a trend for participants with low Openness rating both positively and negatively framed messages higher than participants with high Openness, however, this was not significant. This supports hypothesis (H2a) that peoples' ratings of perceived persuasiveness for framings vary depending on the personality of participants.

Interaction Effects between Personality, Cialdini's Principles and Framing. Figure 4(b) shows that all positively framed messages for all Cialdini's principles were rated higher than the corresponding negatively framed ones. A two-way repeated measures ANOVA on Cialdini's principles and framing showed that there is a significant interaction effect between Cialdini's principles and framing, $F(2.85, 429.91) = 4.24$, $p < 0.01$. This supports hypothesis H3a.

A two-way repeated measures ANOVA on Cialdini's principles and framings with personality traits as Between-Subjects Effects showed that there is a significant 3-way interaction effect between Emotional Stability, Cialdini's principles and framing, $F(3, 444) = 2.74$, $p = 0.04$. Figure 5(c) shows hardly any difference in the messages rated by participants with high and low Emotional Stability on

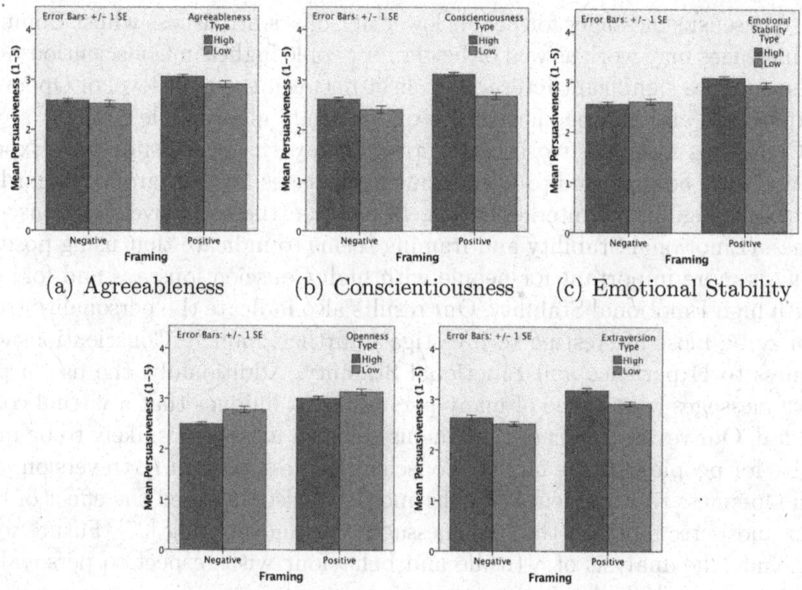

(a) Agreeableness (b) Conscientiousness (c) Emotional Stability

(d) Openness (e) Extraversion

Fig. 6. Mean persuasiveness of framing based on personality traits types

Cialdini's principles, but the statistics indicate that there is a difference when framing is also taken into account. This supports hypothesis H3b.

5 Conclusion

This paper considers how persuasive messages (within the healthy eating domain) should be communicated to individuals with different personality traits and how the persuasive message type (based on Cialdini's principles) and message framing can impact persuasiveness. Our results suggest that messages exploiting the principle of authority are the most effective across a range of personality types. So, if a virtual healthy eating coach is able to present only one message, then an Authority type seems the most likely to persuade. Similarly, it seems best to use a positively framed message, as we found that these are regarded more persuasive, independent of personality type. However, a virtual coach likely needs to use multiple messages over multiple days. Using the same type of message repeatedly may diminish its effectiveness; more investigations are needed to study the persuasiveness of messages when used in a sequence. Additionally, there may only be a limited number of messages per type available, so even if a sequence of Authority messages were the most effective, it is likely that other types may need to be used. The second most effective message types were Commitment and Consensus, and overall, there was no significant difference between them. However, from the results it seems that it may be better to use Consensus messages for people lower in Conscientiousness whilst Commitment messages may work as well or better for people higher in Conscientiousness. We also found a significant interaction effect between the trait level of Openness to Experience and the persuasiveness of the Cialdini principles, and it seems that Consensus messages work better for people with low Openness to Experience, and may be preferred to Commitment messages for that group. Regarding framing, the significant interaction effects between the trait levels of Conscientiousness/Emotional Stability and framing seems to indicate that using positive framing is more important for people with high Conscientiousness and for people with high Emotional Stability. Our results also indicate the personality traits that may be most interesting to investigate further, namely Conscientiousness, Openness to Experience and Emotional Stability. Additionally, the use of persuasive messages is only one of many persuasive techniques that a virtual coach could use. Our results indicate that the use of such messages is likely to be most effective for people who are high in Conscientiousness, high in Extraversion, and low in Openness to Experience. Furthermore, we also analysed the effect of age, gender and personality on the best message type and framing [27]. Future work will include the analysis of attitude and behaviour with respect to personality, message type and framing.

Overall, our work has applications in the area of virtual agents — by identifying the personality of a user they may be interacting with, such agents can tailor their persuasive techniques and messages to improve the outcomes of their interactions, and our work is a first step towards achieving this. We have provided some insights into the process for the selection of persuasive messages. These can then be incorporated into a heuristic to enable an intelligent agent system to deploy these adaptations. We can further develop this approach by taking into account the user's previous records on healthy eating behaviour such as including healthy foods in their diets as well as existing attitudes. For example, a person who regularly eats three portions of fruits and vegetables a day may require a different message in comparison to one who regularly eats only one portion a day. Further work is also needed on testing persuasive messages in the real world, including the longitudinal effects on behaviour (i.e., actual rather than perceived persuasiveness) and the effect of message sequences.

References

1. Churchill, S., Good, A., Pavey, L.: Promoting the avoidance of high-calorie snacks. The role of temporal message framing and eating self-efficacy. Appetite **80**, 131–136 (2014)
2. Lo, S.H., Smith, S.G., Taylor, M., Good, A., von Wagner, C.: The effect of temporal framing on behavioral intentions, expectations, and behavior: the case of healthy eating. J. Appl. Biobehav. Res. **17**(3), 202–213 (2012)
3. Mazzotta, I., de Rosis, F., Carofiglio, V.: Portia: a user-adapted persuasion system in the healthy-eating domain. IEEE Intell. Syst. **22**(6), 42–51 (2007)
4. Berkovsky, S., Freyne, J.: Group-based recipe recommendations: analysis of data aggregation strategies. In: Proceedings of the 4th ACM Conference on Recommender Systems, pp. 111–118. ACM, New York (2010)

5. Cialdini, R.B.: Influence: The Psychology of Persuasion. Harper Collins, New York (2009)
6. Masthoff, J., Grasso, F., Ham, J.: Preface to the special issue on personalization and behavior change. User Model. User-Adap. Inter. **24**(5), 345–350 (2014)
7. Orji, R., Vassileva, J., Mandryk, R.L.: Modeling the efficacy of persuasive strategies for different gamer types in serious games for health. User Model. User-Adap. Inter. **24**(5), 453–498 (2014)
8. Smith, K.A., Dennis, M., Masthoff, J.: Personalizing reminders to personality for melanoma self-checking. In: Proceedings of the 2016 Conference on User Modeling Adaptation and Personalization, pp. 85–93 (2016)
9. Michie, S., Richardson, M., Johnston, M., Abraham, C., Francis, J., Hardeman, W., Eccles, M.P., Cane, J., Wood, C.E.: The behavior change technique taxonomy (v1) of 93 hierarchically clustered techniques: building an international consensus for the reporting of behavior change interventions. Ann. Behav. Med. **46**(1), 81–95 (2013)
10. Miceli, M., De Rosis, F.D., Poggi, I.: Emotional and non-emotional persuasion. Appl. Artif. Intell. **20**(10), 849–879 (2006)
11. Grasso, F.: Rhetorical coding of health promotion dialogues. In: Dojat, M., Keravnou, E.T., Barahona, P. (eds.) AIME 2003. LNCS (LNAI), vol. 2780, pp. 179–188. Springer, Heidelberg (2003). doi:10.1007/978-3-540-39907-0_26
12. Grasso, F., Cawsey, A., Jones, R.: Dialectical argumentation to solve conflicts in advice giving: a case study in the promotion of healthy nutrition. Int. J. Hum. Comput. Stud. **53**(6), 1077–1115 (2000)
13. Kaptein, M., Lacroix, J., Saini, P.: Individual differences in persuadability in the health promotion domain. In: Ploug, T., Hasle, P., Oinas-Kukkonen, H. (eds.) PERSUASIVE 2010. LNCS, vol. 6137, pp. 94–105. Springer, Heidelberg (2010). doi:10.1007/978-3-642-13226-1_11
14. Kaptein, M., Markopoulos, P., de Ruyter, B., Aarts, E.: Personalizing persuasive technologies: explicit and implicit personalization using persuasion profiles. Int. J. Hum. Comput. Stud. **77**, 38–51 (2015)
15. Kaptein, M., van Halteren, A.; Adaptive persuasive messaging to increase service retention: using persuasion profiles to increase the effectiveness of email reminders. Pers. Ubiquit. Comput. **17**(6), 1173–1185 (2013)
16. Kaptein, M.C., Markopoulos, P., de Ruyter, B., Aarts, E.: Persuasion in ambient intelligence. J. Ambient Intell. Hum. Comput. **1**(1), 43–56 (2010)
17. Higgins, E.T.: Beyond pleasure and pain. Am. Psychol. **52**(12), 1280–1300 (1997)
18. Dijkstra, A., Rothman, A., Pietersma, S.: The persuasive effects of framing messages on fruit and vegetable consumption according to regulatory focus theory. Psychol. Health **26**(8), 1036–1048 (2011)
19. de Graaf, A., van den Putte, B., de Bruijn, G.J.: Effects of issue involvement and framing of a responsible drinking message on attitudes, intentions, and behavior. J. Health Commun. **20**(8), 989–994 (2015)
20. Godinho, C.A., Alvarez, M.J., Lima, M.L.: Emphasizing the losses or the gains: comparing situational and individual moderators of framed messages to promote fruit and vegetable intake. Appetite **96**, 416–425 (2016)
21. Amazon: Amazon mechanical turk (2016). www.mturk.com. Accessed 30 May 2016
22. Taylor, W.L.: Cloze procedure: a new tool for measuring readability. Journalism Q. **30**, 415–433 (1953)
23. Randolph, J.J.: Free-marginal multirater kappa (multirater k[free]): an alternative to fleiss fixed-marginal multirater Kappa. In: Joensuu Learning and Instruction Symposium 2005 (2005)

24. Goldberg, L.R.: The structure of phenotypic personality traits. Am. Psychol. **48**(1), 26–34 (1993)
25. Dennis, M., Masthoff, J., Mellish, C.: The quest for validated personality trait stories. In: Proceedings of the 2012 ACM International Conference on Intelligent User Interfaces, pp. 273–276. ACM, New York (2012)
26. Smith, K.A., Dennis, M., Masthoff, J., Tintarev, N.: A method of bootstrapping adapation using personality stories. In progress
27. J. Thomas, R., Masthoff, J., Oren, N.: Personalising healthy eating messages to age, gender and personality: using Cialdini's principle's and framing. In: IUI 2017 Companion. ACM (2017)

Motivations, Facilitators, and Barriers

Computers and People Alike

Investigating the Similarity-Attraction Paradigm in Persuasive Technology

Peter A.M. Ruijten(✉) and Tiange Zhao

Eindhoven University of Technology, Eindhoven, The Netherlands
p.a.m.ruijten@tue.nl, tiagozhao@gmail.com

Abstract. A study is presented that tests the relation between the (perceived) personality of an online interactive system and the personality of its user. We expected a system with a dominant interaction style to be more persuasive than a submissive one. Moreover, we expected people with dominant personalities to be persuaded more by a dominant system, while people with submissive personalities would be persuaded more by a submissive one. These expectations were tested in a study where participants were provided with automated persuasive messages that had either a dominant or a submissive style. Results support our hypotheses and show that the similarity-attraction paradigm can be extended to persuasive technologies. However, findings also show that the dominant system is perceived as less likable. Although it is hard to predict whether these effects occur in real-world settings, the current work could help creating technologies that adapt their persuasive messages to their users.

1 Introduction

Did you ever notice that many interactive technologies have persuasive elements in them? And that these technologies sometimes easily change your mind or behavior and sometimes they do not? The extent to which those persuasive elements have effect on you could be largely determined by your personality. It has been argued that the relation between type of advice you receive and your personality may have the potential to enhance the impact of persuasive technologies [2]. In addition, one of the most powerful persuasion principles is similarity [39]. Would it be possible to apply this similarity principle by adapting a persuasive message to match the personality of an individual user?

In this paper, we present a study in which we test the relation between system personality and user personality and its effect on persuasiveness of an online persuasive system. The current work is largely based on earlier work on people's perceptions of computer personalities and their relation with individual personalities [29]. It aims to go one step further and test the effects on persuasion of an online automated system.

P.W. de Vries et al. (Eds.): PERSUASIVE 2017, LNCS 10171, pp. 135–147, 2017.
DOI: 10.1007/978-3-319-55134-0_11

1.1 Personalized Persuasion

In recent years, research on persuasive technology has shifted from a focus on group persuasion towards individual and personalized persuasion [22]. According to Kaptein and colleagues [22], the effectiveness of persuasive strategies can be improved by making use of personalized strategies. Research has shown that these persuasive strategies are more effective when they are designed for individuals as opposed to a one-size-fits-all design [21,22,34,38].

This means that a strategy that influences the behavior of one type of person may not have the same effect on another type of person. In order to maximize its persuasive effectiveness, it is important to adapt the persuasive technology to its user. One way to do this is by means of personalization. More specifically, by focusing on the individual characteristics [33]. This could lead to more positive evaluations of advertisements [17], and even to changes in (healthy) behavior [38]. These findings suggest that adapting persuasive messages to an individual's personality can increase a persuasive technology's impact.

1.2 Masculinity and Interpersonal Dominance

A commonly used personality trait model is the Five Factor Model of Personality [8,15]. This model allows individual personality traits to be identified and to subsequently be divided into five different dimensions. Based on these trait dimensions, persuasive technologies can be personalized to increase their effectiveness for each individual. The model as presented by Digman [8] consists of the following five personality traits: Openness to experience, Conscientiousness, Neuroticism, Agreeableness, and Extraversion. The extraversion component is also referred to as the power, status, or control factor, and it ranges from submissive to dominance. Dominant personalities are more likely to exert power over others, while submissive personalities tend to avoid such behavior [24].

A trait that is found to be correlated with a person's interpersonal dominance is masculinity [41]. It should therefore not come as a surprise that gender has an influence on a person's interpersonal dominance orientation. Indeed, analyses of conversations by males and females showed that males control topics and interrupt more than females [12,40]. These effects could be related to findings on gender differences in influenceability [10,11]. In sum, males tend to use more dominant language than females, and females appear more influenceable than males. Therefore we could expect a dominant communication style to be more persuasive than a submissive one.

1.3 Similarity as Persuasive Cue

One of the most powerful persuasive principles is similarity [39]. People can be persuaded more easily by others who are similar to themselves, because they like them more [6]. This phenomenon is known as the similarity-attraction paradigm, and can be described as the tendency of people to be attracted to others who are similar to themselves [4,9,23]. This paradigm has been empirically supported by a large number of studies. For example, perceived similarity is found to be the most significant factor in interpersonal attraction in college students [31], and mutual friends among high school and college students tend to have similar personality profiles [19]. Moreover, among the countless studies on spousal relationships, similarity of personalities appears to be a key determinant in marital satisfaction [3,27]. Finally, people are more sensitive to persuasion from similar others or members of their in-groups than from dissimilar others [6,7].

This similarity-attraction effect does not only occur in human-human interactions. The Computers Are Social Actors (CASA) paradigm [30] shows that interactions between people and computers are fundamentally social, and that social responses to computers happen without conscious attention [28]. In one particular study, Nass and colleagues [29] investigated whether the similarity-attraction paradigm has the same effects in human-computer interactions. In this study, people with dominant versus submissive personalities were randomly matched with a computer with either a dominant or a submissive personality. This personality was created by applying several social cues. Results showed that people did not only evaluate the computer with a similar personality more positively, but they were also more satisfied with it [29].

1.4 Research Aims

The current study was designed to investigate effects of different communication styles on perceptions and persuasiveness of an online automated system. Following earlier findings on masculinity and interpersonal dominance [5,41], we expected that the system with a dominant communication style would be more persuasive than the system with a submissive communication style.

Based on earlier work on the similarity-attraction paradigm and perceptions of computer personalities [29], we hypothesized that people with dominant personalities would have more positive perceptions of the dominant system, whereas people with submissive personalities would have more positive perceptions of the submissive system. This same pattern was expected to be found for the persuasiveness of the system. These hypotheses were tested in an online study where participants performed a restaurant ranking task. This task was chosen because it enabled us to naturally provide an automated persuasive message with either a dominant or a submissive style.

2 Methods

2.1 Participants and Design

One hundred and thirty-one students (72 males and 59 females; age M = 19.71, SD = 1.94, Range = 17 to 27) sampled from a first year Bachelor course participated in the experiment. They were randomly assigned to one of two online automated systems that varied in their communication style, being either dominant or submissive. This factor is referred to as Communication Style in the remainder of this paper.

2.2 Materials

Restaurant Ranking Task. Participants performed the restaurant ranking task adopted from Andrews and Manandhar [1]. The task started with a scenario informing participants that they were going to have dinner with one of their friends, paying at most €50 per person. After the scenario was presented, participants were shown a page with information about five imaginary restaurants, differing on various attributes such as food quality, service quality, and whether it matches the friend's favorite cuisine. Participants were asked to rank their top three restaurants out of this list of five options. A screenshot of one of the options is shown in Fig. 1.

Del Posto

This brand new Italian restaurant is certainly not the average pizza & pasta place! The mouth-watering menu, brought to you on a tablet, shows traditional Italian dishes, such as antipasti, parmigiana, pasta and pizza from a wood-burned oven. On the tablet you receive information about the ingredients of the dishes and about the choice and the origin of wines. Del Posto will give you a full Italiana experience!

Food Quality:	★★★★☆ 4 stars
Service Quality:	★★☆☆☆ 2 stars
Average Cost	45 euro/person
Favorite Cuisine:	★★★☆☆ 3 stars
Decor:	★★★★★ 5 stars

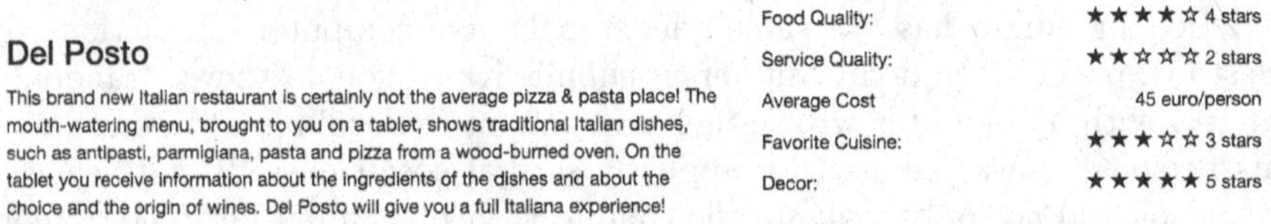

Fig. 1. Screenshot of the description of one of the five options in the restaurant ranking task. Each of the options had different characteristics as indicated by the star ratings.

On the next page, the automated system showed a message that was designed to persuade participants to flip the order of their top three restaurants. Half of the participants received this message with a dominant style, the other half received the submissive one. Participants were told that the message was coming from a chief editor working for a restaurant recommendation website. The messages were created based on earlier work on computer personalities [29]. More specifically, participants in the dominant condition encountered a chief editor named 'Max' who was displaying strong language expressed in the form of assertions and commands with a bold, sans-serif typeface. An example of such a message is shown below. The words between brackets varied based on the choices participants made and the characteristics of those choices.

"You have chosen [Del Posto] as your top preference. However, you will definitely ruin the dinner if you keep it as your best choice! Think about it: [this restaurant serves the worst service]! It is absolutely unacceptable to make such a mistake!"

Participants in the submissive condition encountered a chief editor named 'Linus' who was displaying weaker language expressed in the form of questions and suggestions with an italicized, serif typeface. An example of such a message is shown on the next page.

"You have chosen [Del Posto] as your top preference. However, would you please allow me to remind you that [the service quality in this restaurant is not good]? Perhaps you would like to reconsider your top preference?"

After reading the persuasive message from the chief editor, participants were provided the possibility to change their initial ranking of the restaurants. They could only change the order of their initial top 3 preferences, without adding other restaurants to their ranking.

Persuasiveness of the system was measured by the difference between the initial and final rankings. This difference indicates the number of switches between two options that is needed to change the initial choice into the final choice. When for example a participant's initial choice was A, B, C, and the final choice was B, A, C, persuasiveness would be 1 since there is one switch needed. An overview of the number of switches that is needed between every initial and final choice is presented in Table 1.

Table 1. Persuasiveness as the difference between initial and final order. The number represents the number of switches needed to transform the initial choice into the final choice.

Initial order	Final order	Persuasiveness
A, B, C	A, B, C	0
A, B, C	B, A, C	1
A, B, C	A, C, B	1
A, B, C	B, C, A	2
A, B, C	C, A, B	2
A, B, C	C, B, A	3

Interpersonal Dominance and System Perceptions. Two questionnaires were included in the experiment. The first measured masculinity, a concept highly correlated with interpersonal dominance (see [41]). This questionnaire consisted of 20 personality characteristics adopted from the BSRI masculinity scale [25] on which participants had to indicate on a 7-point scale (ranging from 1 'not at all' to 7 'completely') to which extent those characteristics fitted their personality (α = .86). Responses on the scale were averaged to form an Interpersonal Dominance score.

The second questionnaire measured perceptions of the online system adopted from earlier work on perceptions of computers [29]. This questionnaire consisted of six indices. Four of them focused on user perceptions of the system: Dominance (9 items, α = .76), Submissiveness (5 items, α = .76), Affiliation (4 items, α = .94), and Competence (11 items, α = .93). Two of them focused on perceptions of the interaction with the system: Satisfaction (7 items, α = .91) and Benefits (3 items, α = .89). Answers on all six indices were averaged to form separate cores. In addition, an Index of System Dominance was created by calculating the difference between the Dominance and Submissiveness scores.

2.3 Procedure

The experiment was performed online. Upon opening the website, participants were informed about the goal of the experiment and they gave informed consent by clicking the 'Continue' button. After this, participants indicated their age and gender (which was used for counterbalancing the number of male and female participants in each condition).

Next, they arrived at the page with instructions about the task and the restaurant scenario, and made their initial ranking. After they were confronted with the persuasive message from the chief editor, participants indicated their final ranking and finished the task. Participants then completed the questionnaires and finally were thanked for their contribution in the experiment. They were debriefed about the goal of the experiment and presented with the results in one of the next lectures in the course. Participants were not compensated for their participation.

3 Results

To test whether the manipulation was successful and the dominant system was indeed perceived as more dominant than the submissive system, the Index of System Dominance was submitted to an independent t-test with the two communication styles of the system as groups. Results showed that the manipulation worked sufficiently: the dominant system (M = 2.00, SD = 1.53) was perceived as more dominant than the submissive one (M = 1.46, SD = 1.60), $t(129) = 1.94$, $p = .05$, $d = 0.34$, see Fig. 2.

3.1 Dominant Personality and Perceptions of the System

After checking the manipulation, we explored whether user perceptions differed between the two communication styles. For each of the four indices of user perceptions, we performed an ANOVA with the two communication styles as groups. Results of these analyses indicated a significant effect of Communication Style on Affiliation, $F(1, 129) = 5.63$, $p = .02$, $\eta_p^2 = 0.04$. As can be seen in Fig. 3a, participants indicated that they felt less Affiliation toward the dominant system ($M = 3.63$, $SD = 1.27$) than towards the submissive one ($M = 4.14$, $SD = 1.20$).

In line with earlier findings [29], the dominant system ($M = 4.18$, $SD = 1.03$) was also perceived with a lower Competence than the submissive one ($M = 4.47$, $SD = 0.98$), see Fig. 3b. However, this effect was only significant when a one-tailed test was performed, $F(1, 129) = 2.68$, p(1-tailed) = .05, $\eta_p^2 = 0.02$. For the perceptions of people's interactions with the system, no significant effects of Communication Style were found on Satisfaction with the system and perceived Benefits of the system (both F's < 1, both p's > .60).

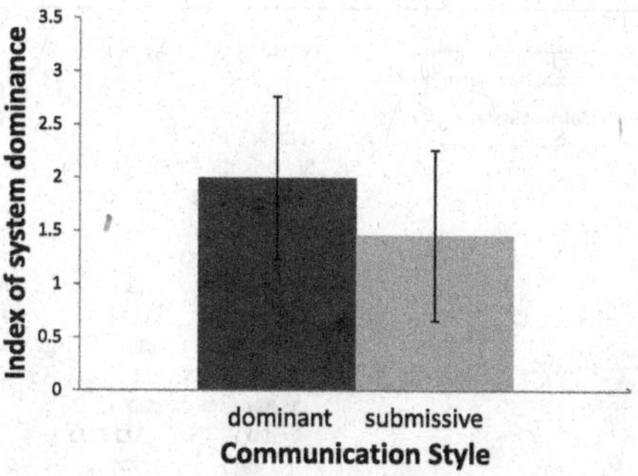

Fig. 2. Visualization of the Index of System Dominance per Communication Style. Whiskers represent standard deviations.

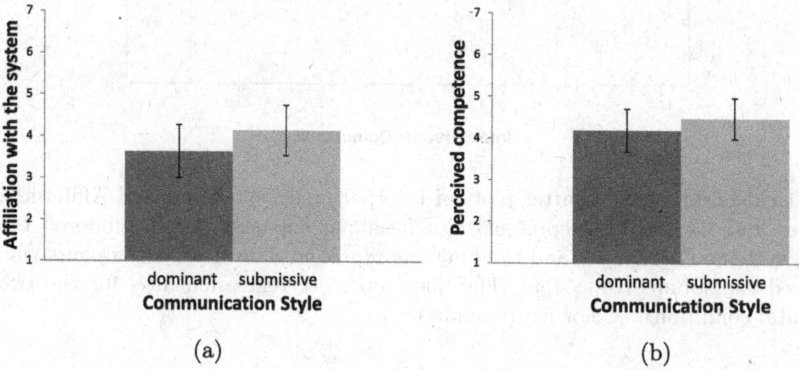

Fig. 3. Visualization of indices of user perceptions with (a) Affiliation and (b) Competence for both experimental conditions. Whiskers represent standard deviations.

3.2 The Similarity-Attraction Paradigm

Next, we tested the hypothesis that people with dominant personalities have more positive perceptions of the dominant system, whereas people with submissive personalities have more positive perceptions of the submissive system. For each of the four indices of user perceptions of the system, we performed a multiple linear regression analysis with Communication Style and Interpersonal Dominance as predictors. Results showed that Interpersonal Dominance did not significantly predict Affiliation ($\beta = .12$, $t(128) = 1.38$, *ns*). However, Communication Style did significantly predict Affiliation, $\beta = .19$, $t(128) = 2.21$, $p = .03$. Moreover, Communication Style and Interpersonal Dominance together explained a significant proportion of variance in Affiliation, $R^2 = .06$, $F(2, 128) = 3.79$, $p = .03$. As can be seen from the regression lines in Fig. 4, participants with low Interpersonal Dominance felt more affiliation towards the submissive system, whereas participants with high Interpersonal Dominance felt more affiliation towards the dominant system. No significant effects of either

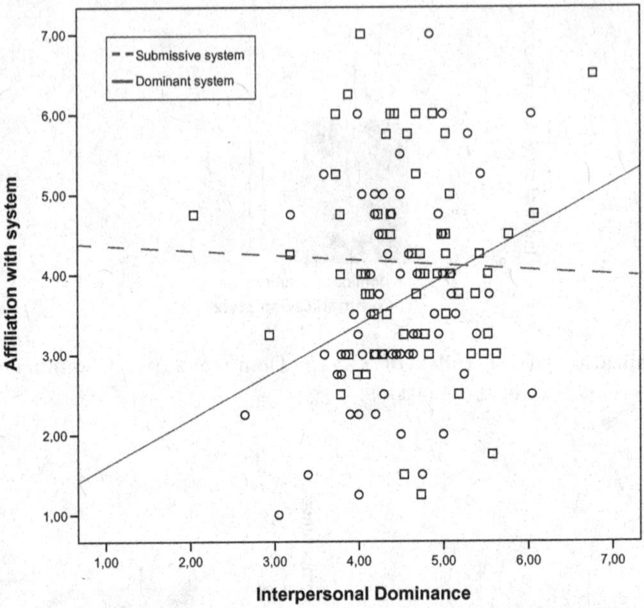

Fig. 4. Visualization of two scatter plots of Interpersonal Dominance and Affiliation. The circles and the red line represent data from participants who encountered the dominant system. The squares and the blue line represent data from participants who encountered the submissive system. The lines represent regression lines for the two experimental conditions. (Color figure online)

Communication Style or Interpersonal Dominance on any of the other three indices of user perceptions were found.

3.3 Persuasion

To test the hypothesis that the dominant system would be more persuasive than the submissive one, we tested the relation between Communication Style and Persuasiveness. This relation was tested with a Kendall's τ because Persuasiveness is an ordinal dependent variable, making it unsuitable to perform a t-test or an ANOVA. Results showed that the relationship was significant, $r_\tau = .20$, $p = .02$. As can be seen in Fig. 5, the dominant system ($M = 0.77$, $SD = 0.97$) was more persuasive than the submissive system ($M = 0.45$, $SD = 0.91$).

To test the hypothesis that participants with dominant personalities would be more persuaded by the dominant system, while participants with submissive personalities would be persuaded more by the submissive system, Persuasiveness was submitted to an ordinal regression analysis. Note that Persuasiveness is an ordinal dependent variable with more than two levels (in our case 4 levels, ranging from 0 to 3). Therefore ordinal regression is the best analytic approach. The model significantly supports the expected relationship between the two predictors and Persuasiveness, $\chi^2(2, 130) = 6.35$, $p = .04$.

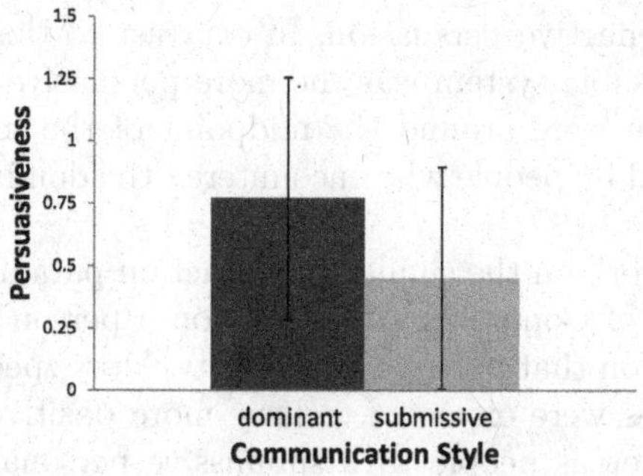

Fig. 5. Visualization of Persuasiveness per Communication Style. Whiskers represent standard deviations.

4 Discussion

The current study was designed to investigate effects of communication style of an online automated feedback system on perceptions and persuasiveness of the system. Based on earlier work on masculinity and interpersonal dominance [5,41], we expected a dominant communication style to be more persuasive than a submissive one. In line with earlier findings [29], we also expected that people would have more positive evaluations of and be more persuaded by a system that matched their personality in terms of dominance. Results of a study in which people performed a restaurant ranking task were in line with these expectations.

We expected the dominant system to be more persuasive than the submissive one. This hypothesis was confirmed, indicating that the system that displayed strong language in the form of assertions was more likely to influence people's behavior than the system that displayed weaker language in the form of questions and suggestions. This effect could also be explained by people's tendency to obey orders from authority figures [26]. Similar to some of the Milgram studies, the dominant system used phrases such as "you will definitely ruin the dinner" or "it is absolutely unacceptable to make such a mistake". This type of language could have made participants more likely to obey and thus change the order of their preferred restaurants.

The rudeness of the messages could also explain why participants felt more affiliation towards the submissive system than towards the dominant one. This finding poses an interesting design challenge. Much work in the domain of persuasive technology is performed with the aim to develop and design systems that can influence a person's behavior, while at the same time providing a pleasant interaction. Our findings show that the most persuasive system is also the least likable one, which is of course an unwanted side-effect.

This lack of likability for the dominant system could potentially even harm the persuasiveness of the system, since source likability is shown to be a determinant of persuasion [37]. Indeed, earlier research did show that computers that flatter their users are liked more than those that do not flatter [13], opening up the way towards effective persuasion. In contrast to this, the current study shows that the less likable system was the more persuasive one. However, both averages on Affiliation were around the midpoint of the scale, indicating that the *dislike* experienced by people who encountered the dominant system was not that strong.

Based on earlier work on the similarity-attraction paradigm [29], we hypothesized that the effect of Communication Style on a person's perceptions of the system would depend on that person's personality. More specifically, people with dominant personalities were expected to have more positive perceptions of the dominant system, whereas people with submissive personalities were expected to have more positive perceptions of the submissive system. This hypothesis was confirmed, replicating earlier findings on the similarity-attraction paradigm.

4.1 Limitations and Future Work

One issue that could have influenced the findings is the change of content between the persuasive messages: whereas the submissive message was phrased as a kind reminder, the dominant one used much stronger (negative) language. The importance of content and language of persuasive messages was already discussed by Fogg [14]. He stated that language can serve multiple purposes, one of which is to elevate the mood of the user to make a request more fun and seem easy. One key issue here is the difference between the two routes of persuasion [35]. When a person carefully reads the content of a persuasive message, they are more likely to be influenced through the central route [32]. This could cause a strong (negative) message to backfire. On the other hand, when a person relies on simple cues, they are more likely to be influenced though the peripheral route [32], making the content of the message less important. The content thus is more relevant for the central route than for the peripheral route. However, in the current study we changed both the content and the typeface of the message. Whereas the content could have persuaded people through the central route, the typeface could have done so through the peripheral route.

This study was performed with a rather heterogeneous group of students, with 90% of them being 22 or younger, and all of them sampled from a single University course. This makes it hard to generalize our findings to other target groups in specific e-coaching or health domains. Despite of this, results clearly show the similarity-attraction paradigm. An important next step is thus to extend our findings into different target groups and to include more personality traits in the system. For example, a study on the relation between a person's personality (measured on the Big Five Inventory, see [20]) and their perceptions of persuasive technology showed a number of interesting correlations [16]. It could thus be beneficial for a persuasive system to adapt its messages to match users' specific personality traits [16].

The reported effects are relatively small. One explanation for this is that the interaction people had with the system was very short. The importance of understanding long-term effects of persuasive technology has been stressed before [18,36], but they are still hardly investigated. Additionally, persuasion was measured with a single decision-making task, and not over a series of decisions made over time. While a single decision shows a direct persuasive effect, a longer series of decisions would be a more realistic representation of a persuasive system. For example, the dominant system might quickly become annoying and lose its advantage. In order to better understand the effectiveness of persuasive technology in the form of personalized persuasion, future studies should be designed to investigate these long-term effects.

4.2 Conclusions

The current study was designed with the aim to extend earlier work on the similarity-attraction paradigm by including effects on persuasion. The paradigm appears to influence the effectiveness of persuasive technology. In other words, a personalized persuasive interface that adapts its communication style to match the user's personality could be more effective in changing that user's behavior. Although it is hard to predict whether these effects also work in real-world settings, we hope that the work in this paper can ultimately help creating technologies that adapt their persuasive messages to their users.

References

1. Andrews, P., Manandhar, S.: Measure of belief change as an evaluation of persuasion. In: Proceedings of the Persuasive Technology and Digital Behaviour Intervention Symposium, pp. 4–9. The Society for the Study of Artificial Intelligence and the Simulation of Behaviour, London (2009)
2. Berkovsky, S., Freyne, J., Oinas-Kukkonen, H.: Influencing individually: fusing personalization and persuasion. ACM Trans. Interact. Intell. Syst. (TIIS) **2**(2), 4153–4157 (2012)
3. Blazer, J.A.: Complementary needs and marital happiness. Marriage Fam. Living **25**(1), 89–95 (1963)
4. Byrne, D.: Attitudes and attraction. In: Advances in Experimental Social Psychology, vol. 4, pp. 35–89 (1969)
5. Carli, L.L.: Gendered communication and social influence. In: Ryan, M.K., Branscombe, N.R. (eds.) The SAGE Handbook of Gender and Psychology, pp. 199–215. SAGE (2013)
6. Cialdini, R.B.: Influence: Science and Practice, vol. 4. Pearson Education, Boston (2009)
7. Cialdini, R.B., Trost, M.R.: Social influence: social norms, conformity and compliance (1998)
8. Digman, J.M.: Personality structure: emergence of the five-factor model. Annu. Rev. Psychol. **41**(1), 417–440 (1990)
9. Duck, S.W., Craig, G.: Personality similarity and the development of friendship: a longitudinal study. Br. J. Soc. Clin. Psychol. **17**(3), 237–242 (1978)
10. Eagly, A.H.: Sex differences in influenceability. Psychol. Bull. **85**(1), 86–116 (1978)
11. Eagly, A.H., Carli, L.L.: Sex of researchers and sex-typed communications as determinants of sex differences in influenceability: a meta-analysis of social influence studies. Psychol. Bull. **90**(1), 1–20 (1981)

12. Fishman, P.: Interaction: the work women do. In: Sociolinguistics: A Reader and Coursebook (1978)
13. Fogg, B.J.: Charismatic computers: creating more likable and persuasive interactive technologies by leveraging principles from social psychology (1998)
14. Fogg, B.J.: Persuasive Technology: Using Computers to Change What We Think and Do. Morgan Kaufmann, New York (2003)
15. Goldberg, L.R.: The structure of phenotypic personality traits. Am. Psychol. **48**(1), 26–34 (1993)
16. Halko, S., Kientz, J.A.: Personality and persuasive technology: an exploratory study on health-promoting mobile applications. In: Ploug, T., Hasle, P., Oinas-Kukkonen, H. (eds.) PERSUASIVE 2010. LNCS, vol. 6137, pp. 150–161. Springer, Heidelberg (2010). doi:10.1007/978-3-642-13226-1_16
17. Hirsh, J.B., Kang, S.K., Bodenhausen, G.V.: Personalized persuasion tailoring persuasive appeals to recipients personality traits. Psychol. Sci. **23**(6), 578–581 (2012)
18. IJsselsteijn, W., Kort, Y., Midden, C., Eggen, B., Hoven, E.: Persuasive technology for human well-being: setting the scene. In: IJsselsteijn, W.A., Kort, Y.A.W., Midden, C., Eggen, B., Hoven, E. (eds.) PERSUASIVE 2006. LNCS, vol. 3962, pp. 1–5. Springer, Heidelberg (2006). doi:10.1007/11755494_1
19. Izard, C.: Personality similarity and friendship. J. Abnorm. Soc. Psychol. **61**(1), 47–51 (1960)
20. John, O.P., Donahue, E.M., Kentle, R.L.: The Big Five Inventory - Versions 4a and 54. University of California, Berkeley (1991)
21. Kaptein, M., Lacroix, J., Saini, P.: Individual differences in persuadability in the health promotion domain. In: Ploug, T., Hasle, P., Oinas-Kukkonen, H. (eds.) PERSUASIVE 2010. LNCS, vol. 6137, pp. 94–105. Springer, Heidelberg (2010). doi:10.1007/978-3-642-13226-1_11
22. Kaptein, M., Markopoulos, P., Ruyter, B., Aarts, E.: Can you be persuaded? Individual differences in susceptibility to persuasion. In: Gross, T., Gulliksen, J., Kotzé, P., Oestreicher, L., Palanque, P., Prates, R.O., Winckler, M. (eds.) INTERACT 2009. LNCS, vol. 5726, pp. 115–118. Springer, Heidelberg (2009). doi:10.1007/978-3-642-03655-2_13
23. Kelly, E.L.: Consistency of the adult personality. Am. Psychol. **10**(11), 659–681 (1955)
24. Kiesler, D.J.: The 1982 interpersonal circle: a taxonomy for complementarity in human transactions. Psychol. Rev. **90**(3), 185–214 (1983)
25. Locke, K.D.: Circumplex scales of interpersonal values: reliability, validity, and applicability to interpersonal problems and personality disorders. J. Pers. Assess. **75**(2), 249–267 (2000)
26. Milgram, S.: Behavioral study of obedience. J. Abnorm. Soc. Psychol. **67**(4), 371–378 (1963)
27. Murstein, B.I.: The complementary need hypothesis in newlyweds and middle-aged married couples. J. Abnorm. Soc. Psychol. **63**(1), 194–197 (1961)
28. Nass, C., Moon, Y.: Machines and mindlessness: social responses to computers. J. Soc. Issues **56**(1), 81–103 (2000)
29. Nass, C., Moon, Y., Fogg, B., Reeves, B., Dryer, C.: Can computer personalities be human personalities? In: Conference Companion on Human Factors in Computing Systems, pp. 228–229. ACM (1995)
30. Nass, C., Steuer, J., Tauber, E.R.: Computers are social actors. In: Proceedings of the SIGCHI Conference on Human Factors in Computing Systems, pp. 72–78. ACM (1994)

31. Newcomb, T.M.: The prediction of interpersonal attraction. Am. Psychol. **11**(11), 575–586 (1956)
32. Oinas-Kukkonen, H., Harjumaa, M.: A systematic framework for designing and evaluating persuasive systems. In: Oinas-Kukkonen, H., Hasle, P., Harjumaa, M., Segerståhl, K., Øhrstrøm, P. (eds.) PERSUASIVE 2008. LNCS, vol. 5033, pp. 164–176. Springer, Heidelberg (2008). doi:10.1007/978-3-540-68504-3_15
33. Orji, R., Mandryk, R.L., Vassileva, J.: Gender, age, and responsiveness to Cialdini's persuasion strategies. In: MacTavish, T., Basapur, S. (eds.) PERSUASIVE 2015. LNCS, vol. 9072, pp. 147–159. Springer, Cham (2015). doi:10.1007/978-3-319-20306-5_14
34. Orji, R.O., Vassileva, J., Mandryk, R.L.: Modeling gender differences in healthy eating determinants for persuasive intervention design. In: Berkovsky, S., Freyne, J. (eds.) PERSUASIVE 2013. LNCS, vol. 7822, pp. 161–173. Springer, Heidelberg (2013). doi:10.1007/978-3-642-37157-8_20
35. Petty, R.E., Cacioppo, J.T.: Communication and Persuasion: Central and Peripheral Routes to Attitude Change. Springer, New York (1986)
36. Reitberger, W., Meschtscherjakov, A., Tscheligi, M., de Ruyter, B., Ham, J.: Measuring (ambient) persuasive technologies. In: Proceedings of Measuring Behavior, pp. 489–490. Noldus (2010)
37. Roskos-Ewoldsen, D.R., Fazio, R.H.: The accessibility of source likability as a determinant of persuasion. Pers. Soc. Psychol. Bull. **18**(1), 19–25 (1992)
38. Sakai, R., Peteghem, S., Sande, L., Banach, P., Kaptein, M.: Personalized persuasion in ambient intelligence: the APStairs system. In: Keyson, D.V., et al. (eds.) AmI 2011. LNCS, vol. 7040, pp. 205–209. Springer, Heidelberg (2011). doi:10.1007/978-3-642-25167-2_26
39. Tajfel, H.: Social psychology of intergroup relations. Annu. Rev. Psychol. **33**(1), 1–39 (1982)
40. West, C., Fenstermaker, S., et al.: Power, inequality, and the accomplishment of gender: an ethnomethodological view. In: Theory on Gender/Feminism on Theory, pp. 151–174 (1993)
41. Wiggins, J.S., Broughton, R.: The interpersonal circle: a structural model for the integration of personality research. Perspect. Pers. **1**, 1–47 (1985)

Office Workers' Perceived Barriers and Facilitators to Taking Regular Micro-breaks at Work: A Diary-Probed Interview Study

Yitong Huang[1(✉)], Steve Benford[1], Hilde Hendrickx[2], Rob Treloar[2], and Holly Blake[1]

[1] The University of Nottingham, Nottingham, UK
{yitong.huang,steve.benford,holly.blake}@nottingham.ac.uk
[2] Unilever R&D, Sharnbrook, UK
{hilde.hendrickx,rob.treloar}@unilever.com

Abstract. Research has suggested regular breaks in sedentary office work are important for health, wellbeing and long-term productivity. Although many computerized break reminders exist, few are based on user needs and requirements as determined by formative research. This paper reports empirical findings from a diary-probed interview study with 20 office workers on their perceived barriers and facilitators to taking regular micro-breaks at work. This work makes two contributions to the Persuasive Technology (PT) community: a diagnosis of the full range of determinants and levers for changing office work break behaviours; a demonstration of applying the Behaviour Change Wheel (BCW), an intervention development framework originating from Health Psychology, to elicit theory-based design recommendations for a potential PT.

Keywords: Workplace sedentary behaviour · Requirement elicitation method

1 Introduction

It is well recognized in health sciences that too much sitting, especially prolonged sitting (>60 min) without breaks, is associated with increased risks for metabolic syndrome, obesity, cardiovascular diseases and a range of other conditions [19], regardless of the amount of exercise [15]. Occupational sitting is a health hazard for sedentary office workers [17], who would benefit from hourly micro-breaks (3–5 min), that involve light physical activities even as simple as walking to the kitchen and refilling a mug, to alleviate the metabolic dysfunction caused by long periods of sitting [14]. In addition to physical health benefits, management literature has suggested micro-breaks are essential in maintaining employees' psychological wellbeing and energy level [9].

There exist many computerized break reminders that come in a variety of forms, ranging from popup windows on workstation screens and browser plug-ins based on the Pomodoro Technique [4], to wearable gadgets with vibrating inactivity alerts (e.g. Jawbone, Apple Watch). Previous Human-Computer Interaction (HCI) work has approached the problem

P.W. de Vries et al. (Eds.): PERSUASIVE 2017, LNCS 10171, pp. 149–161, 2017.
DOI: 10.1007/978-3-319-55134-0_12

from a cognitive perspective, in relation to interruption and attention management in organizational contexts [11, 12]; some has led to the identification of opportune moments for delivering persuasive messages during working hours [21].

However, to date, there is a lack of research looking into the problem through the lens of behaviour change. Just as a doctor needs to diagnose a patient's problem before writing out a prescription, PT designers will also benefit from carrying out a thorough behavioural diagnosis, which should reveal all behavioural facets that require modifications. In view of that, we conducted a study to answer the question of what facilitates and hinders micro-breaks at work from office workers' perspectives. The following section will introduce the methodological framework that guides our study. Afterwards, we present the study method, findings and a recommended selection of intervention functions and behaviour change techniques. The paper concludes with a discussion of limitations and practical recommendations.

2 Framework Guiding the Study: Behaviour Change Wheel

There are numerous theories that predict and explain behaviour change, presenting PT designers with the challenge of selecting theories most appropriate and relevant to the problem under investigation. It is positive that an increasing number of PT studies are underpinned by psychology theories, such as goal-setting theory [5, 10] and Transtheoretical Models [5]. However, the selection of theory is still heavily reliant on the designer's instinctive understanding of the behaviour and existing knowledge of psychology theories, rather than a systematic and theoretically guided process [13]. Such an approach excludes potentially relevant and viable theories and persuasive strategies [7]. For instance, the widely known Transtheoretical Model and Health Belief Model are increasingly questioned for their failure to address automatic motivational factors (e.g. impulses, habits, and emotions) that can be powerful drives for some behaviours [20].

To guide the process of selecting and translating theories into intervention design, several intervention development frameworks have been proposed, although most of these have been judged as not sufficiently coherent, comprehensive or well-linked to a model of behaviour, according to Michie et al. [13], who consequently developed the Behaviour Change Wheel (BCW).

The BCW is underpinned by a behavioural model at the centre called "COM-B", which breaks down behavioural problems in terms of possible deficits in three aspects (with two subcomponents in each aspect), namely Capability (psychological and physical), Opportunity (physical and social) and Motivation (automatic and reflective). The BCW also summarises nine intervention functions (e.g. education, coercion, restriction, environmental restructuring) that address one or more of the six COM-B components. As those intervention functions are defined in very general terms, they are further delineated with 93 Behaviour Change Techniques (BCTs) (e.g. "habit formation", "social reward"), which are irreducible active ingredients within an intervention package.

In the PT community, the Fogg Behaviour Model (FBM) [6] is probably the most well-known model for analysing behaviours. We believe COM-B has at least two advantages over FBM as a behavioural model. First, FBM is merely a model for analysing behaviours, whereas the COM-B is situated within the BCW, which would translate the behavioural analysis into the design of specific intervention features that target those COM-B components to produce the change. Second, developers of the BCW have made deliberate efforts to link each COM-B component with one or more domains under the Theoretical Domain Framework (TDF), an integrative framework that groups behaviour change theories into 14 domains based on overlapping constructs [3]. In a nutshell, the TDF can be seen as a variant of the COM-B model with a more fine-grained classification of facets underlying behaviours. The compatibility with TDF is valuable for two reasons. First, the TDF is a validated model already used by psychologists to elicit and analyse data in behavioural diagnosis [7, 8]. Second, while Fogg also attempts to expand on the FBM with several subcomponents (e.g. pleasure/pain, social deviance), the TDF covers a much wider range of psychological mechanisms in a more systematic manner (i.e. based on overlaps in theoretical constructs). In the next section, we will explain how the COM-B and TDF have been used for the elicitation and analysis of data in our study.

3 Method

3.1 Participants and Recruitment

The study was promoted via posters and news bulletins at the University of Nottingham, and staff mailing lists for two non-profit organisations (NPO). Office workers spending at least 2 days of the week in sedentary (chair-bound most of the time) or semi-sedentary (intermittently chair-bound and moving around but without substantial walking or physical labour) jobs volunteered to participate and were directed to an online screening questionnaire; we excluded office workers who felt that they had no discretion over timing of micro-breaks, because changing those peoples' patterns most likely required organisational/policy change and were thus beyond the scope of this work. As a result, we recruited 20 eligible participants (F = 12, M = 8, mean age = 35.4 ± 11.4 yrs. old), who were employed in a variety of office-based roles including project management, communication, IT support, clinical research admin, filmmaking, teaching and research.

3.2 Procedure and Materials

Data collection consisted of two main stages, a 2-day diary period and a 1-hour interview session. Each participant attended a 15-min briefing session with the researcher, at which they consented to participate in the study, answered demographic questions and were given a diary pad together with verbal and written instructions of the diary protocol. Participants were requested to record any two workdays in the following week as continuous series of sitting and break episodes, and note down the time whenever they left and returned to seat (Fig. 1 left). Participants were told the definition of breaks as any "interruption in sitting". For each break, participants needed to take a photo of the physical context of this break and complete a "work break experience" form, which elicited in-situ responses about the decision and experience relating to the break; for instance, the form asked participants to complete sentences such as the following, "I wish I had taken this break earlier/later (delete where inappropriate), because…".

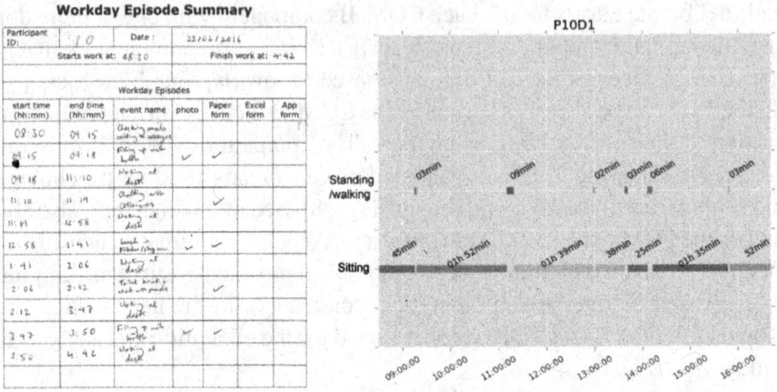

Fig. 1. A sit-break pattern graph produced based on diary entries

When participants expected an unusual day (e.g. fieldwork, conferences or travels that were not part of everyday routines), they were asked to keep the diary the following workday. The diary was collected once 2 full workdays were recorded. The recorded lengths of workdays in the study ranged from 6 h 48 min to 10 h 51 min. Semi-structured interviews based on the TDF (see Table 1) were conducted within a week following the collection of diaries. Prior to each interview, the researcher reviewed all diaries and marked specific events or behavioural patterns pertinent to theoretical constructs [3] for elaboration and clarification in interviews. The researcher also transcribed diary entries onto a spreadsheet, which was then used to produce a visual representation of the sit-break pattern for each participant (Fig. 1). Moreover, the researcher also produced two graphs with dummy data illustrating two disparate workstyles, namely "Workstyle 1" that has breaks every 2–3 h, and "Workstyle 2" that features a micro-break in every 45 min to 60 min of sitting. Those dummy examples were meant to facilitate discussion on the pros and cons of different workstyles. Starting from Participant 8, a ranking of participants based on their accumulative durations of prolonged sitting episodes per day was produced and brought to interviews.

3.3 Data Analysis

To retain links between quotes and individual respondents, the Framework Approach to qualitative analysis [18] was used – the first author read through all interview transcripts and coded relevant quotes onto TDF domains and COM-B components. Coding was then reviewed by two health psychologists familiar with the BCW, after which minor revisions were made. Afterwards, the first author summarised sub-themes on perceived barriers and facilitators emerging under each domain and counted their frequency. The BCW guide [13] that links COM-B/TDF domains to intervention functions was used to select intervention functions and BCTs likely to be effective.

Table 1. Interview topic guide based on COM-B and TDF

COM-B	TDF Domain	Eliciting Questions
Psychological Capability	Knowledge	What do you think would be the ideal work break pattern?
		How did you get to know about it? What do you think it is based on?
	Skills	How easy or difficult would you find it to follow workstyle 2?
	Memory, attention and decision process	If it were not for this study, would you always have an idea of how long you've been sitting for?
		Was that break a conscious decision? What were your thoughts when you decided that?
	Behavioural regulation	Do you set any rules for yourself regarding when you should stand up and move around?
		Do you have a system to help monitor whether you have taken regular breaks on workdays?
Reflective Motivation	Belief about consequences	(Show two work break patterns) Which of the two do you think is better? Why? How convinced are you? (prompts: in terms of health, productivity, social and mood consequences respectively)
	Belief about capabilities	Would you find it helpful to have a piece of technology that 1) monitors and displays your sitting time 2) triggers inactivity alerts 3) gives you feedback on your break pattern at the end of each day?
	Optimism	How confident do you feel about breaking up your sitting with regular micro-breaks?
	Goal	Do you want to change your current sitting patter in any way?
		Compared to the goal of completing your work, to what extent is having a healthier work pattern a priority for you? What about in the long-term?
	Intention	Is taking regular micro-breaks something you intend to do?
	Social/professional role and identity	Is sitting and working at desk for a prolonged period of time consistent with your professional standard?
		To what extent do you see yourself as someone conscious of the health impacts of your own lifestyle choices?
Automatic Motivation	Reinforcement	Would you say that generally you are in the habit of sitting for over 60 minutes/taking regular breaks? If not, what would be helpful in developing/breaking that routine/habit?
		Do you feel your break time experience is rewarding enough at the moment?
	Emotion	Does taking a break evoke an emotional response? Is the decision to take breaks influenced by any emotion?
Physical Opportunity	Environmental context and resources	What break facilities would you like to have access to?
		Are there any other factor that facilitates or hinders micro-breaks? (e.g. nature and structure of work that demands long period of concentration to get into the flow/for consistent outcome)
		How do you like the idea of having a smart cup in the office that prompts breaks?
Social Opportunity	Social influences	What's the culture of taking breaks in your workplace?
		How do your manager/supervisor perceive taking regular breaks?
		Would you feel part of a "crowd" or any social pressure if you follow workstyle 2?
		How do you find the ranking I showed you? Would you be motivated by that?

4 Results

The analysis revealed facilitators and barriers to the target behaviour (i.e. break prolonged sitting with hourly micro-breaks) in 11 associated TDF domains and 5 relevant COM-B components, as summarised in Table 2. The last column presents frequency with which each belief was mentioned among 20 interviewees. A domain was judged to be relevant, if it had prevalent belief(s) stated by over 10 participants (bold), or if several competing beliefs were raised by over 10 participants (underlined).

4.1 Factors Underlying Micro-break Behaviours

In this subsection, we highlight four important high-level themes, summarised from sub-themes in Table 2. For each quote, we report the respondent's gender, job and employer/ work setting(s) in bracket.

Variability of reflective motivation to take breaks among participants. First, we found a high variability across 20 participants in their *beliefs about consequences* of different work break styles. Only 11 participants strongly believed in the health benefits of regular micro-breaks, whereas the rest believed that prolonged sitting only had a limited impact on themselves, because they were still young, or/and that they had sufficient physical activities outside of work (e.g. 10,000 step/day) to outbalance the adverse impact of occupational sitting. When it came to productivity, 13 participants believed micro-breaks had an overall beneficial effect on work, which was a potential motivator for micro-break behaviours; however, there was also the concern that micro-breaks were particularly unproductive for certain tasks (e.g. programming, writing), as mentioned by 14 participants. In relation to beliefs about consequences, limited or uncertain *knowledge* about the optimum break interval or scientific rationale behind the recommendation was another barrier to the target behaviour, e.g. "only because my watch tells me every hour to get up… but until you send a non-Apple-paid doctor in front of me and tell me you absolutely should stand up every hour. Then I would do it" (P16, male, filmmaker, NPO).

This was mirrored by the variability in participants' *intention* to adopt a more regular break pattern. 9 out of 20 participants had clearly made the decision and some efforts to improve their break patterns before participating in the study (e.g. P2 (male, project coordinator, NPO): "I looked at this about two months ago, just apps for the MacBook for reminding you to take breaks."); 10 participants had started contemplating or preparing for the target behaviour change since participating in the study (e.g. P12 (female, clinical research admin, university/healthcare): "I think it wasn't long enough to change my behaviour then. But now I'm perhaps more aware of making sure I get up and make a drink… But even though I know I would be healthier… it's like I need that motivation to actually do it."); 2 participants had expressed no intention to change by the end of the debriefing interview. Admittedly, our study had a self-selected sample of participants who were potentially better-educated and more health-aware than average office workers. However, this variability still suggested that theoretical constructs in the

Table 2. Summary of sub-themes mapped onto COM-B and TDF with frequency counts

COM-B	TDF domains	Sub-themes about perceived facilitators and barriers	FQ. (n=20)
psychological capability	1. Knowledge	unsure about the optimum break interval or scientific rationale behind it	**16**
	2. Memory, attention and decision	forget to take breaks	**18**
		forget having taken breaks and for how many times	**20**
		complex decision process involving multiple factors	**15**
	3. Behavioural regulation	need to break existing habit and apply new "if-then" rules	**13**
		need a system to ease self-monitoring of break behaviours, provide feedback on my behaviours and progress over time	**12**
Reflective Motivation	4. Belief about consequences	**health consequences:** a. believe micro-breaks have independent health benefits	**11**
		b. unconvinced of benefits of micro-breaks	9
		productivity: a. overall speaking, workstyle 2 is more productive than 1	**13**
		b. workstyle 2 is particularly unproductive for some tasks	**14**
	5. Belief about capabilities	a technology with automatic tracking function, prompts/cues and visual feedback would give me confidence in improving my break pattern, despite some difficulty at the beginning	**12**
	6. Goals	Workstyle 2 is a goal with changing priority and accessibility	**11**
	7. Intentions	a. having taken an action towards the target behaviour change	7
		b. contemplating or preparing for the target behaviour change	10
		c. no intention to change even after participating in the study	3
Automatic Motivation	8. Reinforcement	a. existing habits that contribute to regular break behaviours	**12**
		b. existing habits that contribute to prolonged sitting	**12**
	9. Emotion	breaks evoke positive affect or remove negative affect	**12**
		breaks evoke or do not help with negative affect, or negative affect hinders micro-break behaviours	10
Physical Opportunity	10. Environmental context and resources	**the organisational culture and climate** a. encourages micro-breaks and active work culture	10
		b. neither encourages or discourages breaks despite flexibility	7
		c. discourages regular breaks and I feel I am being watched	3
		heavy workload and tight deadlines impel me to sit and work continuously longer than I would like to	**15**
		need prompts/cues; existing reminders have flaws in design	**20**
COM-B	TDF domains	Sub-themes about perceived facilitators and barriers	FQ. (n=20)
Social Opportunity	11. Social influences	**direct social interactions** that a. prompt breaks (e.g. social support, invite each other for breaks, short or walking meetings)	**11**
		b. inhibit breaks (e.g. bring drinks back to seat for each other, prolonged meetings without comfort breaks)	5
		social norm and social pressure: a. other people are good at taking regular micro-breaks and there is no pressure on sitting down to work	10
		b. other people sit quite a lot and prolonged sitting is perceived as hard-working	7

intentions domain, such as stages of change [16], was still relevant to and predictive of the target behaviour to a certain extent.

Notably, participants' motivations to break up their sitting with micro-breaks fluctuated at different times, suggesting the relevance of the *goals* domain, and its distinction from the *intentions* domain. According to the TDF, an intention refers to a conscious decision or resolve to act in a certain way, which is relatively stable, whereas a goal is the mental representation of a desired end state, which can become more or less accessible to the person depending on the context [3]. In our sample, 11 participants perceived the hourly break pattern as a desirable end state (i.e. a goal), but the goal priority can change depending on the work state, e.g. "I had several documents open, trying to match things up, I didn't want to interrupt my train of thought, but I know it's good. Just there are so many other things going on, it is a low priority (to take breaks)" (P7, female, clinical research admin, university/healthcare). For some participants, the goal to take regular breaks pertained to distal health outcomes, which, despite its prominence at the time of interview, might have to give way to more proximal goals that were mostly work-related in an office setting, e.g. "…because that's so long term, whereas you can see the short-term effects and take that on board more" (P13, female, clinical research admin, university/healthcare); "in my mode as I am NOW trying to be good, saying it is a good idea to break every 50 or 60 min, I am going to say if the display is there and it makes you break earlier then that is good. But if I am really into whatever I am doing, I would throw it out of the window or something" (P9, male, academic, university).

Lack of cognitive resources for regulating break behaviours at work. Apart from reflective motivation to adopt a healthier break pattern, a lack of cognitive resources for *memory, attention and decision processes* was another perceived barrier.

Firstly, participants reported the common experience of being "entrenched" or "engrossed" in work and forgetting to take breaks, e.g. "I lose track of time very easily, especially if I'm coding. I know afterwards when I look at the watch and I see that it's been 3 h and I haven't moved" (P1, female, academic, university). Sometimes the concentration on work led to the neglect of physiological triggers for breaks, "if we have got 10 units of attention and 10 units focused on the screen on what we are writing, then we are not going to notice that your foot hurts" (P9). Delaying toilet and water breaks despite physiological needs was a common behaviour reported by participants, which will be discussed in relation to automatic motivation in the next subsection.

Secondly, people felt they lacked the memory capacity to remember how many breaks they had taken or how many episodes of prolonged sitting they had accumulated throughout the day. For instance, P6 (male, clinical research IT support, university/healthcare) said, "It (the diary) made me realise I'd taken a break that perhaps wouldn't normally even register in my head that I'd taken a break." This had implications for the *belief about capability* domain, as participants thought a physical record of breaks like the paper diary offered the reassurance that they were able to cut off potential prolonged sitting with simple micro-breaks, "sometimes my breaks are so short that I didn't consider them as breaks. But I had to write them down. Then I thought, ah, that's a nice break. Even if it was 5 min, that's something" (P1). Visual feedback such as Fig. 1 could also facilitate self-reflection and potentially improve self-efficacy, "I think if the technology would be there, it would make me work much better to that pattern… especially if it would be something where I would look back on what I've done and then just review myself in actually you're improving or not improving on what I want to do" (P7, female, clinical research admin, university/healthcare).

Finally, decisions about whether to take a break or "power through" could become quite complex and dependent on many factors, such as progress in the current task, physical and mental fatigue, expectancy of outcome, next appointment arrangement etc. This was well illustrated by an incident reported by P18 (male, tech support/project management, university), "if I had a task I needed to complete for a meeting for 11:00 a.m., I'd look at that break reminder and go, 'right, am I going to get this stuff done for 11:00 a.m., if I have a cup of tea now?' I'd then either think, 'yes I am, I'll have a cup of tea because I know that the number of minutes after the break will be of a higher quality in terms of production and freshness than just pushing through' or I'd look and go 'you know what? It's going to be pretty damn close for me to finish, what I'll do is I'll keep going and have a cup of tea as a reward when I've done it.' It's a judgement call on an individual basis." However, given the complexity of the decision process and scarcity of cognitive resources at work, many break-related decisions were carried out "naturally" and "as a habit", as described in participants' diaries and interviews.

Habitual and emotional responses towards cues/prompts for breaks. 12 participants mentioned existing habits and routines that contributed to regular breaks. Some (P1, P7, P10, P13, P16) reported the habit of sipping water constantly while seated, so they were prompted to stand up every one to two hours to go to the toilet or/and refill vessels. The only smoker (P17, male, admin, university) in this study had the least accumulative prolonged sitting time, as he felt a strong impulse to take a cigarette break every hour. While smoking is certainly discouraged, these examples illustrated physiological needs could act as an efficient (i.e. require little attentional resources) and powerful (i.e. not easily controlled by intentionality) mechanism for instigating breaks and regulating overall break patterns.

On the other hand, automatic motivation could also work as higher-level mental processes [2] and unconscious self-regulation [1], which was manifested as habitual delay of breaks despite physiological needs for water and toilet breaks. Some described frequent moments when they put off breaks until reaching a natural "break point" in their task. In this case, some (P1, P13, P14, P16) had established a dependent relationship between natural "break points" in working tasks and the response to take a break; indeed, participants commonly described "micro-breaks" as rewards for completing a good amount of work. However, the risk with this contingency was that the scheduled work might take much longer than planned. e.g. "sometimes you plan it and then you don't plan it correctly and it takes a lot longer. It would've been good for a trigger at one hour to say this is when you should have stopped, you haven't, but you should have a break anyway" (P14, female, clinical research admin, university/healthcare). This suggested the need for applying new "if-then" rules in breaking existing habits, and the usefulness of prompts/cues that were triggered after every certain minutes of sitting.

Modifying ingrained work patterns could also involve *emotions*. For instance, P6 reported that his prolonged sitting habit stemmed from 20-year working history at a small private company, where "everything was urgent. If things broke down, they needed them repairing and you had to deal with it immediately", and this sense of urgency continued to influence his current work practice, "I think the biggest thing for me is to not feel so guilty if I'm running behind schedule". In contrast, P7 who took hourly micro-breaks most of the time, said, "because I know it is better for you, so I enjoy taking the breaks and don't feel guilty about it." This suggested affect could influence break behaviours in both directions and could act as a potential lever for change.

Organisational culture and interpersonal influences. While all participants in this study had freedom to take micro-breaks, organisational culture and the level of institutional control ranged from encouraging flexible, active and interactive work practices (e.g. "we are encouraged even, to be active and engage. You judge a person on what they do rather than how they look or where they are" (P18, male, tech support, university)), to no explicit expectation or surveillance (e.g. "I feel alright to take a break whenever, because they don't know what I'm doing as well" (P4, female, researcher, university/healthcare)), and monitoring and discouraging individuals leaving seats during office hours (e.g. "the manager will come and say where's so and so? They're not in a meeting according to my diary… Why aren't they sitting at their desks?" (P20, female, tech support, university)).

While it is challenging for a PT to directly change organisational climates, the study revealed the potential to make use of interpersonal influences on individual attitudes and behaviours. When participants were presented with the ranking based on healthiness of their own sitting patterns against those of others, they started comparing their own data with others', and making comments on both themselves and those at the bottom. The ranking might motivate them to "see how close to the top of this league you can get by having the appropriate number of breaks, never sitting for too long" (P9).

In addition to social comparison, social interactions directly affected break frequency. For example, some offices had the culture of inviting each other to make drinks together (P4, P5, P7, P12–15), whereas others had the practice of one person bringing drinks for the rest, who could remain seated and working for longer (e.g. P8, P11). Social support was another potential facilitator, e.g. "because we are in a caring environment and people do care about their colleague's health… so I think if you felt that somebody else had been sitting there for longer than is healthy then I think you could say something to them" (P19, female, clinical research admin, university/healthcare).

4.2 Selection of Intervention Functions and BCTs

In this section, we present our recommendation on intervention functions and BCTs that designers can consider when designing a PT to encourage regular micro-break behaviours. The selection was based on what the BCW suggests as potentially viable for addressing the 5 COM-B components and 11 TDF domains displayed in Table 2. A full result table can be download from http://dx.doi.org/10.17639/nott.72, which details how the identified barriers lead to recommendations that are exemplified with potential system features. As the BCTs are underpinned by behaviour change theories, PT designers need to refer to [13] for their full definitions when applying them.

Recommended intervention functions: Education, persuasion, training, enablement, environmental reconstructuring, modelling and incentivisation.

Recommended BCTs: "Credible sources", "information about health consequences", "information about social and environmental consequences", "information about emotional consequences", "feedback on behaviour", "focus on past success", "verbal persuasion about capability", "information about other's approval", "social comparison", "behavioural practice", "habit formation", "adding objects to the environment", "prompts and cues", "conserving mental resources", "action planning", "goal setting", "self-monitoring of behaviour", "review behaviour goals", "monitoring of emotional consequences", "demonstration of the behaviour", "social reward", and "reward approximation".

5 Conclusion and Recommendation

In this study, we have presented a systematic diagnosis of office workers' perceived barriers to taking micro-work-breaks. As barriers identified fall into multiple domains, a single-faceted intervention is insufficient to produce the desired behaviour change outcome. Therefore, for PT designers and practitioners whose aim is to implement a PT that is most likely to be effective, we suggest a multi-component intervention that incorporates as many of the recommended BCTs as possible, within the practical constraints imposed on them. That being said, we acknowledge the fact that our study is limited by the small sample size and potential self-selection bias, which means the identified determinants and recommended interventions may not be equally applicable to the whole population of office workers. Hence, we suggest PT designers consider our recommendations as potentially effective options informed by theories, but narrow the list down to design requirements most appropriate and feasible for different local contexts. A heuristic for doing this is to download the full result table, select the identified barriers supported with participants' quotes that seem to be most relevant to your local context, and thoroughly consider the corresponding BCTs and system features.

Methodologically speaking, we have demonstrated to the broader readership in the PT community a systematic approach, using the COM-B/TDF to diagnose a behaviour change problem and the BCW to translate the diagnosis into theory-informed design recommendations. We believe this approach does lend itself to a comprehensive coverage of factors in the exploratory phase of design research and for generation of new research directions and hypotheses; but the extent to which those factors and BCTs impact on the behaviour need to be tested with further experimental studies.

Acknowledgement. We would like to thank Kathryn Morgan and Rachael Travers for helping transcribe interviews, and Anna Roberts for reviewing the coding and intervention mapping. This research was supported by the Horizon Centre for Doctoral Training at the University of Nottingham (RCUK Grant No. EP/L015463/1) and by the RCUK's Horizon Digital Economy Research Institute (RCUK Grant No. EP/G065802/1) and Unilever UK Ltd. The study received ethics approval from School of Computer Science Ethics Committee, University of Nottingham.

References

1. Aarts, H., Custers, R.: Unconscious goal pursuit: nonconscious goal regulation and motivation. In: Ryan, R. (ed.) The Oxford Handbook of Human Motivation, pp. 232–247. Oxford University Press, New York (2012)
2. Bargh, J.A., Ferguson, M.J.: Beyond behaviorism: on the automaticity of higher mental processes. Psychol. Bull. **126**(6), 925–945 (2000)
3. Cane, J., et al.: Validation of the theoretical domains framework for use in behaviour change and implementation research. Implement. Sci. **7**(1), 37 (2012)
4. Cirillo, F.: The Pomodoro Technique. FC Garage, Berlin (2013)
5. Consolvo, S., et al.: Theory-driven design strategies for technologies that support behavior change in everyday life. In: Proceedings of the 27th International Conference on Human Factors in Computing Systems - CHI 2009, pp. 405–414 (2009)
6. Fogg, B.: A behavior model for persuasive design. In: The 4th International Conference on Persuasive Technology, p. 40. ACM Press, New York (2009)
7. Francis, J.J., et al.: Evidence-based selection of theories for designing behaviour change interventions: using methods based on theoretical construct domains to understand clinicians' blood transfusion behaviour. Br. J. Health Psychol. **14**(Pt 4), 625–646 (2009)
8. French, S.D., et al.: Developing theory-informed behaviour change interventions to implement evidence into practice: a systematic approach using the Theoretical Domains Framework. Implement. Sci. **7**(1), 38 (2012)
9. Fritz, C., et al.: It's the little things that matter: an examination of knowledge workers' energy management. Acad. Manag. Perspect. **25**, 28–39 (2011)
10. Herrmanny, K., Ziegler, J., Dogangün, A.: Supporting users in setting effective goals in activity tracking. In: Meschtscherjakov, A., Ruyter, B., Fuchsberger, V., Murer, M., Tscheligi, M. (eds.) PERSUASIVE 2016. LNCS, vol. 9638, pp. 15–26. Springer, Heidelberg (2016). doi: 10.1007/978-3-319-31510-2_2
11. Jett, Q.R., George, J.M.: Work interrupted: a closer look at the role of interruptions in organizational life. Acad. Manage. Rev. **28**, 494–507 (2003)
12. Mark, G., et al.: No task left behind? Examining the nature of fragmented work. In: Proceedings of the SIGCHI Conference on Human Factors in Computing Systems - CHI 2005, pp. 321–330 (2005)
13. Michie, S., et al.: The Behaviour Change Wheel: A Guide to Designing Interventions. Silverback Publishing, London (2014)
14. Owen, N., et al.: Too much sitting: a novel and important predictor of chronic disease risk? Br. J. Sports Med. **43**(2), 81–83 (2009)
15. Pate, R.R., et al.: The evolving definition of "sedentary". Exerc. Sport Sci. Rev. **36**(4), 173–178 (2008)
16. Prochaska, J.O., Velicer, W.F.: The transtheoretical model of health behavior change. Am. J. Heal. Promot. AJHP **12**(1), 38–48 (1997)
17. Ryan, C.G., et al.: Sitting patterns at work: objective measurement of adherence to current recommendations. Ergonomics **54**(6), 531–538 (2011)
18. Srivastava, A., Thomson, S.B.: Framework analysis: a qualitative methodology for applied policy research. J. Adm. Gov. **4**(2), 72–79 (2009)
19. Tremblay, M.S., et al.: Physiological and health implications of a sedentary lifestyle. Appl. Physiol. Nutr. Metab. **35**(6), 725–740 (2010)

20. West, R.: Time for a change: putting the Transtheoretical (Stages of Change) Model to rest. Addiction **100**, 1036–1040 (2005)
21. Züger, M., Fritz, T.: Interruptibility of software developers and its prediction using psycho-physiological sensors. In: Proceedings of the 33rd Annual ACM Conference Human Factors in Computing Systems - CHI 2015, pp. 2981–2990 (2015)

On the Design of *Subly*: Instilling Behavior Change During Web Surfing Through Subliminal Priming

Ana Caraban[1(✉)], Evangelos Karapanos[2], Vítor Teixeira[1], Sean A. Munson[3], and Pedro Campos[1]

[1] Madeira Interactive Technologies Institute, University of Madeira, Funchal, Portugal
ana.caraban@tecnico.ulisboa.pt
[2] Persuasive Technologies Lab, Cyprus University of Technology, Limassol, Cyprus
[3] DUB Group, University of Washington, Seattle, USA

Abstract. With 50% of people spending over 6 h per day surfing the web, web browsers offer a promising platform for the delivery of behavior change interventions. One technique might be *subliminal priming* of behavioral concepts (e.g., walking). This paper presents *Subly*, an open-source plugin for Google's *Chrome* browser that primes behavioral concepts through slight emphasis on words and phrases as people browse the Internet. Such priming interventions might be employed across several domains, such as breaking sedentary activity, promoting safe use of the Internet among minors, promoting civil discourse and breaking undesirable online habits such excessive use of social media. We present two studies with *Subly*: one that identifies the threshold of subliminal perception and one that demonstrates the efficacy of *Subly* in a picture-selection task. We conclude with opportunities and ethical considerations arising from the future use of *Subly* to achieve behavior change.

Keywords: Persuasive technology · Behavior change · Subliminal · Nudging

1 Introduction

Interest in technologies for behavior change has been increasing. Consider, for instance, the case of physical activity trackers. With physical inactivity becoming the fourth leading cause of death worldwide [40], activity trackers are gaining momentum both in research and practice as they can offer many benefits, including increased awareness of one's behaviors, taking agency to manage one's health, and identifying opportunities for self-regulation in daily life [19].

Many behavior change technologies have been designed to support change through *reflection*. For example, the personal informatics [26] model describes a process of collecting, integrating, and reflecting on one's personal data to develop greater knowledge about one's behaviors and, in turn, act accordingly.

One limitation of this reflection-based approach is its reliance on user motivation to explore the data (or at least pay attention to it) and identify opportunities for action. This motivation is not always there, and self-regulation does not work for everyone. Researchers have raised concerns over the long-term adoption and effectiveness of

P.W. de Vries et al. (Eds.): PERSUASIVE 2017, LNCS 10171, pp. 163–174, 2017.
DOI: 10.1007/978-3-319-55134-0_13

activity trackers (e.g., [12, 14, 19, 23, 32]). One study found a third of owners of activity trackers discard the tracker within the first six months [23]. Another study of tracker adoption found 66% of participants used the tracker longer than two days, 38% longer than a week, and only 14% longer than two weeks [14].

In response, researchers have investigated strategies that require little cognitive effort to reflect on habits and encourage behaviors [15, 30]. Implicit suggestions and positive reinforcements, or *nudges*, offer opportunities for shaping judgments through unforced recommendations when decisions take place [24]. For instance, some approaches adjust default values (e.g., step goal) to lead people to meet a higher mark [13] or attempt to encourage certain choices by providing incentives (e.g., rewards or price discounts) when people adopt the intended conduct (e.g., smoking cessation [36] or healthy eating [29]). Another approach is creating friction at moments of decision-making; for example, when someone retrieves a car key from the *Keymoment* key holder, it drops a bike key, forcing the user to interact with it and nudging them towards sustainable modes of transit [22]. Many of these approaches, however, face a notable challenge: they still require conscious attention and ultimately reflection about possible actions during the decision making moment, entailing a risk of *reactance* [9].

In the light of these challenges, how can we design technologies that subtly and instinctively influence behavior without conscious guidance? In this paper, we explore *subliminal priming* as an approach to motivate change and propose *Subly*, an open-source plugin for Google's *Chrome* browser that primes behavioral concepts (e.g., to walk) through slight emphasis on words as people surf through the Internet. Leveraging Kostakos's [21] argument for three types of contributions in HCI – *data*, *tools* and *theory* – this papers contributes an open-source research tool that allows third-party researchers to design and validate their own subliminal behavior change interventions (see Fig. 1). We present two studies that examined the threshold of subliminal perception and demonstrate *Subly's* effectiveness in a picture-selection task. We conclude with ethical considerations arising from the future use of *Subly* and present a possible scenario as an opportunity for *Subly* to encourage behavior change.

Fig. 1. *Subly* is an open-source plugin for *Chrome* browser that enables priming of behavioral concepts through slight changes in the opacity of certain words. In this example, we primed the words drink, water and consume.

2 The Dual Process Theory and the Decision-Making Process

According to Dual Process Theory [18, 33], decision-making happens through two cooperative systems: the automatic and the reflective. The *automatic* system is the principal mode of thinking. Behavioral decisions are made unconsciously and intuitively through associative inferences (i.e., prior experience), hence faster and with less effort. In contrast, the *reflective* is a knowledge-based process. Decisions are made through a conscious, rational process, and therefore, more effortful and slower.

While the two systems cooperate, one tends to dominate [18]. As we have a predisposition to reduce effort and make decisions instinctively, the *reflective system* only comes into action in situations that the *automatic* system cannot handle, overriding unconscious judgments [18]. Thus, most of our daily actions are controlled by the automatic mind, and technologies that take advantage of this can have significant impact on human behavior. Consider, for instance, the popular road sign stripes that are strategically positioned closer and closer together as the road approaches a sharp curve, making drivers believe they are speeding [34]. Such a simple but powerful intervention, tapping on the automatic mind, can decrease car accidents by 36%. Similarly, careful manipulation of plate size and color can result in a 10-15% decrease in calorie intake during meals [1].

While the dominance of the automatic mind in daily behavioral decisions seems evident, current persuasive technologies focus on the reflective mind. Mercer et al. [28] found self-monitoring to be the most commonly adopted strategy in commercial activity trackers. Also, West et al. [39] found knowledge provision and increasing awareness of one's behaviors to be the main strategies adopted by health (40% of 3336 apps). Adams et al. [1] reviewed 176 research prototypes and found only 11(6%) appealed to the automatic mind. In the remainder of the section, we delve into priming, as one way to influence the automatic mind and guide conducts.

2.1 Influencing Behavior Through Priming

Vast literature suggests that designers can leverage the automatic mind by transmitting information in a way that is not accessible to individual awareness (see [8] for a review). The exposure to a stimulus below the threshold of conscious perception is called *subliminal priming* and can be achieved either *visually* (e.g., brief duration, subtle changes in color) or *audibly* (e.g., back masking) [8]. This strategy relies on the idea that, while not affecting individual reasoning, a subliminal stimulus is still processed by the unconscious mind and may trigger action through inferences. For instance, priming the word "water" may trigger the respective mental representation and instigate drinking behavior [7].

2.1.1 Behavior Change

Research has repeatedly highlighted the effectiveness of subliminal priming in influencing people's attitudes and behaviors. For instance, Dijksterhuis [8] found that priming stimuli related with relaxation (e.g., rest and relax) lowered participant heart rates and blood pressures, affecting their cardiovascular activity. Priming “old age” was found to make people walk slower down a corridor [3]. Légal et al. [24] found that priming the word ‘trust’ before reading a persuasive message about tap water consumption led to a more favorable evaluation and increased individuals’ behavioral intentions in accordance with the message, possibly through the non-conscious activation of the goal “to trust.” Priming the logos of certain brands was linked to heightened intention to select a drink from that brand [20]. More interesting, Veltkamp et al. [35] showed that priming certain behavioral concepts (e.g., drinking water) can be successful even in the absence of need deprivation (e.g., thirst).

2.1.2 Intelligent Tutoring Systems

To escape from purely cognitive approaches of teaching that often lead to disengagement [27], subliminal cues have been employed to accelerate the learning process. Chalfoun and Frasson [5], for instance, demonstrated the impact that subliminal priming has on challenging tasks. They asked participants to construct a magic square in an on-screen 2D puzzle that required applying three successive tricks to be solved. By flashing the tricks needed to infer the solution, they reduced mistakes and moves by 44% compared to the control condition. Beyond supporting task execution, subliminal stimuli were found to generate positive emotions such as increased enjoyment of the activity and increased levels of motivation. Jraidi et al. [17] found that the subliminal priming of words that had the capacity to increase individuals’ self-efficacy and self-esteem (e.g., ‘smart’) had a positive impact on their problem-solving abilities and led to positive emotional reactions and higher scores in the quiz.

2.1.3 Task Support

The complexity of digital tasks, combined with people’s lack of motivation to engage with task-support applications, has encouraged researchers to study subliminal cueing as a means of assisting novice users during interactions with computing. Wallace et al. [37], for instance, explored how subliminal cues may be used to support the use of a text-editor by novices, through displaying use instructions. They found users seek help less frequently when they displayed required task-information visibly but not too briefly for participants to notice.

2.1.4 Boosting Creative Processes and Generating Emotional Experiences

Positive priming has further been found to increase individual performance in creative tasks, such as brainstorming and creative writing. Lewis et al. [25] explored the impact of positive, negative, and neutral images, presented subliminally during a creative writing task, on people’s affective state and writing performance. Positive affect images (e.g., a laughing baby) positively influenced the quality of people’s ideas. Through a different path, Dijksterhuis et al. [8] presented pairs of positive and self-related words subliminally and found they increased individual perceptions of self-esteem. Fitzsimons et al. [11] and Wang et al. [38] studied how subliminal priming may affect individuals’ creativity during brainstorming, and found that visually priming the logo of well-known creative brands (e.g., Apple) made participants feel and reason more creatively and boosted the diversity of ideas during the brainstorming task.

3 Study 1: Threshold of Subliminal Perception

In our research, we sought to examine and develop the capability to prime certain messages by manipulating the opacity of certain words in text people are already reading. Our first study aimed to determine the threshold of subliminal perception. In other words, how much can we vary the opacity of primed cues so that the reader does still not perceive the difference?

We recruited a total of 25 participants (mean age = 25, 17 male). All were native English speakers, the language we used for the study. None of the participants had any known form of visual impairment. Participants were brought into a control room without prior awareness of the purpose of the study. They were seated about 70 cm from the screen, a *Retina LED IPS 13.3"* display at resolution *1440 × 900* px.

Following a procedure similar to that of Pfleging et al. [31], we presented two words at the screen (font-style Times New Roman, 16 px). Both words were presented at full opacity. For one of the words, we started decreasing the opacity at a rate of 1% every 0.5 s. Participants were asked to look at the screen and to press the space bar once they noticed a change. This was performed five times. We repeated the same procedure in four conditions to understand how the threshold of subliminal perception would vary based on the *type of vision* (*foveal*, if participants would gaze at the word, or *peripheral*, if participants would gaze at a distance) and the color of the word presented in *webpages* (*black*, for standard text, versus *blue*, for hyperlinks). To simulate the foveal and peripheral conditions, we varied the distance between the two words. In the foveal condition, both words were placed next to each other. In the peripheral condition, words were placed at a distance and participants were asked to gaze at one of the words (the one with constant opacity). Each participant went through all four conditions, in random order, thus performing a total of twenty trials.

3.1 Findings

The visibility thresholds were set based on the value of the opacity at the time the participants pressed the space bar key. We then computed an overall opacity value for each font color, black and blue, taking into account both the foveal and the peripheral vision condition, using the following formula:

$$\alpha_d = \alpha_{\text{foveal}} + \frac{d}{d_{\max}} * (\alpha_{\text{peripheral}} - \alpha_{\text{foveal}})$$

where α_f and α_p are the opacity values for the foveal and the peripheral vision conditions respectively, d is the distance between the two words being displayed (which reflects the distance between the word whose opacity was changing and participants' gaze position, as participants were asked to gaze at the non-changing word), and d_{max} denotes the fixed value of the screen diagonal (=1510).

Across all conditions, we obtained a 3% higher value in the opacity threshold for text displayed in black ($\alpha_d = 0.76$) than for text displayed in blue ($\alpha_d = 0.73$). This difference between blue and black text was not statistically significant.

4 Study 2: Efficacy of Subliminal Cueing

In this study, we wanted to examine the efficacy of subliminal priming while reading text online. The task consisted of reading a short sentence, which referred to at least three concepts (e.g., "A *bird*, a *cat* and a *bear* were standing on a tree"). After reading the sentence, participants would be asked to quickly select one of three images, each reflecting one of the concepts presented in the sentence. We wanted to see whether priming one concept would make individuals more likely to select the relevant image.

4.1 Method

4.1.1 Developing the Material

We first crafted sentence-image pairs so that all three concepts were equally represented. To do so, we selected a total of twenty-seven short sentences from literature, which embraced at least three different ideas, where each idea was easily reflected in one image (e.g. a bear). Out of the twenty-seven sentences, we wanted to select the nine that had the most equal representation of the three concepts and posed the least challenges to participants during reading. We invited a total of 95 participants (mean age = 40, 43 male, 52 female, all native English speakers) through Mechanical Turk, and compensated them $0.30 for their participation. To avoid participant fatigue, we presented a random set of 14 out of the 27 sentences to each participant. During the task, participants were asked to read each sentence, and without re-thinking, to select one of the three pictures presented, horizontally aligned in random order. After completing each selection task, they were asked to rate how difficult it was to make the selection, on a 5-point scale. We also logged the time spent in each task, both for the sentence reading and the image selection. We removed from our sample sentences that participants reported as confusing or difficult (e.g., uncommon or ambiguous), or ones in which participants spent substantially more time in reading them, given their text length (we measured the median time spent and removed the phrases where the values were sorely larger). From the remaining pool, we selected the nine sentences that displayed the most balanced selection distribution, so that all three concepts were equally represented in the text.

4.1.2 Study Procedure

We recruited a total of 30 participants on campus (mean age = 29, 13 male, 17 female, all English native speakers). We used the properties of subliminal priming suggested by the first study (0.76% opacity, font color black). Participants were presented with the nine sentences in random order and were asked to select one of the three pictures counterbalanced presented. The prime of each sentence was randomized. We asked participants to read out loud, so that we better understood their comprehension of the text and any impact the priming had on their reading. We video-recorded users' tasks.

We used a *Tobii Pro TX300* eye tracker to monitor participants' eye gaze while reading the text and logged the time spent in each task, both for the sentence reading and the image selection. At the end of each sentence and image selection task, participants were asked to rate how difficult it was to make the selection, on a 5-point scale, and elaborate verbally on the reasons for any challenge experienced. At the end of the study, we asked participants if they noticed something unusual and if there was anything that disrupted their reading and how.

4.1.3 Conditions

All participants experienced three conditions: *subliminal*, *supraliminal* (referring to a stimulus that can be consciously perceived if attention is directed to it), and *no-cue* (control condition), the order was block randomized. The nine sentences were randomly distributed across the three conditions (three sentences per condition), and the order of the nine sentences was randomized. In the subliminal condition, the primed words were presented at opacity level of 0.76, while the remaining text was presented at 100%. In the supraliminal condition, the text value was further decreased to 0.66 (see Fig. 2) and the primes were presented with full opacity. In the control condition, all words were presented in full opacity.

"Their blades crashed together twice, then slipped past each other only to be blocked by upraised shields, but the bigger man gave ground at the impact."

Fig. 2. Subly priming the words "blocked" and "shields" supraliminally with opacity levels of 66% and 100% (primes).

4.2 Findings

4.2.1 Did Priming Disrupt the Reading Process?

To understand whether priming disrupted the reading process, we looked at two aspects: the total *reading time*, and the *number of pauses* people made while reading. An analysis of variance revealed no significant effect of condition on reading time, $F(2) = 0.04$, $p = 0.96$, and no significant interaction between condition and the sentence, $F(16) = 1.43$, $p = 0.13$. As expected, a significant effect of the sentence on reading time was found ($F(8) = 9.14$, $p < 0.001$).

We then looked for pauses in participant reading. A pause was inferred from a combination of participant gaze behavior and verbal protocol, being defined as *a pause in reading leading to silence, a hesitation, or a repetition in reading*. Two researchers independently reviewed all tasks by replaying the gaze and audio recordings, and were asked to note when a pause was identified in each of the tasks. Inter-rater agreement was strong (Cohen's Kappa = 0.85). Pauses in reading were exceptionally rare across conditions, possibly due to the short length of the sentences, with only two pauses identified in the supraliminal condition, one in the subliminal and one in control. No significant differences existed across conditions (Fisher's exact test, $p = 0.99$).

4.2.2 Did Priming Affect Users' Image Selection?

A single image would have a probability of being selected 33% of the time, as there were three images presented for each selection task. However, this might vary with individuals or the selection task. Participants selected the primed concepts 28% of the time (25 out of 90) in the control condition, where no priming took place.

Subliminal primes were found to lead to a significantly higher chance of selecting the primed word (42%, 38 out of 90) compared to the control condition, χ^2 (1, N = 180) = 4.13, $p < 0.05$. In contrast, supraliminal priming did not have any impact on users' image selection, as compared to the control condition; the primed word was selected only 24% of the time (22 out 90), χ^2 (1, N = 180) = 0.26, $p = 0.61$.

One plausible explanation for the lack of effect in the supraliminal condition could be rooted in the *aversion effect*, which suggests negative responses to unfamiliar stimuli when presented supraliminally (consciously) [4]. All in all, these results are consistent with prior work that suggests mere exposure effects to be stronger when the stimulus is presented below the limits of perception [4].

5 Design Your Own Intervention with *Subly*

These results give reason to believe that subliminal priming during web surfing should be investigated as a possible vector for behavior change interventions in daily life. Building a tool that makes priming interventions available for use can help identify the right application domains for subliminal priming and validate its efficacy across a wider range of tasks and decision environments. *Subly* supports these goals. It consists of a plugin for Google's Chrome browser and an easy to use study setup panel that enables third-party researchers to design their own behavior change interventions and deploy them in the wild. The source code is available at https://github.com/SublyM-ITI/Subly and a detailed tutorial is available at http://subly.m-iti.org/.

Subly includes a *study setup panel* that enables researchers to define the properties of their study such as type of cueing (flashing message or emphasis using opacity), its properties (size, color or other style of the stimulus by inserting the envisioned CSS widespread style), length of exposure, exposure frequency, opacity threshold, and the list of words or phrases to be primed.

Subly distinguishes itself from existing tools that display subliminal messages (e.g., *Free Subliminal Text, SubliminalMessages* and *Subliminal Power*) in four primary ways. First, it embraces an *authentication* system that identifies each user and corresponding data and allows remote customization of cues, content, and properties for each user. Researchers may *define events* that trigger priming, either events in external applications (e.g., data from a personal device such as an activity tracker) or events in web browsing behavior (e.g., mouse overlaying a word of interest) (Fig. 3).

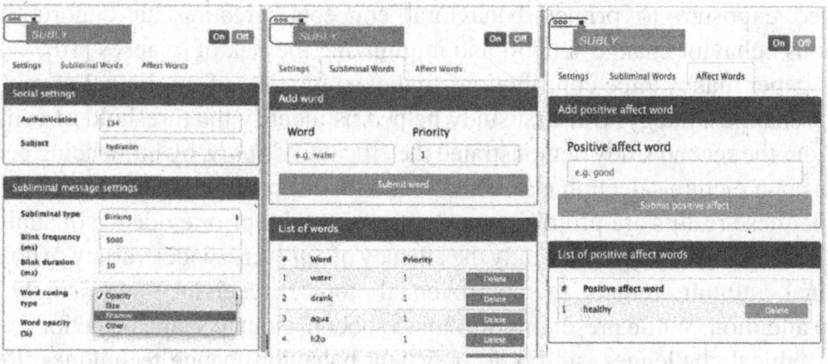

Fig. 3. The study setup panel enables researchers to define all aspects of the intervention as the cues and their subliminal properties (e.g. opacity of the cues).

Second, *Subly* connects to a database and further *logs user interactions with primes and webpages*. The tool records the webpages visited and users' active time on them (a researcher may opt to remove this for privacy reasons), the number and list of primed words and their frequency of occurrence on a web page, user interactions with primed words based on mouse activity (e.g., clicks, selections and *proximal attention* – the duration in milliseconds for which the mouse was in close vicinity to a primed word, assuming a strong relationship between mouse and gaze position [6]). Data can be used for analysis purposes, or in the creation or adjustments of event triggers for priming based on monitoring results.

Further, it allows a prioritization of stimuli (e.g., preferences for certain words or behaviors among others). For instance, in a situation were prolonged sedentary behavior is detected; the preference to prime a goal-related stimulus (e.g., *exercise*) can increase and be intensified to match the user's need.

There are still many opportunities and different techniques to be explored for integrating subliminal primes into web content. Multimedia content (e.g., images or audio) can be added to bias perception while using visual and auditory information to incite multiple senses [16, 25]. Simply manipulating text and images properties, as color, shape, size, texture or reversing the order of words might similarly go unnoticed.

Lastly, as open-source, *Subly* is readily expandable. It can be adapted for other purposes or research questions. For instance, it might be possible adapt it to predict readiness for change based on the content visited (e.g., users might search for opportunities to increase physical activity), to predict user emotional states based on content of pages visited, or to infer when individuals most need or would be receptive to the subliminal incentive.

6 Discussion and Conclusion

With prior research suggesting that subliminal priming effects last from 2 min to more than 24 h [2], and by integrating with a pervasive activity such as browsing, *Subly* enables prolonged exposure to primed behavioral concepts, creating an opportunity for continuous behavior change support and minimizing the risk of relapses [10].

This paper makes three contributions to the exploration of subliminal priming as a behavior change strategy. Our first study helped us identify the threshold of conscious perception; the second study demonstrated the efficacy of *Subly*, by influencing people's decisions in a picture selection task consistent with our priming intervention.

While these results are promising and consonant with prior evidence of subliminal stimuli, we highlight the need to study the efficacy of *Subly* in a wider variety of domains. Subliminal stimuli remain a controversial topic, receiving surprisingly little research attention. While presenting stimuli to subconsciously shape behavior certainly presents ethical challenges, so too do prevalent behavior change techniques. Regular cognitive appeals to change behavior can limit people's attention to other tasks while failing to take advantage of people's primary means for processing information – the automatic system. With this research, we seek to inspire new discussions about the role of subliminal stimuli. We also contribute with *Subly,* a tool that can be readily used and extended by researchers and others to replicate our studies and investigate new questions. Making the tool available for widespread use can help identify and evaluate new application domains for *Subly* and evaluate its efficacy across a wide variety of pro-social domains, such as decreasing sedentary activity, coping with stress at work, reminding patients to take their medications until the behavior becomes habitual, promoting safe use of the Internet among minors, improving search task performances, reducing online bullying, encouraging people to take breaks, or conveying information while performing a task (e.g. while writing an email, subliminal priming might be used to bring users' attention to missing fields or remind them about an attachment). Given that automated processes govern much of our behavior, we envision *Subly* helping change people's attitudes and behaviors, facilitating goal enactment and engaging them during behavior change.

Finally, *Subly* poses significant ethical challenges that need to be considered before and during its use. When priming behavior concepts that may influence wellbeing and health, researchers should adhere to principles like transparent disclosure and ensure that *Subly* supports users' goals and interests and that they are aware of how it functions. Thus, using *Subly* entails signing a *Statement of Use* in which researchers agree that potential risks are minimized and appropriate considering potential benefits stemming from the use of the tool.

Acknowledgement. ARDITI – Agência Regional para o Desenvolvimento e Tecnologia under the scope of the Project M1420-09-5369-000001 PhD Studentship.

References

1. Adams, A., Costa, J., Jung, M., Choudhury, T.: Mindless computing: designing technologies to subtly influence behavior. In: Proceedings of Ubicomp 2015, pp. 719–730 (2015)
2. Americans Internet Access: http://www.pewinternet.org/2015/06/26/americans-internet-access-2000-2015/

3. Bargh, J., Chen, M., Burrows, L.: Automaticity of social behavior: Direct effects of trait construct and stereotype activation on action. JPS **71**(2), 230 (1996)
4. Bornstein, R.: Inhibitory effects of awareness on affective responding: implications for the affect-cognition relationship (1992)
5. Chalfoun, P., Frasson, C.: Subliminal cues while teaching: HCI technique for enhanced learning. Adv. HCI. (2011). Article no: 2
6. Chen, M., Anderson, J., Sohn, M.: What can a mouse cursor tell us more? Correlation of eye/mouse movements on web browsing. In: CHI EA 2001, pp. 281–282 (2001)
7. Custers, R.: How does our unconscious know what we want? The role of affect in goal representations. In: The Psychology of Goals, pp. 179–202 (2009)
8. Dijksterhuis, A.: Unconscious relaxation: the influence of subliminal priming on heart rate (2001)
9. Dillard, J., Shen, L.: On the nature of reactance and its role in persuasive health communication. Commun. Monogr. **72**(2), 144–168 (2005)
10. Epstein, D., Kang, J., Pina, L., Fogarty, J.: Reconsidering the device in the drawer: lapses as a design opportunity in personal informatics. In: Proceedings of Ubicomp 2016, pp. 829–840 (2016)
11. Fitzsimons, G., Chartrand, T.: Automatic effects of brand exposure on motivated behavior: how apple makes you "think different". JC **35**(1), 21–35 (2008)
12. Fritz, T., Huang, E., Murphy, G.: Persuasive technology in the real world: a study of long-term use of activity sensing devices for fitness. In: Proceedings of CHI 2014, pp. 487–496 (2014)
13. Goldhaber-Fiebert, J., Blumenkranz, E., Garber, A.: Committing to exercise: contract design for virtuous habit formation, w16624. NBER (2010)
14. Gouveia, R., Karapanos, E., Hassenzahl, M.: How do we engage with activity trackers? A longitudinal study of Habito. In: Proceedings of Ubicomp 2015, pp. 1305–1316 (2015)
15. Gouveia, R., Pereira, F., Karapanos, E., Munson, S.A., Hassenzahl, M.: Exploring the design space of glanceable feedback for physical activity trackers. In: Proceedings of Ubicomp 2016, pp. 144–155 (2016)
16. Ittersum, K.Van, Wansink, B.: Plate size and color suggestibility: the Delboeuf Illusion's bias on serving and eating behavior. JCR **39**(2), 215–228 (2012)
17. Jraidi, I., Frasson, C.: Subliminally enhancing self-esteem: impact on learner performance and affective state. In: Aleven, V., Kay, J., Mostow, J. (eds.) ITS 2010. LNCS, vol. 6095, pp. 11–20. Springer, Heidelberg (2010). doi:10.1007/978-3-642-13437-1_2
18. Kahneman, D.: Thinking, Fast and Slow (2011)
19. Karapanos, E.: Sustaining user engagement with behavior-change tools. Interactions **22**(4), 48–52 (2015)
20. Karremans, J., Stroebe, W., Claus, J.: Beyond Vicary's fantasies: the impact of subliminal priming and brand choice. JRSP **42**(6), 792–798 (2006)
21. Kostakos, V., Ferreira, D., Pandab, P.: Not what, but HOW to study in HCI: tools, data, theory. In: Adjunct Proceedings of CHI 2015 (2015)
22. Laschke, M., Diefenbach, S., Hassenzahl, M.: Annoying, but in a nice way: an inquiry into the experience of frictional feedback. IJD **9**(2), 129–140 (2015)
23. Ledger, D., McCaffrey, D.: Inside Wearables: How the Science of Human Behavior Change Offers the Secret to Long-term Engagement. Endeavour Partners, Cambridge (2014)
24. Légal, J., Chappé, J., Coiffard, V.: Don't you know that you want to trust me? Subliminal goal priming and persuasion. JESP **48**(1), 358–360 (2012)
25. Lewis, S., Dontcheva, M., Gerber, E.: Affective computational priming and creativity. In: Proceedings of CHI 2011, pp. 735–744 (2011)

26. Li, I., Dey, A., Forlizzi, J.: A stage-based model of personal informatics systems. In: Proceedings of CHI 2010, pp. 557–566 (2010)
27. Malekzadeh, M., Mustafa, M., Lahsasna, A.: A review of emotion regulation in intelligent tutoring systems. Educ. Technol. **18**(4), 435–445 (2015)
28. Mercer, K., Li, M., Giangregorio, L., Burns, C.: Behavior change techniques present in wearable activity trackers: a critical analysis. JMIR mHealth uHealth **4**(2), e40 (2016)
29. Olstad, D., Goonewardene, L., McCargar, L.: Choosing healthier foods in recreational sports settings: a mixed methods investigation of the impact of nudging and an economic incentive. IJBNPA **11**(1), 6 (2014)
30. Ornelas, T., Caraban, A., Gouveia, R., Karapanos, E.: CrowdWalk: leveraging the wisdom of the crowd to inspire walking activities. In: Adjunct Proceedings of Ubicomp 2015, pp. 213–216 (2015)
31. Pfleging, B., Henze, N., Schmidt, A., Rau, D., Reitschuster, B.: Influence of subliminal cueing visual search tasks. In: Proceedings of CHI EA 2013, pp. 1269–1274 (2013)
32. Shih, P., Han, K., Poole, E., Rosson, M., Carroll, J.: Use and adoption challenges of wearable activity trackers. In: iConference 2015 (2015)
33. Strack, F., Deutsch, R.: The reflective-impulsive model. In: Dual-Process Theories of the Social Mind, pp. 92–104 (2014)
34. The road design tricks that make us drive safer (2014). http://www.bbc.com/future/story/20140417-road-designs-that-trick-our-minds
35. Veltkamp, M., Custers, R., Aarts, H.: Motivating consumer behavior by subliminal conditioning in the absence of basic needs: striking even while the iron is cold. JCP **21**(1), 49–56 (2011)
36. Volpp, K., Troxel, A., Pauly, M., Glick, H.: A randomized, controlled trial of financial incentives for smoking cessation. NEJM **7**(360), 699–709 (2009)
37. Wallace, F., Flanery, J., Knezek, G.: The effect of subliminal help presentations on learning a text editor. IPM **27**(2–3), 211–218 (1991)
38. Wang, H., Fussell, S., Cosley, D.: From diversity to creativity: stimulating group brainstorming with cultural differences and conversationally-retrieved pictures. In: Proceedings of CSCW 2011, pp. 265–274 (2011)
39. West, J.H., Hall, P.C., Hanson, C.L., Barnes, M.D., Giraud-Carrier, C., Barrett, J.: There's an app for that: content analysis of paid health and fitness apps. JMIR **14**(3), 72 (2012)
40. WHO|Physical activity. WHO (2015)

Kilowh.at – Increasing Energy Awareness Using an Interactive Energy Comparison Tool

Björn Hedin[1](✉) and Jorge Zapico[2]

[1] KTH Royal Institute of Technology, Stockholm, Sweden
bjornh@kth.se
[2] Linneaus University, Växjö, Sweden
jorgeluis.zapico@lnu.se

Abstract. Reducing the use of energy is important for several reasons, such as saving money and reducing impact on the climate. However, the awareness among non-experts of how much energy is required by different activities is generally low, which can lead to wrong prioritizations. In this study, we have developed an interactive tool to increase "energy awareness". A group of 58 students first did a test to benchmark their current energy awareness, then tried the tool for 10 min, and then did the same test immediately after trying the prototype and one week after trying the prototype. In addition, they answered questions regarding which, if any, of the energy requirement of different activities surprised them, any thoughts about their own energy use aroused after using the prototype and what they thought about using the tool compared to more conventional methods of learning. The results showed a significant learning effect in energy awareness with a very strong effect size of 1.689, that they were most surprised by the energy required to produce a hamburger, 39 of 58 explicitly said they intended to change one or more aspects in order to improve their energy use, where 24 actions involved changing habits and 18 actions was of a one-time investment character. The attitude towards using such a tool instead of more conventional learning was very good and the words most frequently used to describe the tool was good, simple and easy to use, fun, and interesting, but five users also said they were bored after a while. In total the results indicate that using an interactive tool like this even for a limited time is a good way to in an efficient and fun way increase energy awareness.

Keywords: Energy awareness · Technology enhanced learning · Persuasive services · Sustainable HCI

1 Introduction

Energy use is an important topic for sustainability and climate change. While energy efficiencies can be created by optimization and dematerialization, promoting conservation behavior remains an important part of saving energy and avoiding rebound effects [1]. There has been much research during the last decade on using persuasive technologies and eco-visualizations to promote energy conservation. Energy conservation is a dominant topic in using persuasive technologies for sustainability and Sustainable HCI

P.W. de Vries et al. (Eds.): PERSUASIVE 2017, LNCS 10171, pp. 175–185, 2017.
DOI: 10.1007/978-3-319-55134-0_14

research [2]. Technologies such as smart meters allow to measure and provide accurate feedback on energy consumption at both household level and appliance level. Different research activities have used this data with a persuasive intent to promote energy conservation. Many of the smart meters technologies, visualizations and applications created in this research area used quantitative data of energy using watt hour (Wh) or kilowatt hour (kWh) as main unit [3–8].

While kWh is a familiar term for everyone paying electricity bills or selecting an appliance, it is a fairly abstract unit that may be difficult to relate to, as explored in [9]. Even if abstract, it is still a relevant unit, Wood and Newborough have argued that "The kWh is already familiar to most consumers and although few understand this unit, thorough comprehension is not necessary for an effective display" [10]. However, this lack of understanding of qualitative information is a part of a general lack of energy literacy [11, 12]. While users know that having a light bulb on all the time, or charging their mobile phones requires energy, the lack of energy literacy or awareness needed to understand the differences in scale may lead to lack of or suboptimal conservation behavior.

This article explores the understanding of qualitative information of energy consumption and the use of simulation as a way of increasing awareness. A prototype called KiloWhat was created that uses simulation and play to practice and learn about energy consumption, exploring the following research questions:

- RQ1: What is the learning effect of using such a tool?
- RQ2: What energy comparisons surprised the respondents?
- RQ3: What thoughts about personal energy consumption were aroused?
- RQ4: What did they think about the tool compared to more conventional learning approaches?

An experiment was performed with 58 engineering students in four steps (see method chapter for more extensive description):

1. An assessment of their current energy understanding of kWh information.
2. Using and experimenting with the KiloWhat prototype during ten minutes.
3. A second test to evaluate the learning effect, insights and intentions to change as an effect of using the tool.
4. A final test one week later to see how much of knowledge had been forgotten.

This article presents the design of the KiloWhat prototype, the method and results from the test, and discusses the implications of the results for the design of persuasive energy services.

2 Design of the Prototype

The prototype is an online service available at http://kilowh.at/. The main goals of the design are to:

1. Make quantitative kWh information easier to relate to by providing a learning experience where the users can translate kWh into everyday activities

2. Help users to learn differences in scale between the energy consumption of different activities by allowing the users to play and compare between them.

The prototype is based on the code and idea of http://carbon.to which explored the same concept for increasing carbon literacy [13]. The prototype provides the possibility of exploring and comparing the energy use of sixteen different activities such as driving an electric car or having a TV on. The activities are divided in five different categories. The values provided were average values gathered from a search in research databases and should only be seen as rough estimates since there can be large differences within one category.

- *Energy generation:* hours of solar panels, kg of coal, hours running in a treadmill.
- *Home and appliances:* washing machine loads, hours with a LED lamp on, hours with a incandescent light bulb on, hours with a fridge on, hours with Wi-Fi on, mobile phone charges, hours watching television.
- *Transportation:* km driving a gasoline car, km driving an electric car, km driving an electric assisted bike.
- *Heating:* hours heating a house with electric heaters, hours heating a house with geothermal pump.
- *Food:* hamburgers (energy needed to produce).

When accessing the site, it shows by default 1kWh compared to one of the activities selected randomly (Fig. 1a).

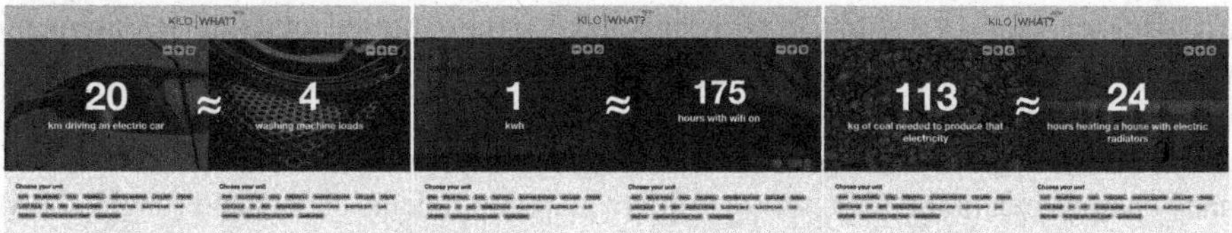

Fig. 1. The user interface of the kilowh.at system

The users can change the unit in the right side for comparing 1kWh to other activities and increase or decrease the amounts using the plus and minus buttons on the top-right corners. The users can also change the unit on the left side to compare different activities against each other (Fig. 1b) or to see what would be needed to generate that amount of energy (Fig. 1c).

The main persuasive strategy used in this prototype is simulation, "enabling users to observe immediately the link between cause and effect" [14]. The intent is that by comparing kilowatt hours against different everyday units, and those units against each other, the users can simulate the energy use of those everyday actions and gain knowledge about them.

The software is a fork of the aforementioned carbon.to. It is developed using Ruby on Rails and JavaScript, the code is available as open source in GitHub (https://github.com/zapico/kilowhat).

As a first design iteration, the prototype was tested with 20 users. The users tested the site and provided feedback. From this feedback, several improvements were added, including a better grouping of activities using colored categories, a unification of time units to hours and the inclusion of several extra activities.

3 Method

In order to evaluate the effect of the prototype, an experiment was designed. The respondents were 58 first-year engineering students who had recently started a five year long educational program within media technology. As part of an introductory course module about "introduction to scientific research" they were required to participate in one scientific experiment. Out of the 58 students, 53 participated fully in the four parts of this experiment. In order to check which students had participated, the participation was not anonymous, but the students were informed that the email addresses collected would not be used to identify individuals. The students were also informed that the results would be used for research, and that they would be presented with the resulting paper with comments as a part of the course module.

The respondents were given instructions by email, and they could do the assignment anywhere and anytime within the given time window. The email told the students about the aim of the prototype, "to increase energy awareness", an introduction to why increasing energy awareness is important, and the setup of the study.

Their first assignment consisted of three tasks that they were required to complete in a sequence within a window of two days, and that the expected total time requirement was 30 min. The first task was to answer a questionnaire where they should give their best guess what 100 kWh corresponded to for 16 different activities, for examples "How many kilometers ride with an electric bike", "how many hours powering at TV" or "how many charges of a mobile phone". In order to help the students understand roughly how much 100 kWh corresponds to, the were informed that the price of 100 kWh electricity is roughly 100 Swedish crowns or about €10. The coincidence that 1 kWh roughly equals 1 Swedish crown makes it easy for users to assess if an estimate is reasonable since money is more easy to relate to. They were instructed not to look up the answer anywhere while answering the questionnaire. The second task was to try out the prototype for exactly 10 min after watching a short screencast introducing the prototype. The third task was then to answer a copy of the first questionnaire but where a three free-text questions were added: Whether they were surprised by any of the comparisons, if using the prototype had given them insights/ideas about their own energy use in a short and long perspective, and how they perceived using the prototype as a learning tool compared to other methods of learning.

Finally, after one week they were required to answer the first questionnaire one final time in order to evaluate how much of the learning effect was lost after one week.

The first part of the analysis was to evaluate the learning effect. After receiving all responses to the questionnaire, the answers were edited [15], where respondents we had good reasons to believe that they had just provided random answers were removed from the analysis. Criteria for removing answers were where it would be obvious for anyone that the answer was not correct. For example, any answer where the students estimated that an electric car would travel longer on 100 kWh than an electric bike. A total of 19 responses were removed in this process.

The answers from the remaining 35 respondents were then analyzed where both p-value and effect size of the learning was calculated. The relevant measure was the order of magnitude the responses were distant from the correct value. I.e. a response 10 times higher than the correct response is as good or bad as a response 10 times lower. Therefore, and also in order to reduce skewness of the data distribution, a transformation was performed on the responses, where the values of the student response were divided by the correct answer if the student answer was lower than the correct answer, and 1/that value if the student answer was higher than the correct answer. This results in a number between 0 and 1 on a ratio scale, where 1 means the answer was exactly correct and 0 means it was infinitely incorrect. For example, if the correct answer was 10 h, then a student answer of 5 or 20 h both would have yielded a value of 0.5, and student answers of 1 or 100 h both would have yielded a value of 0.1. Finally, the results of all 16 questions were averaged for each student. The resulting distributions were tested for normality using the Anderson-Darling normality test and were found to be roughly normally distributed and therefore paired t-tests could be used for analysis.

For analyzing the three free-text answer questions, a-priori codings were used as described in the corresponding sections in the results.

4 Results

The four different research questions were closely associated with specific parts of the study, and the results are accounted for below.

4.1 RQ1: What Is the Learning Effect of Using Such a Tool?

The results from the first questionnaire showed that the energy awareness was low, as expected. The respondents' estimates were most incorrect for the activities that consumed least energy, such as Wi-Fi and charging mobile phones. The median estimation among the 848 data points generated by the 53 students who answered all three questionnaires was 15.7 times off the correct value, and when removing respondents who had given one or several unreasonable responses the remaining 560 data points generated by the remaining 35 students the median value was 11.8 times off the correct value. Their estimates were improved substantially in the questionnaire answered immediately after trying the prototype the median estimate was instead only 1.2 times off the correct value in the 560 data points from the 35 respondents. A paired-samples t-test indicated that scores were significantly higher immediately after trying the prototype ($M = 0.651$, $SD = 0.034$) than before ($M = 0.208$, $SD = 0.017$), $t(34) = -13.688$, $p < .001$, Cohen's $d = 2.775$.

The median results after one more week was 2.89 times off the correct value which was a significant improvement compared to the first questionnaire, but also a significant drop compared to the results of the second questionnaire. A paired-samples t-test indicated that scores were significantly higher one week after trying the prototype

(M = 0.423, SD = 0.025) than before trying the prototype (M = 0.208, SD = 0.017), t(34) = −9.350, p < .001, Cohen's d = 1.698.

Finally, there was a significant decrease of test scores between the second questionnaire and the third questionnaire. A paired-samples t-test indicated that scores were significantly lower one week after trying the prototype (M = 0.423, SD = 0.025) than immediately after trying the prototype (M = 0.651, SD = 0.034), t(34) = 6.366, p < .001, Cohen's d = 1.292. These statistics are presented in Table 1.

Table 1. Results of paired t-tests for the three different questionnaires.

Paired Samples Statistics

Sample	N	Mean	Std. Error Mean	Std. Deviation	Median
Q1	35	0.208	0.017	0.102	0.176
Q2	35	0.651	0.034	0.202	0.704
Q3	35	0.423	0.025	0.147	0.433

Paired Samples Correlations

	N	Correlation	Sig. (2-tailed)
Q1 & Q2	35	0.350	0.040
Q1 & Q3	35	0.452	0.006
Q2 & Q3	35	0.293	0.088

Paired differences

	Mean	Std. Deviation	Std. Error Mean	t	df	Sig. (2-tailed)	Cohen's d
Q1 - Q2	-0.443	0.198	0.032	-13.688	34	0.000	2.775
Q1 - Q3	-0.215	0.136	0.023	-9.350	34	0.000	1.698
Q2 - Q3	-0.228	0.212	0.036	6.366	34	0.000	1.292

4.2 RQ2: What Energy Comparisons Surprised the Respondents?

For RQ2 and RQ3 the results are presented by first showing a number of sample answers, and then presenting the result of the data encoding.

RQ2 was included because we for future work wanted to identify areas where the difference between facts and beliefs were large, since these are promising candidates for behavior change interventions. The question "Did any energy comparisons surprise you? Give examples and describe briefly" was coded using an a-priori coding. The 58 respondents who answered this question generated a total of 78 different coding units. Two answers were "No, I was not surprised" and 14 were unclear or unrelated. The remaining 62 answers were coded as follows using three categories.

Category 1: Much energy required/generated for one activity (absolute value), 28 encoded units

Sample answers: "I had no idea how much energy was required to make a hamburger", "Solar panels were much more effective than I thought", "That so much energy was required to heat a house".

- Hamburger (20 encoded units)

- Solar panels (3)
- Heating (3)
- Washing machine and car (1 each)

Category 2: Little energy required/generated for one activity (absolute value), 14 encoded units
Sample answers: "That charging your phone too often does not make a big difference in the big picture", "That Wi-Fi could be turned on for such a long time", "That you could drive very far with an electric car".

- Mobile charges (6 encoded units)
- Wi-Fi (4)
- Electric car, refrigerator, solar panels and LED (1 each)

Category 3: Two items compared with each other (much energy - little energy), 21 encoded units.
Sample answers: "How big a difference there was between those requiring little [energy] and those requiring more [energy], especially LED vs. light bulb", "I was very surprised by the difference between an electric car and a gasoline car", "How much better geothermal heating was than electric heating".

- Light bulb vs. LED (7 encoded units)
- Car vs. electric car (6)
- Electric heating vs. geothermal heating (3)
- LED vs. Wi-Fi, hamburger vs. mobile charges, hamburger vs. car, hamburger vs. electric bike, refrigerator vs. light bulb (1 each)

4.3 RQ3: What Thoughts About Personal Energy Consumption Were Aroused?

For the analysis of the question "Did you get any thoughts about your own energy consumption? Please describe briefly" we also used an a-priori coding with the categories

1. No, I did not get any thoughts
2. Increased energy awareness
3. Plan to change behavior (in all but one case including explicit increased energy awareness). These were then further categorized as "habits", "one-time investments" and "other".

The 58 answers were categorized using one of the categories above resulting in the following coding of the answers:
Category 1: "No, I did not get any thoughts", 7 respondents
Sample answers: "No", "I don't think I have such a good idea about my own energy consumption so I really, unfortunately, don't know", "No, I have the same thoughts as before. To turn off everything that isn't used and travel as little as possible by car ".

Category 2: Increased energy awareness, but no explicit indication of intention to change, 20 respondents

Category 3: Plan to change, 39 respondents explicitly mentioning energy-related actions of which 38 also indicated increased energy awareness

These 39 respondents indicating they planned to change their behavior generated a total of 45 answers about what they wanted to change, of which 41 were specific and 4 were only general intentions. They 41 specific actions were categorized as follows:

Habits

Sample answers: "Yes, the way I eat. I never knew that meat production could have such a huge energy consumption and therefore affect the environment. That has made me more inclined to review my eating habits", "Yes, I now have more insight in what consumes much electricity at home. I can for example start turning off the TV when I am not watching", "Foremost, I though about how much energy is required for heating homes. Something I will think about is to maybe lower the temperature in the flat and instead put on more clothes"

- Change eating habits (reducing amount of meat) (10)
- Turn things off (light, TV…) (10)
- Decrease temperature at home and put on more clothes (2)
- Reduce use of washing machine, not use tools using much energy unnecessarily (1 each)

One-time investments

Sample answers: "In the short term I think I will exchange more light bulbs to LED lights", "If I ever buy a new car it will probably be an electric car. If Sweden does the same thing as Germany it will also be the only alternative", "As a long term goal, that perhaps not was aroused now but has been there for a while but now received more fuel, was to live and use solar energy and at some time in the future have an electric car - if a car is needed. It was impressive to see how much energy solar panels could generate, and how far an electric car could travel on one charge, especially compared to normal cars"

- Change to LED (8)
- Invest in electric car (5)
- Install solar panels (2)
- Choose more energy efficient products (2)
- Geothermal heating (1)

4.4 RQ4: What Did They Think About the Tool Compared to More Conventional Learning Methods?

Finally, for analyzing the question "What was your experience of using this tool for learning about energy awareness compared to more conventional learning methods", a basic coding was used where the answers were categorized as

Only positive (35), Mixed (16), Only negative (0), Neither (3) and Other/unclear (2).

Next, all descriptive words or units used for describing the tool or its use were collected from the responses, coded so that similar words were grouped and then used for calculating word frequencies. The results were "good" (23), "simple or easy to use"

(17), "fun" (15), "interesting" (7), "clear or good overview" (6), "tiring or boring" (5), "effective" (4), "aesthetic" (3), "difficult to memorize" (2) and one each of "rewarding", "insightful", "interactive", "concrete", "easy to memorize", "unhelpful", "unspecific", "pedagogical", "relevant" and "difficult to understand". Three respondents expressed uncertainty about what "more conventional methods" meant.

5 Conclusion and Discussion

In this paper we have implemented and tested a tool for increasing energy awareness. We set up four main research questions:

5.1 RQ1: What Is the Learning Effect of Using Such a Tool?

The results first confirmed our view that energy awareness in general was very low. However, spending 10 min with the tool developed increases energy awareness significantly, with an effect size after one week of 1.698. Cohen classifies an effect size of at least 0.8 as large, and an effect size of 1.698 should therefore be considered very large. Effect sizes for learning interventions were studied by Hattie [16], who in a study of more than 800 meta analyses about learning the conclusion was that the average learning intervention had an effect size of 0.4.

5.2 RQ2: What Energy Comparisons Surprised the Respondents?

The high amount energy required to produce a hamburger clearly was what most surprised the respondents, whereas they were also surprised by the low amount of energy required by a Wi-Fi router or to charge a mobile phone. All these three were things easy to relate to for students. For pairs of comparisons, they were most surprised by the high difference of pairs with the same functionality such as light bulb vs. led light, gasoline car vs. electric car, and electric heating vs. geothermal heating, indicating that comparing similar activities were easier than comparing apples and oranges.

5.3 RQ3: What Thoughts About Personal Energy Consumption Were Aroused?

When asked about if any thoughts were aroused, 39 of 58 mentioned that they now intended to change something in in their life in order to improve their energy use and of these 38 also indicated they had increased their energy awareness, 20 more indicated they had increased their energy awareness but did not explicitly mention intentions in changing behavior, and 7 said no thoughts were aroused. The most popular habits to change was to eat less meat and to turn of electric equipment when they were not used. It is interesting to note high interest to eat less meat since that is indirect energy, and not something that will save money for the individual, indicating saving money is not the prime concern for these respondents.

5.4 RQ4: What Did They Think About the Tool Compared to More Conventional Learning Approaches?

The general attitude was positive, where 35 respondents gave only positive feedback, 16 gave positive feedback but with some reservation, 5 were neither positive or negative or gave unclear answers and finally 0 were purely negative. Analyzing the free text answers showed the most common words used to describe the tool was good (23), simple or easy to use (17), fun (15), interesting (7), clear or good overview (6) and tiring or boring (5). Of the 82 words identified 72 were positive and 10 were negative.

5.5 Limitations and Further Research

One limitation of this study is that the users were not representative of society in general. They were young and many of them lived with parents and/or did not pay for energy use. Most also run on a tight student budget where major investments are not realistic at the time. They were also relatively well educated and could be expected to have a better knowledge about energy than most other people in their age group.

Further research includes performing another test after 6 months. The respondents will then again answer the questionnaire in order to measure if any learning effect was maintained after 6 months. They will also be reminded of the (possible) actions they said they would take in order to see if the good intentions turned into outcomes.

To sum up, the use of feedback and other persuasive technologies for energy conservation rely on the use of technical units such as kWh which are abstract and not well understood. This study confirms this lack of understanding, both of quantitative energy information and of the ability to compare the energy use of different activities against each other. The results indicate that an interactive tool like the one presented in this article can be an efficient and fun way increase energy awareness.

Acknowledgments. This research was funded by the Swedish Energy Agency.

References

1. Berkhout, P.H., Muskens, J.C., Velthuijsen, J.W.: Defining the rebound effect. Energy Policy **28**(6), 425–432 (2000)
2. DiSalvo, C., Sengers, P., Brynjarsdóttir, H.: Mapping the landscape of sustainable HCI. In: Proceedings of the SIGCHI Conference on Human Factors in Computing Systems, pp. 1975–1984. ACM (2010)
3. Hargreavesn, T., Nye, M., Burgess, J.: Making energy visible: a qualitative field study of how householders interact with feedback from smart energy monitors. Energy Policy **38**(2010), 6111–6119 (2010)
4. Hargreavesn, T., Nye, M., Burgess, J.: Keeping energy visible? Exploring how householders interact with feedback from smart energy monitors in the longer term. Energy Policy **52**(2013), 126–134 (2013)
5. Murtagh, N., Nati, M., Headley, W.R., Gatersleben, B., Gluhak, A., Imran, M.I., Uzzell, D.: Individual energy use and feedback in an office setting: a field trial. Energy Policy **62**, 717–728 (2013)

6. McCalley, T., Kaiser, F., Midden, C., Keser, M., Teunissen, M.: Persuasive appliances: goal priming and behavioral response to product-integrated energy feedback. In: IJsselsteijn, W.A., Kort, Y.A.W., Midden, C., Eggen, B., Hoven, E. (eds.) PERSUASIVE 2006. LNCS, vol. 3962, pp. 45–49. Springer, Heidelberg (2006). doi:10.1007/11755494_7
7. Gamberini, L., Spagnolli, A., Corradi, N., Jacucci, G., Tusa, G., Mikkola, T., Zamboni, L., Hoggan, E.: Tailoring feedback to users' actions in a persuasive game for household electricity conservation. In: Bang, M., Ragnemalm, E.L. (eds.) PERSUASIVE 2012. LNCS, vol. 7284, pp. 100–111. Springer, Heidelberg (2012). doi:10.1007/978-3-642-31037-9_9
8. Yun, R., et al.: Toward the design of a dashboard to promote environmentally sustainable behavior among office workers. In: Berkovsky, S., Freyne, J. (eds.) PERSUASIVE 2013. LNCS, vol. 7822, pp. 246–252. Springer, Heidelberg (2013). doi:10.1007/978-3-642-37157-8_29
9. Karjalainen, S.: Consumer preferences for feedback on household electricity consumption. Energy Build. **43**, 458–467 (2011)
10. Wood, G., Newborough, M.: Energy-use information transfer for intelligent homes: enabling energy conservation with central and local displays. Energy Build. **39**, 495–503 (2007)
11. DeWaters, J.E., Powers, S.E.: Energy literacy of secondary students in New York State (USA): a measure of knowledge, affect, and behavior. Energy Policy **39**, 1699–1710 (2011)
12. Brounen, D., Kok, N., Quigley, J.M.: Energy literacy, awareness, and conservation behavior of residential households. Energy Econ. **38**, 42–50 (2013)
13. Zapico Lamela, J.L., Turpeinen, M., Guath, M.: Kilograms or cups of tea: comparing footprints for better CO2 understanding. PsychNology J. **9**(1), 43–54 (2011)
14. Oinas-Kukkonen, H., Harjumaa, M.: Persuasive systems design: key issues, process model, and system features. Commun. Assoc. Inf. Syst. **24**(1), 28 (2009)
15. Cohen, L., Manion, L., Morrison, K.: Research Methods in Education. Routledge, London (2000)
16. Hattie, J.: Visible Learning: A Synthesis of over 800 Meta-analyses Relating to Achievement. Routledge, London (2009)

Persuasive Technology Against Public Nuisance – Public Urination in the Urban Nightlife District

Randy Bloeme[1(✉)], Peter de Vries[2], Mirjam Galetzka[3], and Paul van Soomeren[1]

[1] DSP-groep, Amsterdam, The Netherlands
{rbloeme,pvansoomeren}@dsp-groep.nl
[2] Psychology of Conflict, Risk, and Safety, University of Twente, Enschede, The Netherlands
p.w.devries@utwente.nl
[3] Communication Science, University of Twente, Enschede, The Netherlands
m.galetzka@utwente.nl

Abstract. Assumptions of the *goal framing theory* are applied to the specific context of a nightlife environment. Focusing on public urination as specific and often occurring antisocial behaviour in nightlife environments, this research explored how choice behaviour of potential public urinators can be influenced in a positive way. One boundary condition was to intervene in choice behaviour without negatively affecting the widely appreciated attractive and stimulating character of nightlife environments. Five experimental forms of nudging and priming are conducted to facilitate alternative social behaviour and to further stimulate potential public urinators to perform social behaviour. This was done by activating positive emotions, presenting visible and accessible alternatives and influencing subjective norms. Facilitating social behaviour reduced public urination by 41%, while additional interventions reduced public urination up to 67%. The results contribute to an extension of *goal framing theory* to specific contexts like nightlife environments.

Keywords: Field experiment · Goal framing theory · Nudging · Priming · Public urination

1 Introduction

Nightlife environments are widely appreciated because of their stimulating and adventurous character. This type of environment offers intense experiences from a wide range of positive and negative emotions ranging from fun and excitement to fear and stress [1, 2]. Although these environments are appreciated, all types of antisocial behaviour occur. Public urination is one of the most common problems in nightlife environments [3, 4]. To discourage this type of behaviour, the main challenge is to take measures by keeping in mind what makes these environments so attractive [4, 5].

In earlier research, it was already pointed out that, in general, strategies to prevent the occurrence of antisocial behaviour are more (cost-)effective than repressive strategies [6]. The most economical procedure is to stimulate the right behaviour by providing response priming instructions (i.e. signs with anti-litter messages) at

P.W. de Vries et al. (Eds.): PERSUASIVE 2017, LNCS 10171, pp. 187–198, 2017.
DOI: 10.1007/978-3-319-55134-0_15

appropriate times, since the effectiveness of general incentives often drops when the incentive stops [7].

Providing response priming instructions to prevent the occurrence of public urination without negatively affecting the attractiveness of the environment can, in our view, be considered persuasive technology [8], i.c. subtle persuasion by the use of technology and environmental interventions.

1.1 Theoretical Approach

One way to explain how persuasive technology works is by looking at the psychological concepts of the *goal framing theory* [9]. The theory states that behavioural goals steer intentions, determine which knowledge and attitudes are activated, how different aspects of a situation are evaluated and which alternatives are considered [10]. Goal framing theory distinguishes three overarching goals which include the most important aspects of human functioning: (1) hedonic goals, focussing on direct need fulfilment, (2) gain goals, focussing on getting and maintaining resources for need fulfilment and (3) normative goals, focussing on conforming to social norms and rules to fit into social contexts. All three overarching goals interact with each other and are all activated to some degree. Due to internal and external cues, a certain goal can become more dominant than others. At that point, behavioural and attentional processes are most strongly framed by this goal (1). The right external cues can be provided at the right moment by using persuasive measures.

Influencing Choice Behaviour. Related to the topic of public urination in nightlife environments, persuasive technology could focus on influencing the choice between prosocial or antisocial behaviour when a person needs to go to the toilet. The prosocial choice is using a toilet and in this case the antisocial behaviour is public urination, for instance in a dark alley.

In order to influence social behaviour via activating specific goals, a basic understanding of the three overarching goals is needed. Direct need fulfilment (hedonic goals) for instance, is explained by theories and models on affect and emotion. When these goals are most strongly activated, emotional bonding from a person with his surrounding environment could for instance be a motivator to behave more prosocially [11].

Getting and maintaining resources for need fulfilment (gain goals) is explained by theories of rational choice. According to these theories, social behaviour depends on the intention of a person and the context of the situation [12]. The intention of a person is formed by attitudes and subjective norms and the perceived behavioural control. Attitudes and subjective norms are formed by expected costs and benefits and the extend to which the person beliefs that a certain behaviour is right or wrong according other people. Perceived behavioural control is formed by the extent to which a person beliefs he or she is able to perform a certain behaviour [12].

The role of conforming to social norms and rules to fit into social contexts (normative goals) is an important concept of the theory of normative conduct, which states that norms systematically influence behaviour once they are activated [13, 14]. The norm activation model (NAM) [15] describes that norms become activated once a person (1) is conscious of the problem and consequences of the certain behaviour, (2) feels responsible for the consequences, (3) is able to identify and (4) perform a prosocial alternative.

The factors described above are requirements in order to stimulate a pro-social choice when a certain goal is activated. These factors already show that there is a certain overlap in the overarching goals. Perceived behavioural control in order to stimulate a pro-social choice when gain goals are activated is for instance closely related to being able to identify and perform a prosocial alternative. These two factors are closely related to the factors needed to activate normative goals. Lindenberg and Steg [9] continue on this overlap by describing how behavioural goals could either strengthen or inhibit each other. In case of public urination, when a person needs to go to the toilet the intention to use a toilet (pro-social choice) will be strengthened once personal norms are activatied. The predictable power of the theory of planned behaviour will grow in that case [16, 17]. However, the opposite is true as well. When a person has the intention to use a toilet but he is not able to easily find one, the personal norm to use a toilet becomes weakened and the chance of public urination grows [9].

The Context of a Nightlife Environment. Knowing which factors could influence the choice between a pro-social of anti-social alternative is important in order to influence this choice behaviour. As described before, the behavioural goal that becomes most strongly activated determines what a person thinks at that moment, which information is processed and which alternatives are considered [10]. However, in this proces the context of a nightlife environment plays an import role. First because of personal norms and views of visitors of a nightlife environment. Visitors often see the nightlife environment as a place where there are different social norms compared to home, school and work [5] and as a place of self-indulgence, with no restrictions [18]. Secondly, attentional processes for instance become restricted with the use of alcohol. Giancola et al. [19] describe this as the *Alcohol Myopia Effect,* which in general leads to a restriction of cognitive processes. One of the problems then is that people are for instance less able to find a suitable pro-social alternative like a toilet or they are less aware of the social norm.

Although cognitive resources are limited in a nightlife environment there are some good opportunities to stimulate pro-social behaviour. One effective way is by providing subtle signals in the direct environment. This is called nudging, small changes in the information that a person is confronted with when making a decision [20]. Nudges respond to automatic processes and are very useful when attention and cognitive processes are limited [21]. An example of a nudge is the use of marked lines on a floor in order to 'guide' people to the stairs instead of a nearby elevator. This could lead to a considerable higher number of people using the stairs [22].

Another opportunity that is closely linked to nudging is *priming*. The concept of priming uses so-called *primes*: subtle incentives that unconsciously activate certain knowledge about a social situation. Often, this knowlegde is about situational norms which represent general accepted beliefs about how to act in a certain situation [23]. A library for instance is a specific environment where it is generally accepted and well known that people should be quiet [24]. Showing a picture of this familiar environment to people could result in these people actually showing this behaviour [23]. Two boundary conditions for the effectiveness of a prime are the extent to which the used prime suits the local context where it is used [25] and the availability of a useful alternative to perform the right behaviour [26].

The Study. In this study several persuasive (technological) interventions are applied in order to stimulate visitors of a nightlife environment to use a toilet instead of choosing for public urination.

Toilet Availability. Having the opportunity to perform the right behaviour is a necessary requirement for the interventions to work [11]. In this case, an opportunity to perform the right behaviour is the availability of a toilet facility (i.e. a public urinal).

Light Projections. As mentioned before, a person's emotional bonding with his/her surrounding environment could be a motivator to behave more prosocial when hedonic goals are active [11]. Well-designed lighting can be a source to stimulate this factor. Where bright light was found to have a positive influence on self-consciousness [27], coloured light was more effective in stimulating pro-social behaviour via affective reactions [28, 29].

Arrows. Earlier studies [22] found that marked lines on the ground steering people to the stairs can lead to a considerable higher number of people using the stairs. The same effect was found when they marked footsteps on the ground leading to a litter bin. The amount of littering in the street was reduced because more people threw their litter in the bin. Following these results, arrows on the street leading to the toilet are used to facilitate perceived control of having a suitable alternative in a context where attention and cognitive processes are often limited.

Graphics. Norms can either be implicitly (e.g. design) or explicitly (e.g. verbal messages) activated [23]. Implicit activation works via unconscious mental representations. For instance by activating the social norm of being silent through the presentation of a graphic of a library environment [23]. In this study, graphics of intimate music concerts are presented in order to represent a situational social norm which suits the local context of a nightlife environment - an important boundary condition as described before.

Arrows and Light Projections Combined. As described before, goal framing theory [9] states that the different overarching goals are always activated to a certain degree and influence each other. In order to explore if different factors activating these goals can strengthen each other, arrows and light projections are combined in this study as well.

Hypotheses. For each intervention it is assumed that it will lead to a decline in the number of public urinators compared to a normal situation where factors leading to a pro-social choice are not actively stimulated.

2 Method

The experiment took place in one of the nightlife areas in Amsterdam, the Rembrandt Square (Rembrandtplein), in eight consecutive weeks in April and May 2016. Two relatively similar sub areas were selected, an experimental and a control location (EL versus CL); in both locations there is a relatively dark alley with recesses facilitating public urination (see Fig. 1). At the EL this alley is app. 55 m long and 3 m wide; the CL measures 70 × 6 m. To reduce public urination, our manipulations were situated in or near the alley at the EL. Average numbers of public urinators on social evenings are 68.57 for the EL and 44.79 for the CL (Fig. 2).

Fig. 1. Experimental (left) and control (right) locations

Fig. 2. Placement of the mobile urinal unit; the entrance to the EL can be seen in the background

2.1 Participants and Design

Our sample consisted of part of the 10 000–15 000 people that on average visit the Rembrandt Square nightlife area during social evenings. The field experiment had a one-factor design, with five manipulations: Toilet Availability, Arrows, Light Projections, Graphics and Arrows and Light Projections Combined. Before the interventions took place, pre-test measurements of public urination were conducted for two weeks at the EL and CL. The interventions were alternated weekly at the EL, each separately for one week, in the subsequent weeks, with the exception of Toilet Availability: this was tested as a stand-alone manipulation, but remained in place during later manipulations. Each week, the same measurements were conducted at the CL.

For each separate condition, measurements at the EL and CL were contrasted with the pre-tests (a standardized difference score of the number of public urinators per hour. between pre-test and intervention week) at both locations before they were contrasted against each other.

Toilet Availability involved placement of a Kros Mobile Urinal Unit to facilitate socially desirable behaviour. Placement was such that visibility and usability were ensured. At about 10 meters distance from the alley in the EL, this urinal unit could be used by four persons at the same time. After placement, the urinal unit remained in place during the entire test period [cf. (2)] (See Fig. 2).

Arrows were used to increase visibility of the urinal arrows. These were taped to the ground with broad yellow tape, accompanied by "WC" (i.e., toilet; white letters printed on a purple sticker), pointing out the urinal to people coming from multiple directions. This manipulation aimed to bring a usable alternative to public urination to the attention of those were about to relieve themselves and who had an alcohol-induced limited attention capacity. See Fig. 3 (left panel).

Light Projections were used to invoke a positive affective response (3). Projectors mounted high up on walls in the alley at the EL projected colourful non-figurative artistic images with round or oval-shaped boundaries onto the floor. See Fig. 3 (right panel).

In the Graphics condition images of intimate musical performances were placed on the walls. Specifically, these images featured singers in the singer-songwriter genre, engaged in passionate performances and in some cases surrounded by a captivated audience. These aimed to activate subjective norms associated with these settings, so as to make people abstain from public urination (which in the portrayed settings can be considered socially undesirable); this is similar to priming manipulations used for instance by (4). Selecting a setting (musical performances) that shares overlap with the festive nightlife atmosphere increases the likelihood of norm activation (5).

In the final condition the Arrows and Light Projections were simultaneously present (Arrows and Light Projections Combined).

2.2 Procedure

Visitors could enter the alley at the EL from four different directions. Because people coming from three of these directions had a better view of the urinal and were closer to it than those coming from the other direction, the former were lumped together in one group.

Fig. 3. Arrows (left-above), Light Projections (right-above) and Graphics (right and left under) manipulations (EL)

Using three cameras, measurements were taken on Fridays and Saturdays between 23:00 and 06:00 h. One camera was aimed at the control location (CL), one on the experimental location (EL), and one on the urinal. After each weekend, the number of public urinators per hour was counted at both the CL and the EL; this variable was our dependent variable. For each public urinator registrations the time was noted, as well as the direction this individual came from. The number of people who used the urinal was also counted. Background variables that were incorporated included the number of passers-by (crowdedness), weather conditions, the number of parked bicycles in the alley, and the number of social interactions (i.e., when people slowed down or stopped to talk to someone else). These latter variables are, however, beyond the scope of this paper, and will not be discussed further. It should only be noted here that these background variables, in general, had little to no influence on the number of public urinators.

Camera footage was made available by the local police department; these could only be scored at the police office precinct, and were erased afterwards.

3 Results

The number of public urinators at the EL during the pre-tests was the highest number counted in this study ($M = 45.00$, $SD = 1.02$). The largest part of this group came from the direction where people had no direct sight at the later placed toilet facility. During the whole study, most public urinators where counted between 00:00 and 04:00 AM.

Toilet availability led to a decrease in the total number of public urinators per night of almost 40% ($M = 28.00$, $SD = 6.11$). The decline of public urinators per hour at the EL ($M = -0.79$, $SD = 0.78$) differs significantly from the variation at the CL ($M = 0.10$, $SD = 0.48$), $t(54) = -3.994$, $p < .001$. Further analyses of both groups entering the EL from different directions show that toilet availability leads to a significant difference in the variation in public urinators with direct view at the toilet ($M = -1.08$, $SD = 0.87$), $t(54) = -5.192$, $p < .001$. There was no effect in the group of people with no direct sight. Facilitating people reduced the total number of public urinators by seventeen people, mainly coming from directions where there was direct sight at the toilet.

Arrows on the street, pointing to the toilet from every direction, led to a decrease of about 51% in the total number of public urinators per night ($M = 23.00$, $SD = 6.11$). The variation in the number of public urinators per hour at the EL ($M = 1.01$, $SD = 1.28$) differs significantly from the variation at the CL ($M = 0.08$, $SD = 0.55$), $t(26) = -2{,}936$, $p = .004$. Comparing the two groups with and without direct sight at the toilet, it turned out that the variation in the number of public urinators per hour only differed significantly in the group with direct view ($M = -1.02$, $SD = 1.41$), $t(26) = -2.738$, $p = .006$. Although the variation in the number of public urinators without direct sight showed a decrease at the CL as well ($M = -.39$, $SD = 1.03$), no significant effect was found using a 95% confidence interval. Arrows on the street reduced the total number of public urinators by twenty-two people, mainly coming from directions where there was direct sight at the toilet.

Light projections led to a decrease of about 51% in the total number of public urinators per night ($M = 23.00$, $SD = 5.19$). The variation in the number of public urinators per hour at the EL ($M = -1.01$, $SD = 1.39$) differs significantly from the CL ($M = -0.36$, $SD = 0.36$), $t(26) = -1.694$, $p = .051$. Further analyses of the both groups showed that the light projections only led to a significant variation in the group with direct sight ($M = -1.33$, $SD = 1.32$), $t(26) = -2.648$, $p = .007$. No effects were found for the group without direct sight. Coloured light projections reduced the total number of public urinators by twenty-two people, mainly coming from directions where there was direct sight at the toilet.

Hanging graphics at both sides of the EL had no effect on the number of public urinators ($M = 42.50$, $SD = 11.93$) compared to the pretest. There was a significant variation in the number of public urinators with direct view on the toilet ($M = -0.87$, $SD = 0.80$), $t(26) = -3.157$, $p = .002$. However, these people were not able to see the graphics before they chose to use the toilet. Therefore, this effect is ascribed to the toilet instead of the graphics.

The combination of arrows and light projections led to a decrease of about 67% in the total number of public urinators ($M = 15.00$, $SD = 4.20$). The variation in the number of public urinators at the EL ($M = -1.39$, $SD = 0.75$) differs significantly from the CL

($M = -0.24$, $SD = 0.40$), $t(26) = -5.026$, $p < .001$. Further analyses of both groups showed again that the intervention only led to a significant variation in the group with direct sight ($M = -1.36$, $SD = 1.14$), differing significantly from the CL, $t = -3.465$, $p = .001$. Although the variation of public urinators without direct sight at the toilet became stronger, following the trend of the variation after using arrows only, no significant effect was found. The combination of arrows and light projections reduced the total number of public urinators by thirty people, mainly coming from directions where there was direct sight at the toilet.

4 Discussion

This study explored how potential public urinators can be stimulated to make a prosocial choice: using a toilet. The study was built on leads mainly from the goal framing theory [9]. Following this theory, it was assumed that the process from a need for a toilet to the choice between using a toilet (pro-social) or public urination (anti-social) is framed by three overarching goals: hedonic, gain and/or normative goals. These goals steer intentions, determine which knowledge and attitudes are activated, how different aspects of a situation are evaluated and which alternatives are considered [10].

First, the study found statistical evidence for the need to have a suitable alternative in order to stimulate pro-social behaviour [1]. All the interventions only had an effect when people were able to directly see the toilet. Toilet availability on its own led to a significant decline in the total number of public urinators with direct sight at the toilet. This could be explained by potential public urinators with an activated gain goal having the possibility to easily find resources for direct need fulfilment [2]. Potential public urinators with active normative goals might have been facilitated in having a suitable alternative to perform social behaviour [3].

The effect of toilet availability can easily be strengthened by adding arrows on the ground pointing towards the toilet facility. It is expected that arrows mainly strengthen the visibility of a useful alternative, an important factor when gain goals are activated [2]. However, this effect is limited to the distance that people have to walk and possibility of having direct sight of the target (toilet facility). The effect of arrows was only found in the group of people having direct sight at the toilet. The effect of adding arrows supports findings from [22] where people easily follow lines on the ground to the stairs. The effect also provides statistical evidence for the assumed thought that potential public urinators are easily led by subtle signal in the direct environment when cost and benefits are limited [4, 5].

The same effect is applicable to coloured light projections focussing on emotional bonding with the direct environment. These light projections were able to further reduce the number of public urinators. Especially when the light projections were combined with arrows. It was striking to see that many people were interacting (e.g. dancing and taking pictures) with the light projections. It is therefore assumed that light projections had a positive effect on emotions [6], but a visible and useful alternative is necessary in order to really influence the choice behaviour of potential public urinators.

No statistical evidence was found for the assumed effect of graphics. On the one hand, this could mean that subjective norms do not influence the choice between a pro- or anti-social choice. For instance, because potential public urinators unconsciously use heuristics about public urinating in nightlife environments instead of moving through a full process of choice where personal norms are being evaluated [4, 5]. On the other hand, it is possible that the intervention did not succeed in activating subjective norms. Lindenberg and Steg [(9] described earlier that subjective norms need the most support in order to be activated while Giancola et al. [19] add that attentional processes might be limited because of the use of alcohol. It is assumed that the latter is true in this case, while it was striking to see during the data collection that people passing by hardly seems to actively look at the graphics on the wall.

Interpreting the results of this study, it should be noted that there are some limitations to the reliability of the CL. Even though the EL and CL had some corresponding physical characteristics there are some important differences as well. First, the CL is in a less crowded part of the nightlife district compared to the EL. Second, the CL has some visible entrances to houses and a hotel, where the EL has fewer and less visible entrances to houses.

Because of the fact that the results of this study are highly contextual, the original plan was to simultaneously perform the experiments at an EL and CL in another large city in the Netherlands. Even though preparations where in an advanced stage, it proved impossible to do that.

Further research could benefit from measuring the degree to which factors activating behavioural goals (e.g. emotion, perceived behavioural control, subjective norms) are already activated. With this information it would be possible as well to explore to what degree these factors should be activated by comparable interventions in order to find people willing to walk further to a toilet when they do not have a direct sight on the toilet.

The same accounts for the use of alcohol. This study did not take into account the number of alcohol that participants used. Measuring levels of alcohol use could provide interesting information about the interaction between alcohol and behavioural goals related to antisocial behaviour in nightlife environments.

4.1 Conclusion

This study shows that there are effective ways to intervene in antisocial behaviour in nightlife environments, without causing any damage to the widely appreciated attraction of this environment. The results offer an extension of the goal framing theory [9] to the specific context of a nightlife environment. In this environment, influencing choice behaviour is most effective by facilitating and stimulating the visibility and usability of a suitable alternative for pro-social behaviour. This could easily be achieved by providing a toilet facility and optionally stimulating extra usage of this toilet by referring to it with arrows on the street. Once these boundary conditions are fulfilled, positively influencing emotional factors could lead to even better results. In that case, the stimulating and adventurous character of the nightlife environment can maintain itself by restraining its own negative effects in a positive way.

References

1. Hubbard, P.: The geographies of 'going out', emotion and embodiment in the evening economy. In: Davidson, J., Bondi, L., Smith, M. (eds.) Emotional Geographies. Ashgate, Aldershot (2005)
2. Jayne, M., Valentine, G., Holloway, S.L.: Alcohol, Drinking, Drunkeness: (Dis)orderly Spaces. Ashgate, Aldershot (2011)
3. Eldridge, A.: Public panics: problematic bodies in social space. Emot. Space Soc. **3**, 40–44 (2010)
4. Van Liempt, I., Van Aalst, I.: Urban surveillance and the struggle between safe and exciting nightlife districts. Surveill. Soc. **93**, 280–292 (2012)
5. Goossens, F.X., et al.: Het Grote Uitgaansonderzoek. Trimbos Instituut, Utrecht (2013)
6. Dwyer, W.O., et al.: Critical review of behavioural interventions to preserve the environment: research since 1980. Environ. Behav. **25**, 275–321 (1993)
7. Burgess, R.L., Clark, R.N., Hendee, J.C.: An experimental analysis of anti-litter procedures. J. Appl. Behav. Anal. **4**, 71–75 (1971)
8. Fogg, B.J.: Persuasive Technology: Using Computers to Change What We Think And Do. Morgan Kaufmann, San Fransisco (2003)
9. Lindenberg, S., Steg, L.: Normative, gain and hedonic goal frames guiding environmental behaviour. J. Soc. Issues **65**, 117–137 (2007)
10. Kruglanski, A.W., Köpetz, C.: What is so special (and non-special) about goals? A view from the cognitive perspective. In: Moskowitz, G.B., Grant, H. (eds.) The Psychology of Goals, pp. 25–55. Guilford Press, New York (2009)
11. Schultz, W., Tabanico, J.: Self, identity, and the natural environment: exploring implicit connections with nature. J. Appl. Soc. Psychol. **37**, 1219–1247 (2007)
12. Ajzen, I.: The theory of planned behaviour. Organ. Behav. Hum. Decis. Processes **50**, 179–211 (1991)
13. Cialdini, R., Trost, M.: Social influence: social norms, conformity and compliance. In: Gilbert, D., Fiske, S., Lindzey, G. (eds.) The Handbook of Social Psychology. pp. 151–192. McGraw-Hill, New York (1998)
14. Kallgren, C.A., Reno, R.R., Cialdini, R.B.: A focus theory of normative conduct: when norms do and do not affect behaviour. Pers. Soc. Psychol. Bull. **26**(8), 1002–1012 (2000)
15. Steg, L., De Groot, J.: Explaining prosocial intentions: testing causal relationships in the norm activation model. Brit. J. Soc. Psychol. **49**, 725–743 (2010)
16. Bamberg, S., Schmidt, P.: Incentives, morality, or habit? Predicting students' car use for university routes with the models of Ajzen, Schwartz and Triandis. Environ. Behav. **35**, 264–285 (2003)
17. Harland, P., Staats, H., Wilke, H.: Explaining proenvironmental intention and behaviour by personal norms and the theory of planned behaviour. J. Appl. Soc. Psychol. **29**, 2505–2528 (1999)
18. Measham, F.: Play space: historical and socio-cultural reflections on drugs, licensed leisure locations, commercialisation and control. Int. J. Drug Policy **15**, 337–345 (2004)
19. Giancola, P.R., et al.: Alcohol myopia revisited: clarifying aggression and other acts of disinhibition through a distorted lens. Perspect. Psychol. Sci. **5**, 265–278 (2010)
20. Momsen, K., Stoerk, T.: from intention to action: can nudges help consumers to choose renewable energy? Energ. Policy **74**, 376–382 (2014)
21. Camerer, C., et al.: Regulation for conservatives: behavioural economics and the case for "asymmetric paternalism''. Univ. Pennsylvania Law Rev. **151**, 1211–1254 (2003)

22. Dijksterhuis, A., van Baaren, R.: Inspiratielijst Voorkomen Zwerfafval In: De Openbare Ruimte. Zwerfafvalcongres (2015)
23. Aarts, H., Dijksterhuis, A.: The silence of the library: environment, situational norm, and social behaviour. J. Pers. Soc. Psychol. **84**, 18–28 (2003)
24. Dijksterhuis, A., Bargh, J.: The perception-behaviour expressway: automatic effects of social perception on social behaviour. Adv. Exp. Soc. Psychol. **33**, 1–40 (2001)
25. Mattila, A.S., Wirtz, J.: Congruency of scent and music as a driver for in-store evaluations and behaviour. J. Retail. **77**, 273–283 (2001)
26. Schultz, P.W., et al.: Littering in context: personal and environmental predictors of littering behaviour. Environ. Behav. **45**, 1–25 (2011)
27. Steidle, A., Werth, L.: In the spotlight: brightness increases self-awareness and reflective self-regulation. J. Environ. Psychol. **39**, 40–50 (2014)
28. Küller, R., et al.: The impact of light and colour on psychological mood: a cross-cultural study of indoor work environments. Ergonomics **49**, 1496–1507 (2006)
29. Lerner, J., Tiedens, L.: Portrait of the angry decision maker: how appraisal tendencies shape anger's influence on cognition. J. Behav. Decis. Making **19**, 115–137 (2006)

Design Principles and Strategies

Commitment Devices as Behavior Change Support Systems: A Study of Users' Perceived Competence and Continuance Intention

Michael Oduor(✉) and Harri Oinas-Kukkonen

Oulu Advanced Research on Services and Information Systems, University of Oulu, P.O. Box 3000, 90014 Oulu, Finland
{michael.oduor,harri.oinas-kukkonen}@oulu.fi

Abstract. The design principles of persuasive systems and the corresponding software features have been shown to have a positive effect on individuals' behavior and systems use. This study continues along the same line by analyzing the effects of these design principles on users' perceived competence and intention to continue to use an online commitment device. A structural equation modelling approach is used to identify the factors that affect the intention to continue to use the system. Data ($N = 173$) collected from the system's users is tested against the proposed research model. The results show that 37% of users' continuance intention is explained by the implemented persuasive software feature categories (computer–human dialogue support, primary task support, perceived credibility, and social support) and perceived competence. Of these categories, primary task support has the strongest effect on perceived competence and continuance intention. The study concludes with a discussion and recommendations for future research.

Keywords: Persuasive systems design · Continuance intention · Perceived competence · Behavior change · Commitment device

1 Introduction

People require self-commitment devices because individuals set goals with varying degrees of formality that in most cases the individuals fail to meet primarily due to a lack of self-control. This draws into doubt the standard model of human behavior: that people are rational beings who when given a set of options always choose the most beneficial one [1]. There are numerous examples of suboptimal health behavior, for example, unwillingness to think about problems when risks are known or data are ambiguous, in which people do not follow through on their (behavior change) goals or make self-beneficial decisions [2].

Failing to meet goals can lead to entering into a commitment contract, a form of a commitment device that is an actual contract between two parties, rather than a unilateral arrangement employed by an individual to restrict his or her own choices [1]. A commitment contract is based on understanding attitude–behavior inconsistency (not doing what one has stated one will do or not acting according to one's own beliefs) and how

P.W. de Vries et al. (Eds.): PERSUASIVE 2017, LNCS 10171, pp. 201–213, 2017.
DOI: 10.1007/978-3-319-55134-0_16

incentives get people to work. As the role of information technology increases in people's day-to-day activities, decision making, and experiences, new opportunities to assist people in making self-beneficial choices have arisen through, for example, the use of persuasive systems to digitally present choices in a way that leverages people's decision-making processes, thus encouraging them to make self-beneficial decisions [3, 4].

Persuasive systems are interactive systems that motivate users and are designed with the intent to change users' attitudes and/or behavior without coercion or deception [3]. These systems and technology, in general, are more effective in persuasion and offer a potential path to long-term behavior change as they leverage technology's capabilities [5] to influence users.

The objective of the present study, thus, is to investigate users' perceptions of an online commitment device. Specifically, the study examines the effect of the persuasive system design (PSD) model's principles [3] on users' perceived competence and how this, in turn, predicts their intention to continue using the commitment device. Data ($N = 173$) collected from the system's users is tested against the proposed research model using partial least squares (PLS), a structural equation modeling approach. Results show that primary task support from the PSD software feature categories is the strongest predictor of users' perceived competence and continuance intention.

2 Theoretical Background

The underlying principle behind persuasive systems is understanding users and integrating this understanding into developing systems (or implementing system features) that encourage users to more actively pursue or focus on their objectives. This could be through designing strategies that help people who want to change their everyday behavior [6], investigating how the design of persuasive systems and intervention characteristics influence users' adoption, continuance intention, and adherence to these systems [7–9], analyzing how the features implemented in a gamified persuasive system are compatible with users' goals and affect system usage, and whether this leads to the desired behavioral outcome [10], among many others.

Many people are aware of the changes they need to make in their behavior but usually have a problem following through on their intentions primarily due to a lack of self-control [11, 12]. That is, people intend to make choices that carefully consider short-term and long-term cost and benefits, "*but at the decision-making moment, they place disproportionate weight on immediate costs and benefits*" [12]. Furthermore, some of the solutions people choose to help them change a behavior or form habits are often easy and do not have an impact because there are no hard commitments. A solution to this problem is the use of commitment devices: "*a way of changing one's incentives to make an otherwise empty threat or promise credible*" [11] and entering into an arrangement that restricts their future choice set by making some choices more expensive or making a present intention for a future action more substantial [1, 13].

As commitment devices (see, e.g., [11] for a list of online commitment devices) also involve persuasive design to motivate users into action, it is essential to empirically analyze the persuasive elements in these systems.

3 Research Model

The PSD model [3] is used as the main conceptual framework in the present study. The model is an integrative framework encompassing the psychological principles behind the design of persuasive systems, the importance of considering the context of use, and the specific features implemented in persuasive systems [3]. In addition to the persuasive features, the study investigated the effect of perceived competence [14, 15] on continuance intention [6, 9]. The measurement items for the constructs can be found in Appendix A. Figure 1 presents the research model.

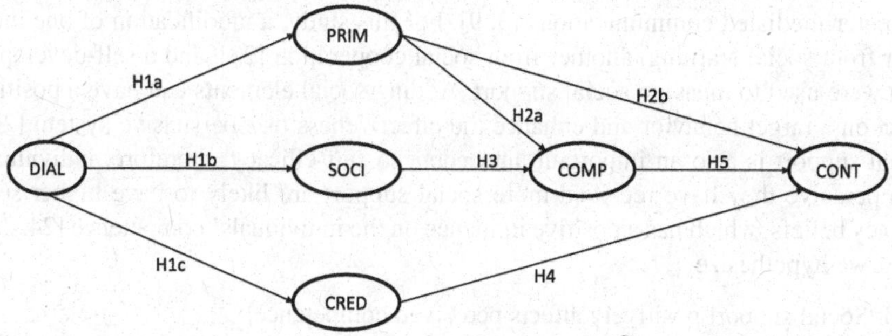

Fig. 1. Research model (Note. DIAL = Dialogue support; PRIM = Primary task support; SOCI = Social support; CRED = Perceived credibility; COMP = Perceived competence; CONT = Continuance intention.)

Computer–Human Dialogue Support. Computer–human dialogue support (henceforth, dialogue support) features facilitate and improve communication between a system and its users, especially in terms of feedback for better guiding users toward their intended goals [3, 17, 18]. Reminding users of their target behavior, for example, will more likely lead to goal achievement, and reminders in various forms [19], such as automated and personalized e-mails and phone calls triggered by technology or a clinician, have been shown to help in task completion [20, 21]. Thus, these features are strongly linked to primary task support. Individuals' interactions with computers are also fundamentally social and from a user's perspective, computers can, in addition to being tools and mediums that increase capability and provide interactive experiences, be social actors that create and enhance relationships [5, 22]. Dialogue support also affects users' positive feelings toward the system, which likely influences the individuals' confidence (credibility) [8]. Therefore, we propose the following hypotheses:

H1a: Dialogue support positively affects primary task support.
H1b: Dialogue support positively affects social support.
H1c: Dialogue support positively affects perceived credibility.

Primary Task Support. Primary task support refers to the way in which a system helps users carry out their primary task (that is, the real-world activity that they would like to complete) [3]. Primary task support aims to enhance users' self-efficacy and reduce the cognitive burden associated with using information systems by providing examples of the correct behavior, enabling reflection, goal-setting, and tracking of one's progress [8, 9]. Therefore, it is likely that the more a system supports users in meeting their primary objective, the more they will feel competent and continue to use the system. We propose the following:

H2a: Primary task support positively affects perceived competence.
H2b: Primary task support positively affects continuance intention.

Social Support. Social support features motivate users by leveraging social influence, especially as social relationships are increasingly maintained and nurtured through computer-mediated communications [3, 9]. For this study, a modification of one indicator from social learning, another from social cooperation [23], and a self-developed item were used to measure social support. Adding social elements can have a positive effect on a target behavior and enhance the effectiveness of a persuasive system [20]. Social support is also an important antecedent to self-efficacy; therefore, individuals who perceive they have received more social support are likely to have higher self-efficacy beliefs, which has a positive influence on the individuals' competence [24, 25]. Thus, we hypothesize:

H3: Social support positively affects perceived competence.

Perceived Credibility. Perceived credibility deals with designing credible and subsequently more persuasive systems. Perceived credibility comprises trustworthiness, expertise, surface credibility, real-world feel, authority, third-party endorsements, and verifiability [3]. Research has shown that understanding how users perceive the credibility of reviews, for example, is important for the survival and development of online discussion forums [26]. Cheung et al. [26] found that a high level of electronic word-of-mouth review credibility enhances users' adoption desires for online recommendations, ultimately leads to continuous use, and attracts repeat visits to other consumer recommendations forums [26]. Therefore, we propose the following:

H4: Perceived credibility positively affects continuance intention.

Perceived Competence. Competence is the desire to feel a sense of satisfaction and effectiveness in attaining important results [15]. Perceived competence is one of the three fundamental needs proposed in the self-determination theory [15]. The theory focuses on motivation and proposes that human beings have basic psychological needs for autonomy, competence, and relatedness [15]. Perceptions of competence are theorized to be important because they facilitate people's goal attainment and provide them with a sense of need satisfaction from engaging in an activity at which they feel effective [14]. Therefore, the more competent one feels, the more likely one is to continue with an activity. Consequently, we propose:

H5: Perceived competence positively affects continuance intention.

4 Research Methodology

Study Context. The artifact examined in the present study is a Web-based Quantified Self (self-tracking data collection and visualization) with commitment contracts system[1]. In the application, an individual creates a contract to spend either less or more time on a particular activity, and if the individual does not stick to the goal(s), then he or she is charged. The system adds real consequences for not meeting planned goals. An individual can set a goal (lose weight, exercise, to-do tasks, etc.), manually input the required results, or connect to external services for automatic reporting. Progress is plotted on a graph that is easily accessible from one's account either through mobile devices or online. If the individual stays on track, then the service is free; however, if the individual goes off track, then he or she pledges money to continue progressing with his or her goal. If the individual goes off track again, he or she is charged.

Table 1. Respondent characteristics ($N = 173$)

Demographics	Value	Frequency	Percent (%)
Gender	Male	131	75.7
	Female	39	22.5
	Unspecified	3	1.7
Age	30 or under	87	50.3
	31–40	53	30.6
	41–50	22	12.7
	50 and older	11	6.4
Marital status	Married	58	33.5
	In a relationship	62	35.8
	Single	53	30.6
Education	High school	20	11.6
	Vocational training	4	2.3
	Bachelor's degree	66	38.2
	Master's degree	49	28.3
	Other advanced degree	9	5.2
	Doctoral degree	25	14.5

Data Collection and Respondent Characteristics. Between March and May 2016, an online survey of the system users was conducted. Data were collected using an online software tool (Webropol) over a 7-week period. Participants were recruited through a public link that was sent to their email addresses. The survey had two main parts. The first part consisted of demographic questions and questions related to the main goal tracked, use history, and use frequency. A total of 173 responses without missing data

[1] https://www.beeminder.com/.

were collected. Overall, about 76% of the respondents were male, 50% were 30 years old or younger, about 70% were either married or in a relationship, and 86% had at least an undergraduate degree. Tables 1 and 2 present the respondents' characteristics.

Table 2. System version, main goals tracked, and use history

	Value	Frequency	Percent (%)
Version	Free Plan	128	74
	Bee Lite	11	6.4
	Plan Bee	27	15.6
	Beemium	7	4
Goals	Fitness and training	34	19.7
	Health	38	22
	Productivity	80	46.2
	Learning	21	12.1
Use history	6 months or less	40	23.1
	6 months to a year	23	13.3
	1 to 2 years	35	20.2
	2 to 3 years	33	19.1
	3 or more years	42	24.3

The second part was the main survey instrument (Appendix A) consisting of seven-point Likert scale items (ranging from strongly disagree to strongly agree).

5 Results and Analysis

The research model was analyzed using the structural equation modeling (SEM) approach, namely, PLS, a technique for simultaneously estimating relations among multiple constructs. SmartPLS software [27] was used for the data analysis. PLS-SEM analysis is composed of two steps: The first step assesses the measurement model, that is, analyzed the relation of each indicator with its corresponding constructs. In the research model, all of the constructs were reflective and measured with at least two indicators. Reflective measurement models are typically assessed in terms of their internal consistency, reliability, and validity. If the measurement model meets the specified criteria, then the second step, the structural model, evaluates the hypothesized relations between the constructs. The evaluation of the predictive capabilities of the structural model is based on the significance of the path coefficients and the relations between the constructs [28].

5.1 Measurement Model

The measurement model represents the relation between constructs and their corresponding measures. Constructs' properties are assessed in terms of their validity and reliability, and measurement items that are not at acceptable levels are removed. Evaluating the measurement model addresses internal consistency (the composite reliability), indicator reliability, convergent validity, and discriminant validity. Reliability and consistency are measured using Cronbach's alpha (CA) and composite reliability (CR)[2] with values ranging between 0 and 1. High values indicate higher levels of reliability, and values higher than the typically applied threshold of 0.7 are acceptable [29]. For composite reliability, values from 0.6 to 0.7 are acceptable in exploratory research, while in more advanced stages of research, values between 0.7 and 0.9 are considered satisfactory. A value of 0.95, however, indicates unnecessary redundancy in the construct items [28], which was the case in the perceived credibility construct. Therefore, CRED 2 (The system provides believable content) was omitted from the construct to decrease the composite reliability to below 0.95.

The inter-construct correlations and the square root of the average variance extracted (AVE) show that all the constructs share more variance with their indicators than with other constructs. This, in addition to the indicators' factor loadings, is a good measure of convergent validity—the extent to which two or more items measure the same construct [30]. High loadings indicate that the measurement items correctly measure the same phenomenon and share a high proportion of variance [28]. Items with outer loadings below 0.6 were omitted (in primary task support and dialogue support). The AVE values for all the constructs were above the suggested minimum of 0.5 [29]. Table 3 shows that all the measurement model criteria were met.

Table 3. Reliability and validity of the constructs

	CA	CR	AVE	COMP	CONT	CRED	DIAL	PRIM	SOCI
COMP	0.88	0.916	0.73	**0.855**					
CONT	0.91	0.938	0.79	0.286	**0.890**				
CRED	0.91	0.945	0.85	0.052	0.137	**0.922**			
DIAL	0.72	0.841	0.63	0.134	0.241	0.266	**0.798**		
PRIM	0.72	0.838	0.64	0.348	0.603	0.268	0.377	**0.800**	
SOCI	0.74	0.848	0.65	0.074	0.190	0.154	0.301	0.178	**0.807**

5.2 Structural Model

Assessing the structural model determines how well the empirical data support the theory, showing the relations between different constructs and specifying how the constructs are related [28]. The key results are the path coefficients and the R^2 values, which represent the hypothesized relations among the constructs and the percentage of the total variance of the dependent variable explained by the independent variables. The complete bootstrapping method with 5,000 resamples and parallel processing with no sign changes was used. Two-tailed bias-corrected and accelerated (Bca) bootstrap was

[2] CA and CR are not commonly used abbreviations for Cronbach's alpha and composite reliability but due to space constraints the abbreviations are used in Table 4.

the confidence interval method used. Figure 2 presents the results of the path model analysis.

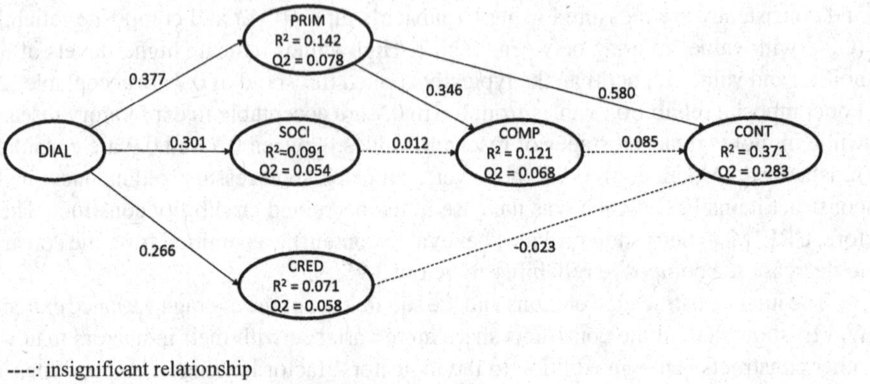

Fig. 2. Structural model results for the full sample

The results of the PLS analysis (Fig. 2) provide moderate support for the research model. Five (H1a, H1b, H1c, H2a, H2b) of the eight hypotheses were supported at $p < .001$, and primary task support and social support explain 12% of the variance in perceived competence. Dialogue support explained 14.2% of the variance in primary task support, 9% of the variance in social support, and 7% of the variance in perceived credibility. A substantial amount of the variance (37%) in intention to continue using the system was explained by primary task support, whereas perceived competence and perceived credibility were found not to be strong predictors of intention to continue to use the system.

Table 4. Total effects and effect sizes (Cohen's f^2 in parentheses)

	COMP	CONT	CRED	DIAL	PRIM	SOCI
COMP		0.085 (0.01)				
CRED		−0.023 (0.001)				
DIAL	0.134	0.224	0.266 (0.076)		0.377 (0.166)	0.301 (0.100)
PRIM	0.346 (0.132)	0.609 (0.436)				
SOCI	0.012	0.001				

Table 4 shows the total effects (direct and indirect effects) and the effect sizes of each construct on the corresponding dependent variables' variance. Effect sizes determine whether the effects shown by path coefficients are small (.02), medium (0.15), or large (.35) [31]. Effect sizes below .02 are considered too weak to be relevant, which is the case for three hypotheses. Dialogue support has the highest total effect (with a medium effect size) on primary task support, whereas primary task support has the highest total effect on perceived competence (a medium effect size) and continuance intention (a large effect size). Dialogue support also has an indirect positive effect on perceived competence and continuance intention.

The predictive relevance of the model was also assessed using Stone-Geisser's cross-validated redundancy measure (the Q^2 value). See Table 5. Values above zero for a certain reflective endogenous construct indicate the path model's predictive relevance for that particular construct [28]. Predictive relevance was demonstrated for all endogenous constructs, with continuance intention exhibiting moderately strong predictive validity.

Table 5. Predictive relevance (Q^2 values)

COMP	CONT	CRED	PRIM	SOCI
0.068	0.283	0.058	0.078	0.054

6 Discussion

The goal of the present study was to examine the applicability of PSD in explaining users' perceived competence and how PSD and competence influence intention to continue using an online commitment device. Continuance intention refers to users' intention to continue using a system after initial acceptance [16]. Continued use instead of initial acceptance explains the success of a system because the psychological motivations that affect users' continuance decisions emerge only after individuals have started using the system [16].

All relations (Fig. 2) except H4 (perceived credibility positively affects continuance intention) were found to be positively related. However, two of the positive relations (H3 and H5) and H4 (which was negative) had very low path coefficients (0.012, 0.085, and –0.023) values meaning the practical significance of the associations is very low. These findings extend the literature on adoption and continuance intentions for persuasive systems [9, 10, 24] by identifying primary task support as an important predictor of perceived competence and continuance intention. This is consistent with findings in Lehto and Oinas-Kukkonen's [9] post-hoc analysis of simpler path models that test the effect of PSD features on use continuance.

The five positive relations (dialogue support predicts primary task support, social support, and perceived credibility, and primary task support predicts perceived competence and continuance intention) demonstrate the important role of PSD in use continuance. In line with previous studies [8, 20, 22] that showed the importance of dialogue support, it was also shown in the present study to be an important factor in evaluating the effectiveness of a system. The latter two relations (primary task support as a predictor of competence and continuance) indicate that supporting a user's primary goal likely leads to him or her feeling more competent and increases the likelihood of continued use.

Interestingly, the direct effect between social support and perceived competence is weak, in contrast to that in [25], for example, who found that social support had an indirect total effect on users' nutrition through self-efficacy and self-regulation. This result may be due to the unique nature of the system compared with other persuasive systems and the social support construct itself. Two measurement items for the construct referred to the perception of support through the system, and the self-developed item was support mediated by the system (Appendix A). On the questionnaire, there was also a question asking users whether they had social supporters (self-appointed peers who can follow up on users' progress as this an optional feature in the system), and 91% (n = 158) replied no. This can help explain the statistically insignificant relation between social support and perceived competence as not many of the respondents perceived or had this type of social support.

The relation between perceived competence and continuance intention, although positive, was statistically insignificant which warrants further investigation. A possibility, as it is one of the three psychological needs [15], is to extend the present model with these other psychological constructs, examine how they correlate with the PSD features, and affect continuance intention. The surprisingly negative but statistically insignificant relation between perceived credibility and continuance intention in contrast to Lehto and Oinas-Kukkonen [10] and Cheung et al. [26] also warrants further attention.

The study had several limitations. First, some of the relations in the model, even those shown to be strong in previous research, have lower or statistically insignificant values. Future research should examine the weaker links more closely. A second limitation is that the research model was applied to only one system. Thus, caution should be exercised about the generalization of the findings to other contexts. Nonetheless, the results should be applicable to other similarly designed systems or those with similar objectives to the one investigated. Last, the effects of demographic attributes and use characteristics were not analyzed in the present study. Future research may employ different procedures, such as subgroup analysis and application of constraints for data selection, to better explain the effects of the various PSD strategies on user behavior.

7 Conclusion

In summary, in accordance with previous research, this study provides evidence that persuasive system design features are associated to varying degrees with users' perceived competence and continuance intention of a commitment device. Further, as in several previous studies, dialogue support was found to have a crucial role in the effectiveness of the system as this feature strongly supports users' interactions with the system and facilitates completion of primary tasks. Primary task support also strongly predicts users' confidence in their ability to carry out a task and their intention to continue using a system. Understanding the place of persuasive systems design in the context of fulfilling users' needs during user engagement with the systems could provide a means to counterbalance factors that lead to the non-adoption or failure of digital interventions, especially considering the nature of commitment devices in which something (usually monetary) is at stake if specific objectives are not fulfilled. As the system investigated supported the pursuance of different goals, how carrying out these different goals is supported by PSD and how this can help in changing long-term behavior should be addressed in future research.

Acknowledgments. We would like to thank Daniel Reeves, the Co-founder/CEO of Beeminder for helping with the distribution of the survey to the system's users.

Appendix A: Measurement Instrument

Constructs	Indicators	Loadings
Dialogue support	The system rewards me	0.778
	The system provides me with appropriate feedback	0.787
	The system provides me with reminders for reaching my goals[a]	***0.570***
	The system encourages me	0.743
Primary task support	They system makes it easier for me to reach my goals	0.825
	The system helps me gradually reach my goals	0.865
	The system helps me keep track of my progress	0.621
	The system offers me personalized content[a]	***0.460***
Perceived credibility	The system provides trustworthy content	0.949
	The system provides believable content[a]	***0.961***
	The system provides accurate content	0.954
	The system provides professional information	0.837
Social support	The system enables me to share with others	0.772
	The system enables me to learn from others	0.883
	The system enables someone (chosen by me) to check on my commitments	0.764
Perceived competence	I feel confident in my ability to achieve my goal	0.830
	I am capable of doing what it takes to achieve my goal	0.899
	I am able to achieve my goal	0.803
	I feel able to meet the challenge of fulfilling my goal	0.885
Continuance intention	I intend to continue using the system	0.943
	I will be using the system in the future	0.936
	I am considering discontinuing using the system	0.870
	I am not going to use the system from now on	0.805

[a]Removed items

References

1. Bryan, G., Karlan, D., Nelson, S.: Commitment devices. Annu. Rev. Econ. **2**(1), 671–698 (2010)
2. Mogler, B.K., Shu, S.B., Fox, C.R., Goldstein, N.J., Victor, R.G., Escarce, J.J., Shapiro, M.F.: Using insights from behavioral economics and social psychology to help patients manage chronic diseases. J. Gen. Intern. Med. **28**(5), 711–718 (2013)
3. Oinas-Kukkonen, H., Harjumaa, M.: Persuasive systems design: key issues, process model, and system features. Commun. Assoc. Inf. Syst. **24**(1), 28 (2009)
4. Lee, M.K., Kiesler, S., Forlizzi, J.: Mining behavioral economics to design persuasive technology for healthy choices. In: Proceedings of the SIGCHI Conference on Human Factors in Computing Systems, pp. 325–334 (2011)
5. Fogg, B.J.: Persuasive Technology: Using Computers to Change What We Think and Do. Morgan Kauffman, San Francisco (2003)
6. Consolvo, S., McDonald, D.W., Landay, J.A.: Theory-driven design strategies for technologies that support behavior change in everyday life. In: Proceedings of the SIGCHI Conference on Human Factors in Computing Systems, pp. 405–414 (2009)
7. Kelders, S.M., Kok, R.N., Ossebaard, H.C., Van Gemert-Pijnen, J.E.W.C.: Persuasive system design does matter: a systematic review of adherence to web-based interventions. J. Med. Internet Res. **14**(6), e152 (2012)
8. Lehto, T., Oinas-Kukkonen, H., Pätiälä, T., Saarelma, O.: Consumers' perceptions of a virtual health check: an empirical investigation. In: Proceedings of ECIS 2012 (2012)
9. Lehto, T., Oinas-Kukkonen, H.: Explaining and predicting perceived effectiveness and use continuance intention of a behaviour change support system for weight loss. Behav. Inf. Technol. **34**(2), 176–189 (2013)
10. Ebermann, C., Brauer, B.: The role of goal frames regarding the impact of gamified persuasive systems on sustainable mobility behavior. In: Proceedings of ECIS 2016 (2016)
11. Reeves, D.: How to do what you want: Akrasia and self-binding, Beeminder BLog (2011). http://blog.beeminder.com/akrasia/. Accessed 10 Feb 2016
12. Beshears, J.L., Choi, J.J., Laibson, D., Madrian, B.C., Sakong, J.: Self control and liquidity: how to design a commitment contract. RAND Working Paper Series WR-895-SSA (2011)
13. Moraveji, N., Akasaka, N., Pea, R., Fogg, B.J.: The role of commitment devices and self-shaping in persuasive technology. In: Extended Abstracts on Human Factors in Computing Systems, CHI 2011, pp. 1591–1596 (2011)
14. Selfdeterminationtheory.org: Perceived competence scales. http://selfdeterminationtheory.org/perceived-competence-scales/. Accessed 22 Sep 2015
15. Deci, E.L., Ryan, R.M.: The 'what' and 'why' of goal pursuits: human needs and the self-determination of behavior. Psychol. Inq. **11**(4), 227–268 (2000)
16. Bhattacherjee, A.: Understanding information systems continuance: an expectation-confirmation model. MIS Q. **25**(3), 351–370 (2001)
17. Lehto, T., Oinas-Kukkonen, H.: Persuasive features in web-based alcohol and smoking interventions: a systematic review of the literature. J. Med. Internet Res. **13**(3), e46 (2011)
18. Langrial, S., Lehto, T., Oinas-Kukkonen, H., Harjumaa, M., Karppinen, P.: Native mobile applications for personal well-being: a persuasive systems design evaluation. In: Proceedings of PACIS 2012 (2012)
19. Ritterband, L.M., Thorndike, F.P., Cox, D.J., Kovatchev, B.P., Gonder-Frederick, L.A.: A behavior change model for internet interventions. Ann. Behav. Med. **38**(1), 18–27 (2009)

20. Chiu, M.-C., Chang, S.-P., Chang, Y.-C., Chu, H.-H., Chen, C.C.-H., Hsiao, F.-H., Ko, J.-C.: Playful bottle: a mobile social persuasion system to motivate healthy water intake. In: Proceedings of the 11th International Conference on Ubiquitous Computing, pp. 185–194 (2009)
21. Langrial, S., Oinas-Kukkonen, H., Lappalainen, P., Lappalainen, R.: Rehearsing to control depressive symptoms through a behavior change support system. In: Extended Abstracts on Human Factors in Computing Systems, CHI 2013, pp. 385–390 (2013)
22. Nass, C., Moon, Y.: Machines and mindlessness: social responses to computers. J. Soc. Issues **56**(1), 81–103 (2000)
23. Stibe, A., Oinas-Kukkonen, H.: Using social influence for motivating customers to generate and share feedback. In: Spagnolli, A., Chittaro, L., Gamberini, L. (eds.) PERSUASIVE 2014. LNCS, vol. 8462, pp. 224–235. Springer, Heidelberg (2014). doi:10.1007/978-3-319-07127-5_19
24. Bandura, A.: Health promotion by social cognitive means. Heal. Educ. Behav. **31**(2), 143–164 (2004)
25. Anderson, E., Winett, R., Wojcik, J.: Self-regulation, self-efficacy, outcome expectations, and social support: social cognitive theory and nutrition behavior. Ann. Behav. Med. **34**(3), 304–312 (2007)
26. Cheung, M.Y., Luo, C., Sia, C.L., Chen, H.: Credibility of electronic word-of-mouth: informational and normative determinants of on-line consumer recommendations. Int. J. Electron. Commer. **13**(4), 9–38 (2009)
27. Ringle, C.M., Wende, S., Becker, J.M.: SmartPLS 3. Boenningstedt: SmartPLS GmbH (2015). https://www.smartpls.com/
28. Hair Jr., J.F., Hult, G.T.M., Ringle, C., Sarstedt, M.: A primer on Partial Least Squares Structural Equation Modeling (PLS-SEM). Sage Publications, Thousand Oaks (2013)
29. Fornell, C., Larcker, D.F.: Evaluating structural equation models with unobservable variables and measurement error. J. Mark. Res. **18**(1), 39–50 (1981)
30. Bagozzi, R.P., Phillips, L.W.: Representing and testing organizational theories: a holistic construal. Adm. Sci. Q. **27**(3), 459–489 (1982)
31. Cohen, J.: Statistical Power Analysis for the Behavioral Sciences, 2nd edn. Routledge, New York (1988)

"Don't Say That!"

A Survey of Persuasive Systems in the Wild

Emma Twersky and Janet Davis(✉)

Department of Mathematics and Computer Science, Whitman College, Walla Walla, WA 99362, USA
{twersker,davisj}@whitman.edu

Abstract. Language use is a type of behavior not yet addressed by the academic persuasive technology community. Yet, many existing applications seek to change users' word choices or writing style. This paper catalogues 32 such applications in common usage or reported in the popular media. We use Schwartz's Theory of Basic Human Values to understand what motivates each attempt to persuade; we use the Persuasive Systems Design (PSD) model to understand contexts and techniques of persuasion. While motivations span the full range of human values, most applications serve values of Achievement, Conformity, or Universalism. Many are autogenous in intent, using reduction, suggestion, and self-monitoring strategies to support behavior change. However, the corpus also includes many endogenous applications that seek to change others' attitudes.

Keywords: Persuasive technology · Persuasive Systems Design (PSD) model · Natural language · Human values

1 Introduction

The popular media is rife with reports of applications that aim to influence the use of language. Consider these recent headlines:

- "Siri corrects people who use the wrong name for Caitlyn Jenner";
- "Watch your tone: Watson can detect attitude";
- "Software Makes Cyberbullies Think Twice Before Sending Mean Messages."

We also encounter such applications in everyday life. This study was inspired when a colleague shared a puzzling interaction with a performance review system. The colleague wrote that she had purged and organized a number of old files during the previous year. The system highlighted the word "old" and suggested a number of alternatives: seasoned, mature, outdated, inactive, and so on. Apparently, the system is designed to defend employers against age discrimination lawsuits by discouraging employees from using words that refer to age.

Like persuasive technologies more broadly, such systems are morally ambiguous. Twitterbots promote what some see as stifling "political correctness" and

P.W. de Vries et al. (Eds.): PERSUASIVE 2017, LNCS 10171, pp. 215–226, 2017.
DOI: 10.1007/978-3-319-55134-0_17

others see as basic respect and human decency. Persuasive tools that influence language use can help us communicate more effectively, and could even help us change how we see ourselves (Patrick and Hagtvedt 2012). At the same time, technology that restricts the use of certain words raises spectres of censorship and "doublethink" (Orwell 1949). Whether technology influences public utterances (e.g., on Twitter) or private utterances (e.g., using Siri), there is potential for both benefits and harms.

Despite the prevalence and significance of applications designed to influence language use, we were unable to find any academic literature on the topic. We are confident there is no discussion of this topic in the proceedings Persuasive Technology Conference, in particular. In this initial survey of existing work "in the wild," we seek to address two questions. First, why deploy technology to influence human language? Second, how (using what mechanisms) do these existing applications influence behavior?

This paper catalogs 32 existing applications already in common use or discussed in the popular media. We analyze this corpus from two distinct points of view. First, inspired by the Value Sensitive Design framework (Friedman et al. 2006), we infer the human values underlying attempts to influence. Second, we analyze each application as a persuasive technology, using the Persuasive Systems Design (PSD) model (Oinas-Kukkonen and Harjumaa 2009).

In the next section, we briefly explain the models that ground our analysis. Then, we explain our method for developing and analyzing the corpus. Next, we present the corpus and our analysis. Finally, we reflect upon what makes these applications persuasive technology and conclude with directions for future work.

2 Background

2.1 Human Values

Our analysis of the motivations underlying persuasive technology applications is inspired by Value Sensitive Design (VSD), a framework for addressing human values throughout the application design process (Friedman et al. 2006). VSD has been proposed as an approach to addressing ethical issues in persuasive technology (Davis 2009). While Friedman et al. (2006) identify values of moral import commonly implicated by information technologies, they do not claim to provide a complete typology of values.

Our choice of Schwartz's (1994) Theory of Basic Human Values as an analytic framework is inspired by Knowles' (2013) analysis of the values underlying pro-environmental persuasive technologies. The Theory of Basic Human Values identifies values recognized across all human cultures. Values are classified in ten motivationally distinct categories: Power, Achievement, Hedonism, Stimulation, Self-Direction, Universalism, Benevolence, Tradition, Conformity, and Security.

2.2 The Persuasive Systems Design Model

The Persuasive Systems Design (PSD) model (Oinas-Kukkonen and Harjumaa 2009) is a comprehensive theoretical framework for describing persuasive technologies. Prior studies have used the PSD model to analyze a corpus of existing applications. For example, Kelders et al. (2011) use the PSD model to analyze web-based health and weight interventions, while Lehto and Oinas-Kukkonen (2011) use it to analyze web-based interventions for substance abuse.

The PSD model considers three phases of persuasive systems design: the Intent, the Event, and the Strategy (Oinas-Kukkonen and Harjumaa 2009).

The Intent includes the identity of the persuader and user, the source of intention (endogenous, exogenous, or autogenous), and the change type (behavior, attitude, or both). Endogenous intent arises from "those who create or produce the interactive technology," exogenous from "those who give access to or distribute the interactive technology to others," and autogenous from "the very person adopting or using the interactive technology."

The Event includes use context and user context. Use context concerns "features arising from the problem domain," while user context concerns "individual differences which influence user's information processing." Because these considerations are not easy to summarize, we adopted the Fogg Behavior Model (Fogg 2009) to further analyze the Event. This model identifies three principal factors necessary for behavior change: motivation, ability, and trigger. A persuasive technology can change behavior by enhancing the user's abilities, by motivating the user, or by providing a trigger to perform the target behavior. Thus, the Fogg Behavior Model captures aspects of the interaction between use context and user context. No prior studies use this model to analyze existing applications.

The Strategy includes the persuasive message and the route of persuasion. The message includes not only content, but also the use of rational or symbolic strategies. The route of persuasion can be direct, in which the user is able to discern and identify the content of the persuasive message, or indirect. The Strategy also includes whether the system is intended primarily for the purpose of persuasion or whether persuasive features are secondary to some other purpose.

Finally, the PSD model catalogs 27 distinct design features, categorized according to whether they provide primary task support, dialogue support, system credibility support, or social support.

3 Methods

We first developed a set of candidate applications, then analyzed the design of each application, and finally inferred motivating values. We analyzed persuasive design before considering values so that we could eliminate from our corpus any applications that lack persuasive intent or do not address language use.

3.1 Building the Corpus

Our corpus was seeded with an initial set of 18 popular media reports of applications that appeared to influence language use collected opportunistically from the second author's Facebook feed over the year June 2015–May 2016.

From this initial corpus, we identified recurring news sources: *Wired*, *Mental Floss*, *Fast Company*, *Google News*, and *LifeHacker*. By searching these sources for the keywords "technology," "word choice", and "word change," we added 3 more applications to our corpus. We also used Google to search for additional articles about known applications. For example, we found listicles with titles such as "Tired Of Getting Offended On The Internet? There's A Web Hack For That" and "5 Free Apps That Make You Seem Smart." We found 11 additional applications in articles in sources including *The New York Times*, *The Washington Post*, *Huffington Post*, *Vox*, *Business Insider*, *Slate*, *Medium*, and *Ars Technica*. Searching for other work by article authors and application creators and resulted in no further applications.

We also searched the *ACM Digital Library* and *Web of Science* for the keywords "persuasive technology," "behavior change," "word choice," and "word change." This method resulted in no relevant applications. We browsed the complete *Proceedings of the Persuasive Technology Conference* and found three tangentially related articles, but none that proposed specific applications. Reviewing citations of these three articles identified no relevant applications. We found articles on applications to help users choose stronger passwords, but excluded these because passwords do not serve as expressive language.

In our analysis of each application, we examined the application itself, its web site, and news stories concerning the application and its creators. We eliminated applications that did not seem to embody persuasive intentions, notably several web-based tools designed to facilitate plagiarism, such as the Article Rewriter[1]. We also eliminated the Trumpweb[2] as it was clearly intended to influence attitudes about a person and not language use. We included iCorrect, a satirical proposal for an iOS feature that helps parents require their children to text with correct spelling and grammar, because the proposal is detailed and plausible.

3.2 PSD Model Analysis

To structure our analysis, we created a spreadsheet of analytic criteria based on the four elements of the PSD model.

For the Intent, we identified the persuader and the users, which let us categorize the source of intention as endogenous, exogenous, or autogenous. We categorized each application as intended to change attitudes, behavior, or both.

For the Event, we interpreted the use context and user context as free text. We further determined whether each application seeks to enhance ability, increase motivation, or provide a trigger.

For the Strategy, we captured the message of each application as free text. We classified each application as using rational strategies, symbolic strategies, or both, and as using primarily direct or indirect persuasion. We determined

---

[1] http://smallseotools.com/article-rewriter/.

[2] https://chrome.google.com/webstore/detail/the-trumpweb/fjkehfaokpmcbigmbgdhmjblecgfkedg.

whether each application was a primarily persuasive system or a persuasive feature secondary to a larger system.

Finally, we assessed whether each application (or its web site) deliberately employed each System Feature.

To begin our analysis, we identified ten applications spanning the corpus. After independently coding these application, we discussed our coding to form a joint interpretation of the criteria. We used this revised understanding to independently code the remaining applications. We compared our independent analyses and discussed them to reach an agreement. Finally, we applied descriptive statistics to understand the corpus as a whole.

3.3 Values Analysis

We assessed whether each application supported each of the 10 categories in the Theory of Basic Human Values. Each of the two authors independently coded these applications, and then we discussed our coding to form a joint interpretation of the categories. Finally, we computed descriptive statistics and performed an informal clustering of applications that implicate similar values.

4 Results

Our corpus comprises 32 persuasive technology applications listed in Table 1. We first introduce the ies clustered by relevant values, and then present our PSD model analysis. We provide raw coding data online.[3]

4.1 Values Analysis

Figure 1 shows that every category of values is implicated by at least one application. Achievement (personal success) and conformity (restraint from violating norms) are addressed by more than half of the applications. The third most common category is universalism (respect for the welfare of all people). The fewest applications appeal to tradition.

We identify six clusters of applications according to common values. Table 1 presents the full corpus, organized according to the following clusters.

1. Conformity-Achievement. These applications are mostly style and grammar checkers that persuade the user to conform to a set of rules for good writing. Within this cluster, we identified four sub-clusters.
   (a) Conformity-Achievement-Universalism includes style checkers that value readability for readers across a broad range of reading skills.
   (b) Conformity-Achievement-Power includes style checkers that help people strengthen the impact of their writing or "avoid embarrassing mistakes."

[3] Link to Google Spreadsheet: http://bit.ly/2f1nzqX.

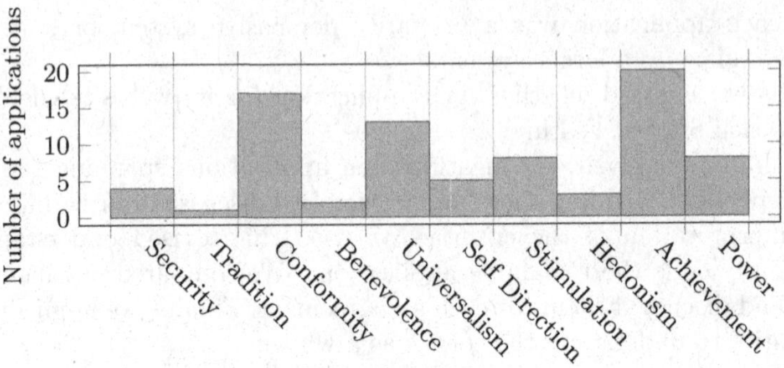

Fig. 1. Value categories present in corpus

   (c) Conformity-Achievement-Tradition includes only iCorrect, which values standard English over slang, abbreviations, and emoji commonly used in text messages.
   (d) The fourth sub-cluster includes applications with no third common value.
2. Universalism-Achievement: These applications focus on improving writing to include a broader range of readers.
3. Stimulation-Hedonism-Achievement: These applications encourage users to do something novel and fun, while learning and challenging themselves.
4. Universalism: These applications focus on treating people with respect.
5. Security-Conformity: These applications seek to keep people safe by enforcing established laws or norms.
6. Outliers: These applications did not fit into other clusters.

4.2 PSD Analysis

The Intent. For each application, we classify the source of intentions as endogenous, exogenous, or autogenous. We find that 13 (40.625%) of the applications are endogenous, 4 (12.5%) are exogenous, and 15 (46.875%) are autogenous. An example of an endogenous application is Common Sans, created for the non-profit Solvatten, which seeks to promote human rights. The clearest example of exogenous intent is iCorrect, because parents adopt it for use by their children. Another example is Textio, which HR managers may adopt on behalf of their team. By contrast, most of the grammar and style tools in our corpus are adopted by the user for their own use.

We classify the change type as attitude, behavior, or both. We find that the majority (23, 71.875%) of applications promote behavior change, 6 (18.75%) promote both attitude and behavior change, and just 3 (9.375%) promote attitude change alone. For example, Grammarly teaches writers to strengthen their tone, grammar, and sentence structure, and thus focuses on behavior change. @DropTheIBot seeks to change both behavior and attitudes: to persuade Tweeters to stop using the phrase "illegal immigrants" and to reconsider the morality

Table 1. The Corpus

Conformity-Achievement		
Conformity-Achievement-Universalism		
Readability Score	Web application that measures readability	www.Readability-Score.com
Microsoft Word	Grammar checker suggests changing "mankind" to "humankind," and recognizes "they" as a singular personal pronoun	products.office.com/en-us/word
Conformity-Achievement-Power		
1checker	Web-based grammar checker	www.1checker.com
Ginger	Web-based grammar checker with sentence rephraser and "personal trainer"	www.gingersoftware.com
Grammarly	Grammar checker as a web browser plugin	www.grammarly.com
Spam Analyse	Web application shows users how to rewrite email newsletters to avoid being flagged as spam	www.SpamAnalyse.com
White Smoke	Grammar checker as a web browser plugin	www.whitesmoke.com
Conformity-Achievement-Tradition		
iCorrect	Satirical proposal to force kids to use correct grammar and punctuation in text messages	www.michaelweisburd.com/icorrect
Conformity-Achievement		
Style and Diction	Command-line tools identify wordy and commonly misused phrases; measure readability	www.gnu.org/software/diction/diction.html
The Passivator	Web browser plugin flags use of passive voice	www.ftrain.com/ThePassivator.html
Universalism-Achievement		
ChangeMyView	Reddit site lets users share and get feedback on persuasive writing	www.reddit.com/r/changemyview
Expresso	Web browser plugin helps users edit their text to improve readability metrics	www.expresso-app.org
Hemingway	Web-based text editor highlights complex sentences and promotes "bold" language	www.hemingwayapp.com
Textio	Web platform helps HR departments eliminate unintended gender or racial bias in job postings	www.textio.com
Unitive	Web application provides structure for writing unbiased job ads	www.unitive.works
Stimulation-Hedonism-Achievement		
Wonder Keyboard	Virtual keyboard for expanding vocabulary and learning English as a second language	www.typewithwonder.com
XKCD Simple Writer	Web-based text editor highlights words not among the 1000 most commonly used	www.xkcd.com/simplewriter
Universalism		
PC2Respect	Web browser plugin changes "politically correct" to "treating people with respect"	www.twitter.com/hashtag/PC2Respect?src=hash
Halogen	HR software suggests alternatives to words such as "old," "pretty," and "short"	www.halogensoftware.com
Honest	Chrome extension changes "skinny," "slim," and "thin" to "fit," "toned," and "healthy"	untitledscience.github.io/HonestChrome
@DropTheIBot	Twitterbot responds to users of the term "illegal immigrant" and exhorts them to use other terms such as "undocumented immigrant"	Account suspended
Common Sans	Typeface that strikes through the phrase "refugee" and adds "human"	www.commonsans.com
Siri	Siri answers questions about "Bruce Jenner" using the name "Caitlyn Jenner" and feminine gender pronouns	www.apple.com/ios/siri

(*continued*)

Table 1. (*continued*)

Security-Conformity		
ReThink	Web browser plugin asks users if they want to send messages that use bullying language	www.rethinkwords.org
Google/Bing	Search term autocomplete algorithms exclude some sexually explicit words	www.Google.com, www.Bing.com
YikYak	YikYak asks users to rethink messages flagged as containing threatening language	www.yikyak.com
Outliers		
Zero Trollerence	Twitter bot "enrolls" users in an "online course" to show them how to change their habits	www.zerotrollerance.guru
Seen	Typeface redacts phrases tracked by the NSA and GCHQ, based on the Snowden papers	www.projectseen.com
Emojimo	Virtual keyboard replaces text with emojis	itunes.apple.com/us/app/emojimo-keyboard/id918318362?mt=8
Just Not Sorry	Gmail plugin identifies phrases that undermine the writer's authority and justifies how and why to rephrase	chrome.google.com/webstore/detail/just-not-sorry-the-gmail/fmegmibednnlgojepmidhlhpjbppmlci
Toneapi	Web application gives feedback on the emotional tone of a text and provides tools for revision	www.toneapi.com
Cliche Finder	Web-based text editor highlights cliches	cliche.theinfo.org

of immigration. By contrast, Common Sans seems unlikely to stop anyone from using the word "refugee," but rather promotes new attitudes towards refugees.

The Event. We find that 24 (75%) of the applications aim to enhance user's abilities, 12 (37.5%) motivate change, and 19 (59.375%) trigger performance of the behavior. Percentages do not add up to 100% since each application could influence user behavior or attitudes in multiple ways.

An example of an application that falls into all three categories is Just Not Sorry. Since users learn what phrases to avoid to strengthen their authority, the application enhances the user's abilities. Since the application gives reasons why users should change their writing style, it provides motivation. Finally, since users are prompted to change their behavior as they are writing in Gmail, the application triggers change. Therefore, Just Not Sorry enhances ability and motivation, as well as providing a trigger.

The Strategy. We find that 13 (40.625%) of the applications use symbolic strategies, 6 (18.75%) use rational strategies, and 13 (40.625%) use both. For example, we classify Common Sans as primarily symbolic. By striking through the word "refugee" and replacing it with "human being," Common Sans conveys the message that refugees are human beings first. By contrast, Readability Score employs rational strategies. This tool presents numeric feedback and recommends improvements based on research findings. An example of an application that uses both rational and symbolic strategies is Toneapi, which gives numeric feedback on readability and tone, but also uses green to represent positive tone and red to represent negative tone.

Next, we find that 30 (93.75%) of the applications use direct persuasion and just 2 (6.25%) use indirect persuasion. One of these is Google's search bar autocomplete algorithm, which does not suggest sexually explicit words as search terms, but also does not point out that suggestions are being withheld. It indirectly discourages the user from searching for certain terms without making its persuasive intent apparent.

Finally, we find that the majority (26, 81.2%) of the applications are intended primarily to persuade, while just 6 (18.75%) constituted persuasive features of a system with some other primary purpose. Siri and Word are examples of applications with secondary persuasive features.

System Features. To complete our application of the PSD model, we analyze system features. We find a total of 59 examples of Primary Task Support features, 40 examples of Dialogue Support features, 112 examples of System Credibility Support features, and 21 examples of Social Support features (Fig. 2). System Credibility Support features, the most common type of features, are found mainly in the applications' websites or in their use of supporting platforms such as Twitter or Google Chrome.

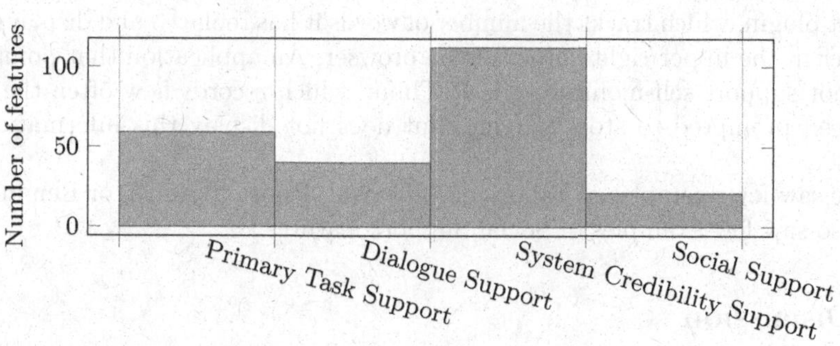

Fig. 2. Number of system features identified in each category across all applications

Excluding System Credibility Support features, the three most common features are Reduction, Self-Monitoring, and Suggestion (Fig. 3).

We find that a majority (25, 78.125%) of the applications include reduction features to make the desired behavior easier to perform. One application that does not employ reduction is Zero Trollerance, which actually complicates users' interactions with Twitter by annoying them.

More specifically, we find that 22 (68.75%) of the applications provide suggestions. For example, Toneapi suggests particular words for the user to change to alter the tone of their writing. It also suggests synonyms that are more positive, more negative, or neutral. Many other style checking tools suggest word changes, including Microsoft Word, Grammarly, Hemingway, 1checker, and so on. One application that does not make suggestions is the Passivator, which flags use of the passive voice but does not suggest alternative phrasing.

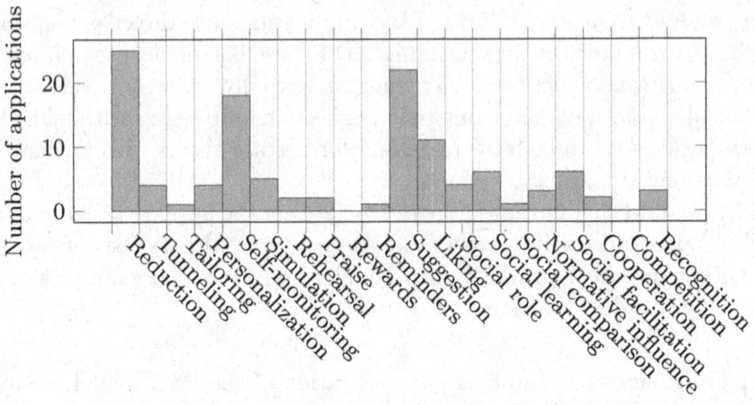

Fig. 3. Number of applications employing each system feature

We find that 18 (56.25%) of the applications include self-monitoring features. For example, Toneapi supports self-monitoring by displaying the prevalence of each emotional tone in a piece of writing. Another example is found in the Honest plugin, which tracks the number of words it has replaced and displays this number in the upper right corner of the browser. An application that pointedly does not support self-monitoring is ReThink, which records how often the user has been prompted to stop bullying, but does not display this information to the user.

We saw few examples of Tailoring, Rehearsal, Praise, Reward, or Reminders. We also saw few examples of Social Support features.

5 Discussion

What defines persuasive technology? We began this research with the assumption that we would not consider grammar or style checkers. They are old applications: the Unix diction tool dates to the 1970s, and Word's grammar checker is familiar to everyone who writes on a computer. Their designs were not informed by persuasive system design principles. However, users adopt these tools with the intent to change their own behavior, specifically their writing habits. These tools do not automatically fix "problems" based on mechanical rules, like autocorrect, but instead make suggestions that the user can adopt or ignore based on their own judgment. Within the PSD model, we can see these tools as autogenous persuasive applications, or behavior change support systems. We find a range of persuasive features, including suggestion, self-monitoring, and personalization. Today, such tools are moving beyond word processors and the Unix command-line. With text messaging and social media, people communicate through writing more than ever. Browser plug-ins such as Ginger and Grammarly can provide pervasive feedback on all our writing, formal and informal. Wonder Keyboard and iCorrect take this to an extreme by integrating with text messaging.

While many applications in our corpus are autogenous in intent, even more are endogenous or exogenous: we can clearly identify the application creator, or a third party, who seeks to others to change or reconsider their language use. Consider, for example, Halogen, Common Sans, or ReThink.

Web search applications particularly challenged our understanding of persuasive technology. Since users are used to having search engines suggest autocompletions of search terms, when sexually explicit terms are not autocompleted, the user is forced to purposefully finish typing them. The absence of suggestion is the point of persuasion. This strategy might be considered "anti-reduction" as users must do more work to perform undesired behaviors.

6 Conclusion

We offer three main contributions to the persuasive technology community:

- we introduce influence of language use as a new domain for design;
- we show how Schwartz's Theory of Basic Human Values and the PSD Model can be used in tandem to analyze the content and form of existing persuasive applications;
- we provide a corpus of applications that can be further studied from a variety of perspectives, not just persuasive technology but also other disciplines such as rhetoric or linguistic anthropology.

By now, this corpus could be expanded with reports of new applications.

We find that motivations span the full range of human values, although most applications serve values of Achievement, Conformity, or Universalism. Most are autogenous in intent; many employ reduction, suggestion, and self-monitoring strategies to support behavior change. However, the corpus also includes endogenous applications that seek to change attitudes. Applications do not only seek to enhance users' abilities, but also provide motivation and triggers.

This first cut of an analysis suggests further opportunities for analysis and theory building. There is more work to be done to understand the relationships between underlying values, persuasive system design, and language use as a kind of behavior. For instance, why are Social Support features comparatively rare in this corpus? Are they ill-suited to influencing language use, or merely underutilized? Furthermore, our corpus includes few persuasive features in systems that serve some other primary purpose. However, many applications in our corpus alter user experiences of existing platforms (e.g., Twitterbots and Web browser plugins). Should the PSD model be extended to explicitly account for such roles? Moreover, is there a greater role for secondary persuasive features in tools for writing and speech? What is the moral dimension in this context of source of intent? Finally, our analysis of the values implicated by these applications was necessarily cursory. Still another direction is to explore in greater depth the moral implications of existing applications using VSD theories and methodology.

This work also suggests opportunities for design. Our PSD model analysis suggests that many of the applications in our corpus could employ persuasive system features more effectively. Hence, one direction for future design work is to iteratively evaluate and improve the design of applications in the corpus, informed by persuasive systems design theory. Further, work in cognitive psychology (e.g., Patrick and Hagtvedt 2012) suggests new opportunities for behavior change support systems to support self-efficacy, growth mindset, or other kinds of personal development. Finally, existing tools such as ChangeMyView suggest opportunities for new tools to help individuals write more persuasively. At the same time, what moral hazards are risked by efforts to influence language use through technology? Development of such behavior change support systems should be informed not only by PSD theory but by consideration of human values.

Acknowledgments. We thank Whitman College and Janice M. and Kim T. Abraham for their generous support of this research. We also thank Heather Hayes, Rachel George, Mercer Hanau, and Justin Lincoln for fruitful discussions of our work in progress.

References

Davis, J.: Design methods for ethical persuasive computing. In: Proceedings of the 4th International Conference on Persuasive Technology, Persuasive 2009, pp. 6:1–6:8. ACM, New York (2009)

Fogg, B.J.: A behavior model for persuasive design. In: Proceedings of the 4th International Conference on Persuasive Technology, Persuasive 2009, pp. 40:1–40:7. ACM, New York (2009)

Friedman, B., Kahn Jr., P.H., Borning, A.: Value sensitive design and information systems. In: Zhang, P., Galetta, D. (eds.) Human-Computer Interaction and Management Information Systems: Foundations. Advances in Management Information Systems Series, pp. 348–372. M.E. Sharpe, Armonk (2006)

Kelders, S.M., Kok, R.N., Van Gemert-Pijnen, J.: A systematic review. In: Proceedings of the 6th International Conference on Persuasive Technology: Persuasive Technology and Design: Enhancing Sustainability and Health, Persuasive 2011, pp. 3:1–3:8. ACM, New York (2011)

Knowles, B.: Re-imagining persuasion: designing for self-transcendence. In: CHI 2013 Extended Abstracts on Human Factors in Computing Systems (CHI EA 2013), pp. 2713–2718. ACM, New York (2013)

Lehto, T., Oinas-Kukkonen, H.: Persuasive features in web-based alcohol and smoking interventions: a systematic review of the literature. J. Med. Internet Res. **13**(3), e46 (2011)

Oinas-Kukkonen, H., Harjumaa, M.: Persuasive systems design: key issues, process model, and system features. Commun. Assoc. Inf. Syst. **24**(1), 28 (2009)

Orwell, G.: Nineteen Eighty-Four. Secker & Warburg, London (1949)

Patrick, V.M., Hagtvedt, H.: "I don't" versus "i can't": when empowered refusal motivates goal-directed behavior. J. Consum. Res. **39**(2), 371–381 (2012)

Schwartz, S.H.: Are there universal aspects in the structure and contents of human values? J. Soc. Issues **50**(4), 19–45 (1994)

Using Argumentation to Persuade Students in an Educational Recommender System

Stella Heras[1(✉)], Paula Rodríguez[2], Javier Palanca[1], Néstor Duque[3], and Vicente Julián[1]

[1] Universitat Politècnica de València, Valencia, Spain
{sheras,jpalanca,vinglada}@dsic.upv.es
[2] Universidad Nacional de Colombia, Medellín, Colombia
parodriguezma@unal.edu.co
[3] Universidad Nacional de Colombia, Manizales, Colombia
ndduqueme@unal.edu.co

Abstract. This paper explores the use of argumentation-based recommendation techniques as persuasive technologies. Concretely, in this work we evaluate how arguments can be used as explanations to influence the behaviour of users towards the use of certain items. The proposed system has been implemented as an educational recommender system for the Federation of Learning Objects Repositories of Colombia that recommends learning objects for students taking into account students profile, preferences, and learning needs. Moreover, the persuasion capacity of the proposed system has been tested over a set of real students.

Keywords: Argumentation · Educational Recommender Systems

1 Introduction

In recent years, educational and learning communities around the world have focus their attention on the huge horizon of opportunities that Internet-based technologies can offer to achieve their goals. With the current proliferation of online learning resources, the outlook for facilitating the access to learning in all kinds of subjects, to all kinds of people anywhere in the world had never been so favourable. For instance, many academic institutions offer Massive Online Open Courses (MOOCs), and most new methodologies for teaching-learning (e.g. flip teaching) rely on digital entities involving educational design characteristics (i.e. Learning Objects, LOs), such as videos, PDF tutorials, etc. that are shared online. While few decades ago the mainstream problem in education was the lack of access to educational resources, nowadays, however, many students have much more information available than they can consume in an efficient way.

Consequently, research in Technology Enhanced Learning (TEL) systems that aim to design, develop, and evaluate socio-technical innovations for various kinds of learning and education is actually of global interest [6]. Among them, there is a growing interest on TEL recommender systems (or Educational Recommender Systems, ERS), which help people to find the more suitable resources for

P.W. de Vries et al. (Eds.): PERSUASIVE 2017, LNCS 10171, pp. 227–239, 2017.
DOI: 10.1007/978-3-319-55134-0_18

their learning objectives, educational level, and learning style (auditory, kinaesthetic, reader, or visual).

In [16], authors proposed an ERS for the Federation of Learning Objects Repositories of Colombia that recommends LOs for students taking into account their learning profile (i.e. personal information, language, topic and format preferences, educational level, and learning style), learning history (i.e. LOs that they already used), and their similarity with other users of the system. The system consists of a recommendation engine that combines several recommendation techniques. Among them, the argumentation-based technique, which selects those LOs for which it is able to generate a greater number of arguments that justify its suitability, achieved the best recommendation results.

An additional advantage of the argumentation-based recommendation technique is that arguments can also be presented to the student as a *persuasion device*. Therefore, in this work, we analyse this ERS as a persuasive technology, and evaluate how arguments can be used as explanations to influence the behaviour of the students towards the use of certain LOs. As pointed out in [20], by applying a particular algorithm in a recommender system, certain types of explanations (explanation styles) may be easier to generate since the algorithm can produce the type of information that the explanation style uses. Our argumentation-based recommendation algorithm uses rules that capture different types of knowledge, which give rise to different explanation styles (content-based, knowledge-based, and collaborative).

Persuasion in ERS is a new area of research with few contributions to date. Section 2 reviews related work in the area. An overview of our ERS is presented in Sect. 3. The persuasive power of the system was evaluated by testing the system with a set of real users and analysing how arguments influence on their decisions. Results are shown in Sect. 4. Finally, Sect. 5 summarises the contributions of the paper and proposes future work.

2 Related Work

Traditional persuasion literature has demonstrated that people are more willing to accept suggestions if they provide from a credible source [8]. Consequently, there is a growing interest in recommender systems research to promote the credibility of the system, and hence the probability of accepting its recommendations. However, as pointed out in [21], the question of how to actually translate credibility into system characteristics in the context of recommender systems is still being investigated. Among the different techniques to improve the system's credibility, providing *explanations* (i.e. arguments) that help users to better understand the good intentions and efforts of the system has proven to be very effective [1]. On the one hand, people trust more on recommendations when they can inspect the operation of the system, and when they can understand the reasons why these recommendations are appropriate to them. Even when users already know the recommendations presented, they prefer recommender systems that are able to justify their suggestions [19]. On the other hand, explanations may avoid bad choices. Therefore, trying to persuade people by means of explanations pursues several important goals in recommender systems: convince users to try or buy (*persuasiveness*); explain how the system works (*transparency*), and help people to make better decisions (*effectiveness*) [20].

In recent years, the literature on ERS has reported a boost in the area [7,17,18,22]. Within ERS, the use of techniques to provide learners and/or educators with insights and explanations in the recommendation process is an important line of research [6]. In educational domains, argumentation theory and tools have a large history of successful applications, specially to teach critical thinking skills in law courses [11]. Some argument-based recommender systems and recommendation techniques have been proposed to recommend music [2], news [4], movies [14], or restaurants [10], or to perform content-based web search [5]. Among them, we share the approach of the movie recommender system based on defeasible logic programming proposed in [3]. In this work, authors defined a preset preference criteria between rules to resolve argument attacks, while we use a probabilistic method to compute the likelihood that an argument prevails over another, which makes the system more adaptive. However, the application of argumentation theory to generate arguments in ERS, remains underexplored. Although the impact of explanations on the performance of recommender systems have already been investigated [21], they do not focus on ERS.

Furthermore, recommending LOs presents some challenges to be able design a recommender system suitable for TEL environments [12]. Each learner uses her own learning process, based on her own tools, methods and paths, and the recommender system must take them all into account. Our work involves a contribution in these areas by proposing an ERS that implements an argumentation framework that allows the system to produce several types of arguments. These arguments justify the recommendations of the system in terms of the students' learning profile (i.e. personal information, language, topic and format preferences, educational level, and learning style), learning history (i.e. LOs that they already used), and their similarity with other students.

3 Persuasive Educational Recommender System Based on Argumentation

This section provides an overview of our persuasive ERS, which consists of several modules. Here, we focus on the argumentation-based recommendation module, which implements the argumentation skills of the system, and on the argument visualisation module, which provides it with persuasion tools. A detailed description on the characteristics of the recommendation algorithms used and the components of the system can be found in [16].

3.1 Recommendation Module

Our ERS implements a recommendation technique based on argumentation theory that uses *metadata* about the LOs to recommend and information about

Table 1. Example program

user_type(*Bob*, *visual*)
resource_type(LO_k, *slides*)
structure(LO_k, *atomic*)
state(LO_k, *final*)
interactivity_type(LO_k, *low*) ← resource_type(LO_k, *slides*)
appropriate_resource(*Bob*, LO_k) ← user_type(*Bob*, *visual*) ∧ resource_type(LO_k, *slides*)
appropriate_interactivity(*Bob*, LO_k) ← user_type(*Bob*, *visual*) ∧ interactivity_type(LO_k, *low*)
educationally_appropriate(*Bob*, LO_k) ← appropriate_resource(*Bob*, LO_k) ∧
appropriate_interactivity(*Bob*, LO_k)
generally_appropriate(LO_k) ← structure(LO_k, *atomic*) ∧ state(LO_k, *final*)
recommend(*Bob*, LO_k) ← educationally_appropriate(*Bob*, LO_k) ∧ generally_appropriate(LO_k)

the *student profile*, and provides the students with those LOs that are better backed by arguments. This is a hybrid technique that combines different types of recommendation styles [15]:

- *content-based*: uses information about the LOs that the user has already assessed in the past, and about the student profile to recommend similar LOs.
- *collaborative*: uses information about the student profile to compute a similarity degree among students. Then, the system recommends LOs that were appropriate for similar students.
- *knowledge-based*: uses information about the LOs that the user has already assessed in the past to recommend similar LOs.

To represent LOs in the system, we follow the *IEEE-LOM* standard[1], a hierarchical data model that defines around 50 metadata fields clustered into 9 categories. Also, student profiles, include their personal information, language, topic and LO's format preferences, educational level, and learning style (auditory, kinaesthetic, reader, or visual).

To represent the logic of the recommendation system, we use a defeasible argumentation formalism based on logic programming (*DeLP* [9]). A defeasible logic program (DeLP) $P = (\Pi, \Delta)$, models strict (Π) and defeasible (Δ) knowledge in the application domain. In our system, the set Π represent *facts* (i.e. strict inference rules with empty body). For instance, in Table 1 $user\_type(Bob, visual)$ is a fact that represents a student named *'Bob'* who has a visual learning style (prefers materials with images, graphs, slides, and visual formats such as png, jpeg, ppt, etc.).

Correspondingly, the set Δ includes defeasible rules of the form $P \leftarrow Q_1, ..., Q_k$. These rules represent the defeasible inference that literals $Q_1, ..., Q_k$ may provide reasons to believe P. There are 4 types of defeasible rules in our ERS, 3 to represent the underlying logic each of the system's recommendation approaches (content-based, collaborative or knowledge-based), and 1 to represent general domain knowledge. Table 2 shows a compendium of the rules[2].

[1] 1484.12.1-2002 - IEEE Standard for Learning Object Metadata: https://standards.ieee.org/findstds/standard/1484.12.1-2002.html.

[2] The complete rule set is not provided due to space restrictions.

Table 2. Defeasible rules

GENERAL RULES
G1: ~recommend(user, LO) ← cost(LO) > 0 *G2: ~recommend(user, LO) ← quality_metric(LO) < 0.7*
CONTENT-BASED RULES
C1: recommend(user, LO) ← educationally_appropriate(user, LO) ∧ generally_appropriate(LO)
C1.1: educationally_appropriate(user, LO) ← appropriate_resource(user, LO) ∧ appropriate_interactivity(user, LO) *C1.1.1: appropriate_resource(user, LO) ← user_type(user, type) ∧ resource_type(LO, type)* *C1.1.2: appropriate_interactivity(user, LO) ← user_type(user, type) ∧ interactivity_type(LO, type)* *C1.2: generally_appropriate(LO) ← structure(LO, atomic) ∧ state(LO, final)*
C2: recommend(user, LO)← educationally_appropriate(user, LO) ∧ generally_appropriate(LO)) ∧ technically_appropriate(user, LO)
C2.1: technically_appropriate(user, LO) ← appropriate_language(user, LO) ∧ appropriate_format(LO) *C2.1.1: appropriate_language(user, LO) ← language_preference(user, language) ∧ object_language(LO, language)* *C2.1.2: appropriate_format(LO) ← format_preference(user, format) ∧ object_format(LO, format)*
C3: recommend(user, LO) ← educationally_appropriate(user, LO) ∧ generally_appropriate (LO) ∧ updated(LO)
C3.1: updated(LO) ← date(LO, date) < 5 years
C4: recommend(user, LO) ← educationally_appropriate(user, LO) ∧ generally_appropriate(LO) ∧ learning_time_appropriate(LO)
C4.1: learning_time_appropriate(LO) ← hours(LO) < γ
COLLABORATIVE RULES
O1: recommend(user1, LO) ← similarity(user1, user2) > α ∧ vote(user2, LO) ≥ 4
KNOWLEDGE-BASED RULES
K1: recommend(user1, LO)← similarity(LO1, LO2) > β ∧ vote(user1, LO2) ≥ 4

For instance, in Table 1 *appropriate_interactivity(Bob,* LO_k*) ← user_type(Bob, visual) ∧ interactivity_type(*LO_k*, low)* is a rule that the system can use to infer that a specific learning object LO_k is appropriate for the student *Bob*, since *Bob* has a visual learning style and LO_k requires a low interaction with student, which is suitable for the visual learning style. Note that the system cannot infer contradictory facts. Thus, provided that $\sim$ represents default logic negation, $visual(bob)$ and $\sim visual(bob)$ cannot be inferred by the same set of rules.

A DeLP program can be queried to resolve if a ground literal can be derived from the program. When the ERS is requested to recommend LOs to a specific user, it tries to derive all possible *recommend(user, LO)* defeasible rules. Also, by backward chaining facts and defeasible rules and following a similar mechanism to the *Selective Linear Definite (SLD)* derivation of standard logic programming, our ERS generates arguments to support the recommendation of specific LOs. Therefore, from each defeasible rule the system can generate an argument to support the literal that can be derived from the rule. Arguments in this framework are defined as follows:

Definition 1 (Argument). *An* argument $\mathcal{A}$ *for* h *(represented as a pair* $\langle h, \mathcal{A}\rangle$*) is a minimal non-contradictory set of facts and defeasible rules that can be chained to derive the literal (or conclusion)* h.

From the program of Table 1 we can derive one argument that supports the recommendation of LO_k to the student *Bob* (⟨recommend(*Bob*, LO_k), {educationally_appropriate(*Bob*, LO_k), generally_appropriate(LO_k)}⟩). Moreover, other arguments to support that LO_k is appropriate for the student *Bob* can also be derived (e.g. ⟨educationally_appropriate(*Bob*, LO_k), {appropriate_resource (*Bob*, LO_k), appropriate_interactivity(*Bob*, LO_k)}⟩). Arguments can be *attacked* by other arguments that *rebut* them (i.e. propose the opposite conclusion) or *undercut* them (i.e. attack clauses of their body).

Definition 2 (Attack). *An argument* $\langle q, \mathcal{B}\rangle$ attacks *argument* $\langle h, \mathcal{A}\rangle$ *if we can derive* $\sim h$ *from* $\mathcal{B}$ *or if* q *implies that one of the clauses of* $\mathcal{A}$ *does no longer hold (there is another argument* $\langle h_1, \mathcal{A}_1\rangle$ *used to derive* $\langle h, \mathcal{A}\rangle$ *such that* $\Pi \cup \{h_1, q\}$ *is contradictory).*

Attacks between arguments are resolved by using a probability measure. This measure estimates the probability that an argument succeeds based on the aggregated probability of the facts and clauses that form the body of the rules used to generate the argument. Thus, our ERS uses a simplified *probabilistic argumentation* framework [13] that assigns probability values to arguments and aggregates these probabilities to compute a *suitability value* to rank and recommend LOs.

Definition 3 (Argumentation Framework). *In our ERS, an argumentation framework is a tuple* (Arg, P_{Arg}, D) *where* Arg *is a set of arguments,* $D \subseteq Arg \times Arg$ *is a defeat relation, and* $P_{Arg} :\rightarrow [0:1]$ *is the probability that an argument holds.*

The probability of an argument $Arg = \langle h, \mathcal{A}\rangle$ is calculated as follows:

$$P_{Arg} = \begin{cases} 1 & \text{if } \mathcal{A} \subseteq \Pi \\ \dfrac{\sum_{i=1}^{k} P_{Q_i}}{k} & \text{if } \mathcal{A} \subseteq \Delta \mid h \leftarrow Q_1, ..., Q_k \end{cases} \tag{1}$$

Facts have probability 1. The probability of defeasible rules is computed as the average of the probabilities of the literals $Q_1, ..., Q_k$ that form their body (i.e. 1 if they are facts, 0 if they cannot be resolved, or P_{Q_i} if they are derived from other defeasible rules). Then, the *suitability* value of a recommendation is computed as the product of the probabilities of all of its supporting arguments.

Definition 4 (Defeat). *In our ERS, an argument* $\langle q, \mathcal{B}\rangle$ defeats *another argument* $\langle h, \mathcal{A}\rangle$ *if* $\mathcal{B}$ *attacks* $\mathcal{A}$ *and* $P_{\mathcal{B}} > P_{\mathcal{A}}$.

For instance, let us assume that in the program of Table 1 the fact *state(*LO_k*, final)* does not exist. Despite that, we can still generate the defeasible argument $\mathcal{A} = \langle$*generally_appropriate(*LO_k*)*, {*structure(*LO_k*, atomic)*}$\rangle$, but with a probability of 0.5. Then, if new information that demonstrates that LO_k is still a

Table 3. Explanation schemes.

ID	Rule	Explanation style	Description
*E*1	*C*1	Content-based	The learning object *LO* fits the topic *T*, is suitable for your *LS* learning style, and it is atomic and stable
*E*2	*C*2	Content-based	The learning object *LO* fits the topic *T*, is suitable for your *LS* learning style, and fits your *L* language and *F* format preferences
*E*3	*C*3	Content-based	The learning object *LO* fits the topic *T*, is suitable for your *LS* learning style, fits your *L* language and *F* format preferences, and it is updated
*E*4	*C*4	Content-based	The learning object *LO* fits the topic *T*, is suitable for your *LS* learning style, and fits your *L* language, *F* format preferences and learning time $< T$ preferences
*E*5	*O*1	Collaborative	The system has found a user with profile similar to yours who used and enjoyed the learning object *LO*
*E*6	*K*1	Knowledge-based	The system has found that you previously used and enjoyed *LOx*, which is similar to this learning object *LOy*

draft is added to the program (e.g. the fact ∼state(LO_k, *final)* is added), its associated argument $\mathcal{B} = \langle \sim state(LO_k, final), \{\sim state(LO_k, final)\}\rangle$ would be generated with probability 1. Then, the argument $\mathcal{B}$ undercuts the argument $\mathcal{A}$, and since its probability is greater, $\mathcal{B}$ defeats $\mathcal{A}$. Furthermore, since the argument $\mathcal{A}$ is invalidated, the argument that supports the recommendation of LO_k to *Bob* does not longer hold and the ERS would not provide this recommendation. Thus, at the end of the recommendation process, those LOs whose supporting arguments hold are sorted from higher to lower suitability and provided to the student.

3.2 Argument Visualisation Module

We have also designed a module for constructing explanations based on the arguments of the system. In the literature, we can find an ambiguous use of the term 'explanation' and 'argument'. Here, we use the term argument to denote the formal representation of the justification for each recommendation (as it is understood by argumentation theory). The term explanation is used to denote the textual representation of these arguments, as it is presented in the user interface). Since the number of rules that we use to generate arguments is finite and small, currently this is a simple module that associates each rule with a scheme of explanation (see Table 3 for some examples). Therefore, in addition to the LO recommended, the ERS is able to show a text with an explanation for this recommendation. Explanations can follow different *explanation styles*, depending on the recommendation approach followed by the system [20]. Our ERS is an hybrid argumentation-based system whose rules defeasible rules implement different recommendation approaches and hence, its arguments have different explanation styles (i.e. *content-based*, *collaborative*, and *knowledge-based*). As pointed out in Sect. 2, with these arguments we pursue to convince students to try specific LOs (improve *persuasiveness*). Also, we want to explain why a student may be interested or not to try a LO (improve *effectiveness*). Furthermore, since our arguments come directly from the rules that we hired to generate the recommendation, we pursue to improve the *transparency* of the system (we want to explain the way the recommendation engine works).

4 Evaluation

In this section, we evaluate the use of arguments in our ERS from several perspectives: to help students to select the objects that best suit their objectives and preferences (*effectiveness*); to persuade the students to try certain LOs (*persuasiveness*), and to explain the students how the system works (*transparency*).

4.1 Data

To run the evaluation tests, we used a database of 75 LOs of different areas (e.g. systems, database, programming, algorithms, etc.) from the Federation of Learning Objects Repositories of Colombia (FROAC). Each object included metadata describing its title, keywords, description, interactive level, resource type, format, and language, among others. Also, 40 students of a computer systems management course of the Universidad Nacional de Colombia were registered as users in the recommendation system. To register in the ERS, students have also to fill out educational information (e.g. educational level, topics of interest (keywords)) and answered a test to automatically determine their learning style. To identify the characteristics of the student profile, the following features were selected:

- Personal information: ID, Name, Surname, Birth Date, Email, Sex, Language, Password, Country, Department, City, Address, Phone.
- Preferences:
  - Level of Interactivity: students were asked if they prefer LOs that allow interaction or just presentation of content. We got a distribution of 8% of students who preferred LOs with very low interactivity level, 25% preferred LOs with low interactivity level, 35% preferred medium interactivity level, 30% preferred high interactivity level, and 3% preferred LOs with very high interactivity level.
  - Language: 95% of students preferred Spanish LOs (as was expected by their nationality), and only 5% preferred LOs in English.
  - Format: 25% of the students selected *jpeg* as they preferred format, 28% *mp4*, and 48% *pdf*.
- Learning Style: we used the VARK learning styles model to determine the learning style of each student. We got a distribution of 50% visual students, 15% auditory, 15% reader, and 20% kinesthetic students.
- Usage History: ID of each LO evaluated, Rating, and Date of use.

4.2 Methodology

Once the students were registered, the evaluation tests were organised into two iterations. First, the system presented to each student a set of the 10 LOs. The LOs 1 to 6 were those with highest suitability value (the aggregated probability of their associated arguments, see Sect. 3.1) sorted in a decreasing order. The other 4 objects were those with lowest suitability value, also sorted in a decreasing order. Explanations for a random set of 5 out of these 10 LOs were shown to the student. In this way we ensure a balanced sample distribution among potentially good and bad recommendations. With this design decision we intend to avoid that the recommendations are so good or obvious that the probability that users like them is so high that showing any explanation is almost irrelevant. Also, if the ERS was able to generate several arguments for the same recommendation, the system randomly selected one of them to show an explanation.

In this 1st iteration, students were requested to provide a rating for the 10 objects on a scale of 1 to 5 (from 1 -dislikes or not relevant- to 5 -likes a lot or more relevant-). After that, we ran a 2nd iteration of the evaluation tests. Here, the system presented explanations (also selected randomly from the set of all arguments of each recommendation) for those LOs that received the lower ratings (from 1 to 3) and were not accompanied by explanations in the 1st iteration. Once again, the students were requested to provide a new rating for these LOs. With these experiments, we obtained a database of 663 ratings, with an average number of 16.5 ratings per student.

4.3 Results

To test the effectiveness of arguments in our ERS, we evaluated the average rating that students gave to the LOs that were presented with and without explanations. Results shown in Fig. 1 demonstrate that those LOs that were presented with explanations got better evaluations. The boundaries of the boxes denote the 25th and 75th percentiles, and the overall average of the rating values are marked by the dots inside the boxes. Therefore, although the improvement is slight in the average, the quantity of objects that got high ratings (from 3 to 5) was greater when explanations were included. This means that explanations, and their underlying arguments, are effective to attract the students' attention towards those LOs that best meet their expectations.

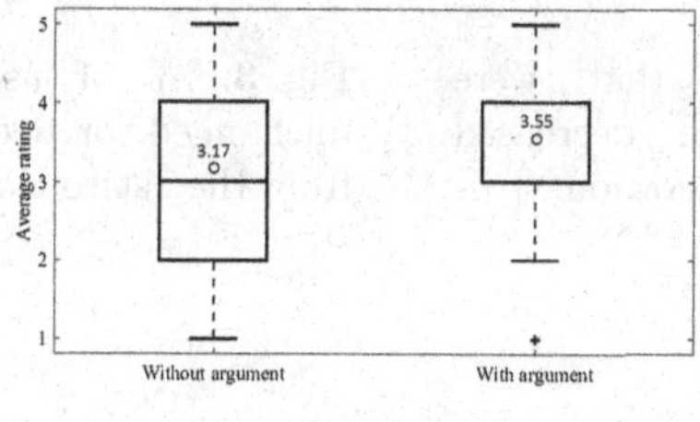

Fig. 1. Average rating in the 1st iteration (with and without arguments).

To test the persuasive power of arguments, in Fig. 2 we show the percentage of ratings that were *improved*, *not changed*, or *decreased* for the same LOs with no further explanation (in the 1st iteration) and when explanations were presented (in the 2nd iteration). 72% of ratings are improved with the explanations. When asked why they increased their ratings, many students claimed that thanks to the explanations they could understand why the system recommended specific LOs to them, and thought they were better. Therefore, explanations proved to enhance transparency and students' understanding of the system's operation. In Fig. 3 we show the percentage of students that changed their ratings (improved, not changed, or decreased) from one iteration to the other (on average over the total LOs evaluated per student). 92% of students improved their ratings when explanations were provided. In both tests, no student decreased his ratings. Then, we can conclude that arguments are useful to persuade students and have a positive influence on their LOs assessment.

Furthermore, in the Figs. 4 and 5, we show the average ratings obtained in each iteration for the different types of arguments and their associated explanation styles. With this experiment, we tried to evaluate if there is any specific type of argument that could be more effective or persuasive. In view of the results, students seem slightly prefer those explanations that are based on their profile and preferences (content-based) or on the opinions of similar users (collaborative). In addition, we can also appreciate an increase in the average scores that students gave to LOs on the 2nd iteration, which can be attributed to the emergence of persuasive explanations from the 1st to the 2nd iteration. Finally, we got very few data for the knowledge-based explanations. This was expected, since this type of explanations are based on arguments that make use of the history of ratings of the students. Few arguments of this type could be generated as the students that participated in the tests were new in the system and the probability to find similar LOs that they previously evaluated was very low.

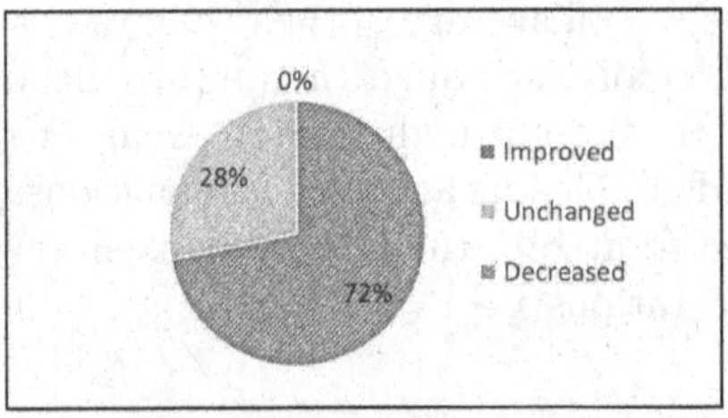

Fig. 2. % of ratings that were improved, unchanged, or decreased from the 1st to the 2nd iteration.

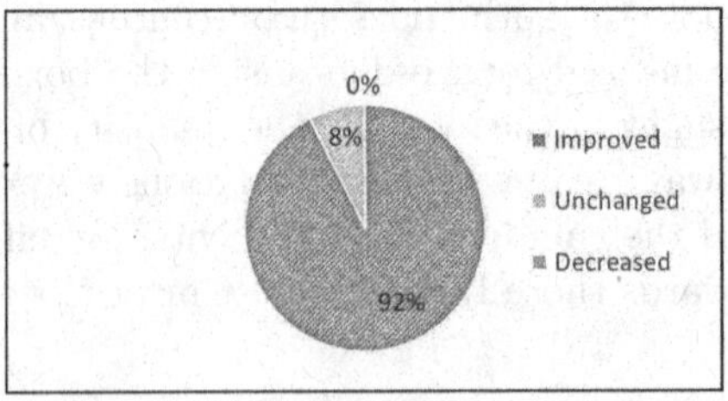

Fig. 3. % of users that improved, unchanged, or decreased their ratings from the 1st to the 2nd iteration.

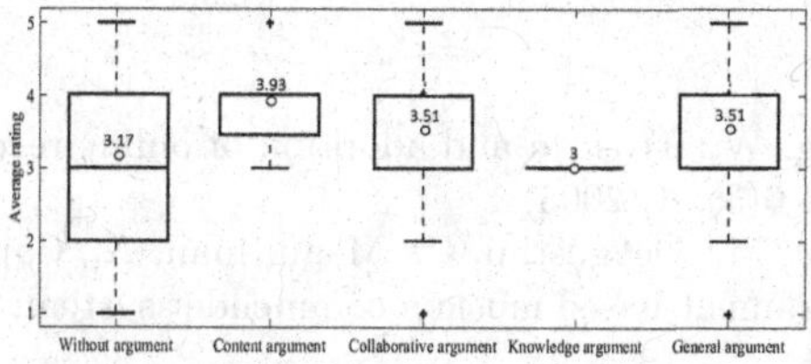

Fig. 4. Average ratings in the 1st iteration for each type of arguments.

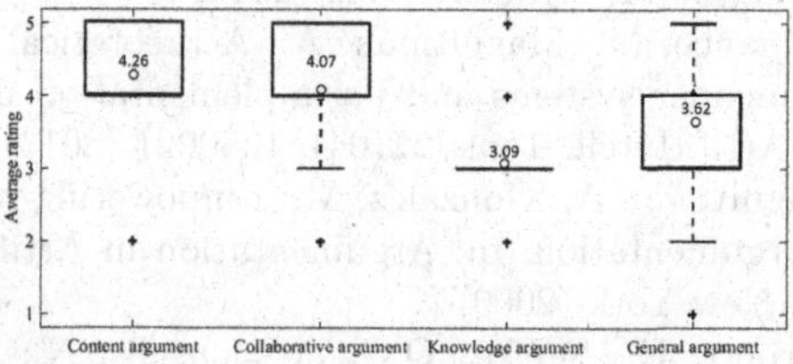

Fig. 5. Average ratings in the 2nd iteration for each type of arguments.

5 Conclusions

This paper has presented an argumentation-based educational recommender system. The system is used a persuasive technology that helps students to find appropriate learning objects for their profile and learning objectives. Our hybrid recommendation algorithm uses defeasible rules that capture different types of knowledge, which give rise to different arguments, and generates explanations based on them. In this work, we have analysed how arguments can be used as explanations to improve the efficiency, persuasive power, and transparency of the system. The good results achieved open an interesting panorama for the application of argumentation as a persuasion technique for e-learning environments. As future work, we plan to test the system with a larger number of students, and analyse how arguments improve the confidence of the users. Moreover, we want to evaluate the role that persuasive arguments play to improve the actual learning results that students achieve with the LOs that they use.

Acknowledgements. This work was funded by the 'Programa Nacional de Formación de Investigadores COLCIENCIAS', and by the COLCIENCIAS project 1119-569-34172 from the Universidad Nacional de Colombia. It was also supported by the projects TIN2015-65515-C4-1-R and TIN2014-55206-R of the Spanish government, and by the grant program for the recruitment of doctors for the Spanish system of science and technology (PAID-10-14) of the Universitat Politècnica de València.

References

1. Benbasat, I., Wang, W.: Trust in and adoption of online recommendation agents. J. Assoc. Inf. Syst. **6**(3), 4 (2005)
2. Briguez, C., Budán, M., Deagustini, C., Maguitman, A., Capobianco, M., Simari, G.: Towards an argument-based music recommender system. COMMA **245**, 83–90 (2012)
3. Briguez, C., Budán, M., Deagustini, C., Maguitman, A., Capobianco, M., Simari, G.: Argument-based mixed recommenders and their application to movie suggestion. Expert Syst. Appl. **41**(14), 6467–6482 (2014)
4. Briguez, C., Capobianco, M., Maguitman, A.: A theoretical framework for trust-based news recommender systems and its implementation using defeasible argumentation. Int. J. Artif. Intell. Tools **22**(04), 1350021 (2013)
5. Chesñevar, C., Maguitman, A., González, M.: Empowering recommendation technologies through argumentation. In: Argumentation in Artificial Intelligence, pp. 403–422. Springer, New York (2009)
6. Drachsler, H., Verbert, K., Santos, O.C., Manouselis, N.: Panorama of recommender systems to support learning. In: Recommender Systems Handbook, pp. 421–451 (2015)
7. Dwivedi, P., Bharadwaj, K.: e-Learning recommender system for a group of learners based on the unified learner profile approach. Expert Syst. **32**(2), 264–276 (2015)
8. Fogg, B.: Persuasive technology: using computers to change what we think and do. Ubiquity **2002**(December), 5 (2002)
9. García, A., Simari, G.: Defeasible logic programming: an argumentative approach. Theory Pract. Logic Programm. **4**(1+2), 95–138 (2004)
10. Heras, S., Rebollo, M., Julián, V.: A dialogue game protocol for recommendation in social networks. In: Corchado, E., Abraham, A., Pedrycz, W. (eds.) HAIS 2008. LNCS (LNAI), vol. 5271, pp. 515–522. Springer, Heidelberg (2008). doi:10.1007/978-3-540-87656-4_64
11. Kirschner, P., Buckingham-Shum, S., Carr, C.: Visualizing Argumentation: Software Tools for Collaborative and Educational Sense-Making. Springer Science & Business Media, Heidelberg (2012)
12. Klašnja-Milićević, A., Ivanović, M., Nanopoulos, A.: Recommender systems in e-learning environments: a survey of the state-of-the-art and possible extensions. Artif. Intell. Rev. **44**(4), 571–604 (2015)
13. Li, H., Oren, N., Norman, T.J.: Probabilistic argumentation frameworks. In: Modgil, S., Oren, N., Toni, F. (eds.) TAFA 2011. LNCS (LNAI), vol. 7132, pp. 1–16. Springer, Heidelberg (2012). doi:10.1007/978-3-642-29184-5_1
14. Recio-García, J., Quijano, L., Díaz-Agudo, B.: Including social factors in an argumentative model for group decision support systems. Decis. Support Syst. **56**, 48–55 (2013)
15. Ricci, F., Rokach, L., Shapira, B.: Recommender Systems Handbook. Springer, New York (2015)
16. Rodríguez, P., Heras, S., Palanca, J., Duque, N., Julián, V.: Argumentation-based hybrid recommender system for recommending learning objects. In: Rovatsos, M., Vouros, G., Julian, V. (eds.) EUMAS/AT -2015. LNCS (LNAI), vol. 9571, pp. 234–248. Springer, Cham (2016). doi:10.1007/978-3-319-33509-4_19
17. Salehi, M., Pourzaferani, M., Razavi, S.: Hybrid attribute-based recommender system for learning material using genetic algorithm and a multidimensional information model. Egypt. Inf. J. **14**(1), 67–78 (2013)

18. Sikka, R., Dhankhar, A., Rana, C.: A survey paper on e-learning recommender systems. Intl. J. Comput. Appl. **47**(9), 27–30 (2012)
19. Sinha, R., Swearingen, K.: The role of transparency in recommender systems. In: Conference on Human Factors in Computing Systems, pp. 830–831. ACM (2002)
20. Tintarev, N., Masthoff, J.: Explaining recommendations: design and evaluation. In: Recommender Systems Handbook, pp. 353–382. Springer, US (2015)
21. Yoo, K.H., Gretzel, U., Zanker, M.: Source factors in recommender system credibility evaluation. In: Recommender Systems Handbook, pp. 689–714. Springer, New York (2015)
22. Zapata, A., Menendez, V., Prieto, M., Romero, C.: A hybrid recommender method for learning objects. In: IJCA Proceedings on Design and Evaluation of Digital Content for Education (DEDCE), vol. 1, pp. 1–7 (2011)

Pokémon WALK: Persuasive Effects of Pokémon GO Game-Design Elements

Alexander Meschtscherjakov(✉), Sandra Trösterer, Artur Lupp, and Manfred Tscheligi

Center for HCI, University of Salzburg, Jakob-Haringer-Straße 8, 5020 Salzburg, Austria
{alexander.meschtscherjakov,sandra.trosterer,artur.lupp, manfred.tscheligi}@sbg.ac.at

Abstract. Pokémon GO is a location-based game that caused hype around the globe in 2016. Its primary objective is to "Catch'em All", meaning to catch all available Pokémon. These are virtual creatures distributed in the real world and the player has to walk around and catch them. Various game-design elements such as fighting against other Pokémon or hatching eggs by covering a predefined distance promote physical activity. In this paper, we present the results of an online survey (N = 124) that investigated whether Pokémon GO persuaded players to walk more and in particular which game-design elements had the highest influence on walking behavior. Results show that Pokémon GO persuades people to leave the home and increase exercise. Game-design elements such as catching as many different Pokémon and completing the Pokédex are more persuasive than fighting and competition challenges.

Keywords: Location-based game · Pokémon GO · Walking behavior

1 Introduction

Mobile location-based games (LBGs) use the player's movement in the real-world to evolve gameplay. Typically, the player has to walk around, search for specific locations, and solve tasks. In order to synchronize the real world movement of the player and the (often) virtual gameplay environment, the player's location is tracked through a GPS signal. With the advent of smartphones that include GPS and motion sensors, a plethora of such mobile games has been presented over the last decade. Since it is an inherent feature of the gameplay to move around in the real world, such games can be regarded as persuasive technology to nudge players to leave the home and walk around.

Pokémon GO is such a LBG, which quickly became a global phenomenon. It is a gaming app for iOS and Android smartphones released in July 2016 around the globe. The player is represented as an avatar, whose goal is to catch, train, and battle Pokémon. A Pokémon is a virtual creature that is located at random places in the real world. Currently, 145 Pokémon can be caught. The app

P.W. de Vries et al. (Eds.): PERSUASIVE 2017, LNCS 10171, pp. 241–252, 2017.
DOI: 10.1007/978-3-319-55134-0_19

uses a customized version of Google Maps to visualize the current geographical location of the avatar on a digital map in real time (Fig. 1-1). Pokémon appear and disappear randomly on the map only if the avatar (a.k.a. the player) is in close vicinity of the respective Pokémon. After tapping on a Pokémon on the map, it is visualized as an overlay on top of the smartphone camera image creating an augmented reality (AR) and the player may catch it by throwing a PokéBall at the it with a flick gesture on the smartphone. Whenever a new Pokémon is caught, it is registered in the player's Pokédex (Fig. 1-2). In order to retrieve PokéBalls and other items, the player needs to visit specific places called PokéStops that are scattered over the map. The game has many game elements that create a diversified gaming experience. For different activities the player earns Experience Points (XP), needed to level-up the players avatar allowing him/her to own and catch stronger Pokémon. During the game the player might use various in-game items (e.g., Candy to evolve new Pokémon) that can be gained by visiting PokéStops for different purposes. Players also may hatch eggs to retrieve new Pokémon by walking for a certain distance (Fig. 1-3).

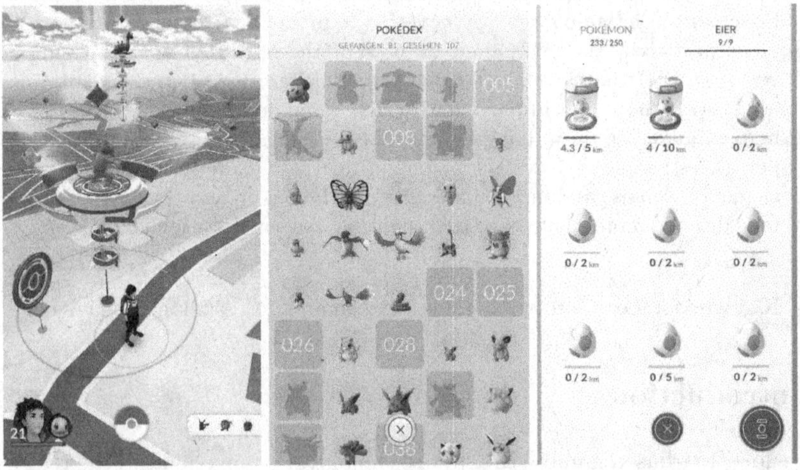

Fig. 1. From left to right: (1) map visualizing the avatar near a PokéStop (purple landmark) and next to a gym (white and blue tower); in the bottom right three close Pokémon are visualized (2) Pokédex with captured (colored) and observed (silhouettes and numbers) Pokémon (3) Eggs deck with a 5 km egg incubated for 4.3 km; after walking for 5 km a new Pokémon hatches. (Color figure online)

Pokémon GO has received a huge media echo calling it a social media phenomenon. It was used for commercial purposes to allure players to visit certain sites. It was also advertised as a game that fosters physical activity and current research suggests that the number of steps a day on average may increase by more than 25% [1]. In this paper, we want to focus on one specific aspect of Pokémon GO, namely its effect on the amount of physical activity with regard to the different game-design elements. Our aim is to identify those game-design elements that are successful in nudging people to exercise more often. Our findings can be used to inform the design of other persuasive games. In particular we were interested in the following three research questions:

RQ1. Does Pokémon GO persuade people (i.e. Pokémon GO player) to be more physically active?
RQ2. Which game-design elements that foster exercising, are important to players and how successful are they in persuading people to walk more?
RQ3. What are the motivations to play or stop playing Pokémon GO and how do players foresee how long they will play Pokémon GO in the future?

In order to answer these research questions, we have conducted an online survey. In the following we report on related work on Pokémon GO and LBGs. Then, we describe the specific game-design elements of Pokémon GO that foster physical activity, which we identified in a structured analysis. After describing study setup and results we discuss our findings and describe which game-design elements were successful in nudging physical activity.

2 Related Work

A good persuasive game [6] has to be effective and compelling in persuading people to actually do something [11]. This can be achieved by using various game-design elements, e.g., by accumulating points, steps, or items to encourage physical activity. Other motivators can be random awards [2] or nostalgia [13]. Pokémon GO combines game-design elements from location-based and role-playing games. LBGs use the movement of the player as a game input modality, thus, persuading people to walk. For that purpose, movement sensors and GPS of the smartphone are used to capture player movements in the wild. Some well-known examples are "Zombies, Run!", "Ingress", and "Geocaching", which, apart from being fun to play, offer a decent persuasive environment [9].

Pokémon GO itself is based on "Ingress", an augmented-reality LBG with a science fiction backstory. Ingress heavily relies on the players physical movement from one portal (i.e., a special physical location like PokéStops) to another in order to interact with them. Due to relying on physical movement as well, Pokémon GO can lead to a significant increase in physical activity for a period of time as shown by Althoff et al. [1]. They claimed that Pokemon Go can significantly increase physical activity on the first 30 days of playing. They used search engine query logs to identify Pokémon GO players and wearable sensor data for a change in movement behavior. Williamson [14] states that the most intriguing aspect of Pokémon GO from public health perspective is that people have to engage in physical activity to be playing the game. It is noted that physical activity is a by-product of the game. Dorward et al. [8] claim that Pokémon GO raises the awareness and engagement with the real-world nature. On the other hand, player's eyes seem to be glued to the screen and not looking at the real environment already has led to various accidents.

Role-playing and goal-setting are two important game-design elements when it comes to motivate players to engage in physical activities [5]. Bartley et al. [4] used role-playing to design a game in which the avatar can be strengthened by physical exercises in real life (e.g., push-ups to increase the avatar's strength). Pokémon GO focuses not only on development of an avatar, but on the development of various Pokémon. Goal-setting is an effective way to encourage physical activities [7,10]. Especially role-playing games enable players to set their own goals with certain pre-set goals in mind, which have to be met in order to advance in the game. Pokémon offers pre-set goals called achievements like the "Kanto" achievement, where the player needs to fill his Pokédex. Apart from pre-set goals, it is free for the player to set own goals to progress in the game.

Another base for the design of persuasive games are gamer types [12]. Bartle et al. [3] categorize four different classes: *achievers*, who want to accomplish certain goals in order to obtain rewards; *explorers*, who want to find out as much as possible about the game; *socializers*, who use the game as a platform to interact with other players and built relationships; and *killers*, who enjoy the competition with other players. Pokémon GO supports all types: achievers level their avatar and fill their Pokédex; explorers visit real world areas and discover new Pokémon; socializers compare their levels and caught Pokémon with others; killers fight in gyms.

3 Persuasive Game-Design Elements

Pokémon GO offers various game-design elements. We were interested in exploring which game-design elements are especially important for players and successful in nudging them to walk more often. In order to identify game-design elements that foster walking, we analyzed the Pokémon GO app by systematically evaluating each possibility to interact with the app. We present the game-design elements we believe to persuade players to be more physical active. We do not claim that this list is comprehensive, since the different game-design elements overlap each other and Pokémon GO is constantly updated with new features. Nonetheless, the chosen game-design elements are sufficiently distinct from each other and have unique features. All game-design items require the player either to explicitly move to a certain place or to move around.

Reaching a Higher Level. The first game-design element is know from many games. It is the overall aim of a game to reach higher levels. The maximum level one can reach in Pokémon GO is currently level 40. A new level is reached by collecting XP, which may be earned by various activities that require walking around such as capturing Pokémon, visiting PokéStops, or fighting in gyms. With each new level the player receives extra items and some items can only be collected after reaching a certain level (e.g., "Hyperballs" above level 20).

Catching New Pokémon. The theme of the traditional Pokémon game is "Gotta Catch'Em All". This theme is also reflected in Pokémon GO. The goal of this game-design element is to collect as many different Pokémon as possible and to fill-up the Pokédex. The maximum number of catchable different Pokémon is currently 145. One possibility to fill-up the Pokédex is to catch new Pokémon in the real world. Since specific Pokémon repeatedly appear at certain locations player may walk deliberately to these locations. Nearby Pokémon are visualized in an extra window in the bottom right corner of the screen. Pokémon that have not been captured yet are visualized as a silhouette (Fig. 1-1).

Evolving New Pokémon. Another possibility to get to new Pokémon is to evolve them from already caught ones. For example, a Pidgey Pokémon may then be evolved into a Pidgeotto, which may be evolved into a Pidgeot. In order to do so the player has to collect Pidgey Candy. By catching a Pokémon, the player not only receives the Pokémon but also three Candies of the correspondent Pokémon type. Thus, in order to evolve new Pokémon, the player has to be physical active catching as many Pokémon as possible and to collect Candy.

Visting PokéStops. A vital game-design element are PokéStops. They are dedicated landmarks scattered all over the real world and visualized as a platform with a blue dice hovering above it (Fig. 1-1). When a player approaches such a PokéStop in the real world the visualization changes and the player may receive a set of items at random (e.g., PokéBalls), which then are stored in the inventory bag. Then the player has to wait some time to be able to use the PokéStop again. Since PokéBalls are needed to proceed in the game, it is crucial for a player to move around and visit PokéStops.

Exploring PokéStops. Pictures of landmarks representing PokéStops (e.g., statues or sculptures) are shown when tapping on a PokéStop on the map. This game-design element is closest to a traditional treasure hunt or geocaching game. Since the landmark is shown as a picture on the map it might be fun for the player not only to get close to the PokéStop and collect items, but to actually search for the shown landmarks in the real world.

Hatching Eggs. Some Pokémon can be achieved by hatching eggs. Pokémon eggs are items that can be gained by visiting PokéStops or by reaching new levels. Currently three different types of eggs are available. They differ in the amount the player has to walk while hatching the egg in an incubator (2 km, 5 km, 10 km). It is known what kind of Pokémon can hatch from which type of egg, but the player does not exactly know, which Pokémon he/she will get until it actually hatches. Egg hatching nudges the player to walk for a certain distance which is shown in a separate window (Fig. 1-3).

Fighting in Gyms. Apart from PokéStops, the map also shows gyms. Such gyms can be sieged by a team (i.e., blue team mystic, red team valor, yellow team instinct). A player can fight against Pokémon of other teams by walking towards a gym and starting a fight. Fighting is done by swipe and tap gestures on the screen. When a player has captured a gym, an own Pokémon may be placed in that gym. Apart from the pride of "owning" a gym, the player is rewarded with stardust and coins once a day as long as he/she is owning the gym. Thus, walking to gyms and fighting is rewarding.

Collaborative Fighting. A gym might not only be occupied by one but several Pokémon from one team. Thus, it might be beneficial to built groups and approach a gym together. By doing so the gym is easier to occupy and also easier to hold against attacks from other teams. Apart from that, it might be a social experience to go for a walk together and capture gyms.

Exploiting Special Events. In October 2016, a Halloween Special event was presented as a new update for Pokémon GO. During that time players were able to experience benefits, such as earning more Candy when capturing a Pokémon. During that time, also psychic and ghost types Pokémon appeared more frequently. Such special events could also lead players to be more physical active during that time, since they could proceed in the game faster.

These game-design elements were then investigated with regard to their importance and effect to increase physical activity in an online study.

4 Study Setup

To answer our research questions, we setup an online questionnaire with LimeSurvey (www.limesurvey.org). It consisted of questions regarding demographics and smartphone gaming behavior (5 items), as well as Pokémon GO playing behavior (10 items) including game statistics (8 items), which could be obtained from the Pokémon GO app. It included 20 statements regarding the importance of specific game-design elements (outlined in the previous section), and whether particular game elements had an impact on the player's walking behavior (e.g., "In order to hatch eggs, I deliberately walk around."). Statements had to be rated on a 7-point Likert scale from 1 (=I totally disagree) to 7 (=I totally agree). Furthermore, we asked participants to rate their overall mobility behavior with respect to Pokémon GO and its dangerous aspects (9 items). Finally, we asked whether participants would play it in the future and their reasons for that with open-ended questions (5 items). Items were self construed and iterated with colleagues to ensure intelligibility.

The questionnaire was offered in German and English. Participants could deliberately take part in a drawing of ten 15 Euro vouchers (Amazon, Google Play Store, iOS App Store). Participants were recruited through various email lists of our university, via announcements on Facebook and on reddit (www.reddit.com), a community platform to share, vote, and discuss latest digital trends. In order to participate in the study, participants need to have reached at least Level 3 to make sure that players have at least some experience in playing Pokémon GO. Level 3 is reached easily within a few hours of playing it.

5 Study Results

The survey was online for 10 days. From the received filled-in questionnaires, we used 124 for analysis, which was done with IBM SPSS Statistics (Version 24). It took on average 11 min to fill in the questionnaire. Five questionnaires were filled-in in German, 119 in English. Of the 124 participants, 27% were female and 73% were male, aged between 13 and 52 years (M = 26.76, SD = 8.36). Participants owned either an Android smartphone (61%) or an iPhone (39%). Most participants played games on their smartphone almost daily (80.6%), or several times a week (13.7%)—the rest less often. All participants were active Pokémon GO players, and only 5.6% played other location-based games.

5.1 Game Statistics and Mobility Behavior

Participants played Pokémon GO almost daily (86.3%), or several times a week (13.7%). On average they actively played Pokémon GO for 14 weeks (SD = 3.65) with a mean duration of 2.24 hours per day (SD = 2.01). More than half of participants stated that they now play Pokémon GO considerably less (12.9%) or less (38.7%) than in the beginning. For 29% of participants playing frequency did not change and 15.3% stated to play more or even considerably more (4%). The reasons for playing less was that it was cumbersome to reach new levels (50%), high battery power consumption (31.3%), less fun (23.4%), less time (15.7%) or technical malfunctions (14.1%). The reasons for playing more was more fun (33.3%), to compare with others (33.3%), by habit (29.2%), or the Halloween Special event (25%).

Most participants were experienced players. Their player level ranged from 10 to 33 (M = 25.48, SD = 3.86) and 88.7% started paying Pokémon GO in July 2016 (release month). Regarding teams, 41.9% belonged to the blue team mystic, 31.5% to the red team valor, 25.8% to the yellow team instinct and 0.8% to no team at all. The number of *different* Pokémon registered in the Pokédex ranged between 44 and 145 (i.e., current maximum) with a mean of 115.64 (SD = 23.39). The overall, number of Pokémon captured ranged between 94 and 9,761 (M = 2,883.17, SD = 2,190.64). Participants stated to have visited between 122 and 14,019 PokéStops (M = 3,502.72, SD = 2,882.58). Distance covered while playing ranged between 25 km and 3,442 km (M = 383.49, SD = 361.43), with one outlier stating to have walked 3,442 km and the rest below 1,121 km.

Regarding the effect of Pokémon GO on their mobility behavior, participants had to rate statements on a scale from 1 (=I totally disagree) to 7 (=I totally agree). Participants strongly agreed that because of Pokémon GO they deliberately take a detour (M = 6.05, SD = 1.17), go out more often (M = 5.86, SD = 1.48), and walk for longer distances (M = 5.86, SD = 1.42). They stated not to exercise less than before, indicated by a low mean rating of this aspect (M = 1.70, SD = 1.37). Participants rather neglected that Pokémon GO is dangerous because of distraction issues (M = 3.07, SD = 1.86) or addiction risks (M = 2.26, SD = 1.69) as the mean ratings were rather low with respect to these aspects.

5.2 Importance and Persuasiveness of Game Elements

We now compare the ratings of game-design elements regarding their importance and to what extend these elements nudge players to increase their

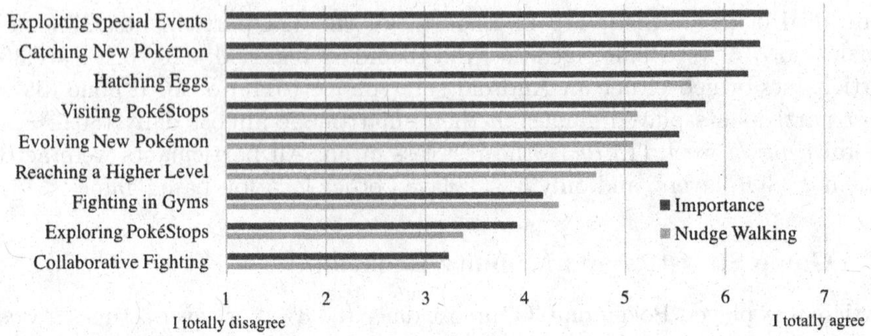

Fig. 2. Mean ratings of importance and the influence on walking behavior of different game-design elements.

physical activity (Fig. 2). Ratings were given on a scale from 1 to 7. Exploiting Special Events was rated highest in terms of importance (M = 6.44, SD = 1.21) and nudging players to walk around more often (M = 6.19, SD = 1.52). Also rated very high regarding importance and nudging physical activity was Catching New Pokémon (importance: M = 6.36, SD = 1.29; nudge walking: M = 5.89, SD = 1.57) and Hatching Eggs (importance: M = 6.23, SD = 1.13; nudge walking: M = 5.66, SD = 1.72). This shows that for most players catching different Pokémon and filling the Pokédex was the motivator to go out more often.

Highly rated, but less than those reported above, was Visiting PokéStops (importance: M = 5.80, SD = 1.30; nudge walking: M = 5.13, SD = 1.86), Evolving New Pokémon with Candy (importance: M = 5.54, SD = 1.56; nudge walking: M = 5.08, SD = 1.74), and Reaching a Higher Level in the overall game (importance: M = 5.43, SD = 1.69; nudge walking: M = 4.71, SD = 1.89). Fighting in Gyms was rated less important (M = 4.18, SD = 2.01) being the only item rated on average lower in the importance scale than in the nudge walking scale (M = 4.34, SD = 2.03). Finally, two items were rated considerably lower then the others. Exploring PokéStops was considered to be neither unimportant nor important (M = 3.92, SD = 2.06) and also the ratings regarding walking were rather low (M = 3.38, SD = 1.93). Collaborative Fighting as a team in gyms was rated low in importance (M = 3.23, SD = 2.24) as well as on the walking scale (M = 3.15, SD = 2.26). The rather high standard deviation in both game-design elements reflects a broad distribution among participant answers.

In order to find out whether the ratings of game-design elements with regard to nudge walking differed significantly from each other, we calculated a one-way ANOVA for repeated measures for these items. We could find a significant result ($F(8,98) = 48.380$, $p < .001$, partial $ETA^2 = .331$). Post-tests to compare the different items were calculated with a Bonferroni-corrected level of significance. Table 1 provides an overview of the pairwise comparisons. We could find significant differences among the highest rated items (i.e. Exploiting Special Events significantly differs from Hatching Eggs, Catching New Pokémon significantly differs from Visiting PokéStops) and, a significantly lower rated group of elements regarding Evolving of New Pokémon and Reaching a Higher Level, with an overlap to the higher rated item Visiting PokéStop to Collect Items and the lower rated Fighting in Gyms. The game elements Exploring PokéStop and Collaborative Fighting received significantly lower ratings compared to other elements, while not significantly differing from each other. The results indicated that there is a relation between the importance of elements and their potential to nudge walking behavior. Indeed, overall we found a positive high correlation (Pearson r = .767, $p < .001$), i.e., the higher the importance rating, the higher was also the rating regarding the walking behavior.

Table 1. Pairwise comparisons of game-design elements regarding nudge walking behavior with Bonferroni-corrected level of significance.

Nr	Game-Design Element	1	2	3	4	5	6	7	8	9
1	Exploiting Special Events		n.s.	*	***	***	***	***	***	***
2	Catching New Pokémon	n.s.		n.s.	**	**	***	***	***	***
3	Hatching Eggs	*	n.s.		n.s.	**	***	***	***	***
4	Visiting PokéStops	***	**	n.s.		n.s.	n.s.	*	***	***
5	Evolving New Pokémon	***	**	**	n.s.		n.s.	n.s.	***	***
6	Reaching a Higher Level	***	***	***	n.s.	n.s.		n.s.	***	***
7	Fighting in Gyms	***	***	***	*	n.s.	n.s.		*	***
8	Exploring PokéStops	***	***	***	***	***	***	*		n.s.
9	Collaborative Fighting	***	***	***	***	***	***	***	n.s.	

*$p<.05$, **$p<.01$, ***$p<.001$

5.3 Motivations Continue or Stop Playing

Regarding the question whether participants will (continue to) play Pokémon GO in the future 88.7% agreed and 11.3% answered that they did not know. None stated to want to stop playing it. Of those stating to play it in the future, 4.5% would play it the next few days, 13.6% in the next few weeks, 36.4% in the next few months, and the majority anticipated to play it in a year (45.5%).

We asked participants to state reasons for keep on playing Pokémon GO. For the analysis of the qualitative data we went through answers and built thematic categories following a grounded theory approach. Then we counted participants' statements belonging to each category. Most mentioned reasons for keeping on playing Pokémon GO in the future were likability and fun (51.82%) and the urge to complete the Pokédex (33,64%). Interestingly, 12,73% stated to continue playing for health reasons and for 2.72% Pokémon GO became a habit. Other reasons mentioned were social aspects (10.91%), nostalgia (6.36%), and competition (6.36%).

6 Discussion

In the following we discuss how our findings contribute to answering our research questions (RQ1–3) along with a discussion on the limitations of our study.

RQ1 Persuadability of Pokémon GO. Pokémon GO is not an exergame, whose purpose is to make exercising entertaining, but it is a fun game that nudges people to go out more often. Our results supports the claim that Pokémon GO is persuading players to be more physically active, which is in line with [1]. Participants walked on average 383 km actively playing for 14 weeks. They agree to go out more often, to walk for longer distances, and deliberately take a detour in order to proceed in the game. Especially catching new Pokémon is motivating people to be more physically active. Different game-design elements contribute nudging people to be more physical active. Most successful are those that directly relate to the ability to catch new Pokémon and fill the Pokédex.

Participant's statements support this claim, when answering why they would continue playing Pokémon GO: "*I'm recovering so it motivates me to exercise or at least have some extra purpose in exercising and I have to walk my dog anyway and he loves the extra long walks or when we go biking to a Pokéstop nearby.*" Another participant explained why she is playing: "*The exercise; I'm not a person who will walk around for no reason or even for a healthy stroll, but this game changed that. My boyfriend and I travel about* 45 min *to go to Battery Park in Delaware, and we spend the whole day there walking around and catching Pokémon and fighting in gyms, and collecting PokéStops. Our first day we ever went we walked just over* 7 miles. *If someone told me I'd walk even* 2 miles *for fun I'd say they're crazy*".

RQ2 Effect of Game-Design Elements. Pokémon GO players act strategical. They rated the exploitation of special events as very important to them. This event nudged them to go out even more often than they already did to play Pokémon GO. For the design of persuasive games we argue that including such special events can be a boost for the motivation of people. Then again, players got more benefits while the event was taking place, leading to more physical activity throughout this time, which might lead to less physical activity after the event. Thus, a balance for such special events needs to be found.

Interestingly, reaching a higher level, which is often the utmost goal of a game is perceived of significant lower importance than catching new Pokémon. This also holds true for very common game-design elements such as the collection of experience points. For Pokémon GO, it is clear that the most important game element is to fill the Pokédex, which is also reflected in the game's theme "Gotta Catch'Em All". Thus, game-design elements that enabled players to find and catch new Pokémon (e.g., visualize the new silhouette of not-yet-caught Pokémon), to acquire or evolve Pokémon (e.g., by hatching eggs or evolving new Pokémon with Candy) were important ones. They were also successful nudging people to do an extra walk. More specifically, game-design elements that required to walk a certain distance (e.g., hatching eggs) persuaded players to do an extra walk. In contrast, other game-design elements such as Fighting in Gyms or elements that resembled geocaching designs (i.e. visit real world equivalence of PokéStops) were rated as not as important. Also, the overall level of the avatar was not considered as being as important as catch all different

Pokémon. Regarding Bartle's player types [3] elements of Pokémon GO that support achievers seem to be more successful in nudging physical activity than killers. Exploring and socializing seems to play a moderate role.

RQ3 Motivations to Keep on Playing. Most participants stated that they will continue to play Pokémon GO in the future with 45.5% foreseeing that they will play within a year's time. For the majority of players fun is the main reason. A major motivator is the goal to fill-up the Pokédex. Even players, who already had a rather full Pokédex stated that they keep on playing to gain XP and reach new levels to be prepared when new catchable Pokémon are presented in the future. Thus, improving the game by making the software more stable and reducing power consumption, keeping players engaged, e.g., through the means of special events, and above all introducing new Pokémon will be necessary for a sustainable success of Pokémon GO. Although most players do not play Pokémon GO primarily for health reasons, for 12,73% a major reason to continue playing is an increased physical activity. To play it together with friends or fight against others is also motivating. Additionally, the nostalgic factor is decisive as many participants stated.

Limitations. There are some shortcoming with our study. Our sample primarily includes active and experienced players. We did not include players, who followed the hype in the first weeks of the release but stopped playing it in the meanwhile. Thus, our findings suggest that Pokémon GO may persuade people to increase exercising that are involved in the game for some time. Participants had a general positive attitude towards Pokémon GO. One could argue that a hype like the one with Pokémon GO could backfire, since many people might follow the hype, but then abandon playing it jumping on the next hype. We did not actually measure walking behavior as [1], but relied on the participants subjective assessment. Having access to game data (e.g., when and for how long would players interact with what game-design element and how this would influence walking behavior) would be beneficial for a deeper analysis. We cannot predict the long-term persuasive effects of Pokémon GO on walking behavior. Although most players seem to be willing to keep on playing especially when the game is being updated in the future, a prediction whether the game has an impact on the future walking behavior is difficult to make. While writing these lines, it was leaked that there will be 100 new Pokémon in future updates. Regarding game-design elements, it is difficult to separate them, since they are all interwoven with each other. E.g., reaching a higher level has an influence on which Pokémon can be caught. Thus, our findings may not be generalizable.

7 Conclusion

We have presented an online survey on the persuadability of Pokémon GO. We have shown that players are more physical active when playing it. Despite fun being the main motivator for people to play it, game-design elements that allows players to collect new Pokémon and filling-up the Pokédex nudges people to go out more often and exercise. The longterm effect on mobile behavior will be dependent on constantly updating the game and the introduction of new catchable Pokémon. Further work, hast to target at the long-term persuasive effect of Pokémon GO and quantifying the effect of game-design elements.

References

1. Althoff, T., White, R.W., Eric, H.: Influence of Pokémon go on physical activity: study and implications. J. Med. Internet Res. **18**(12), e315 (2016)
2. Anselme, P., Robinson, M.J.F.: What motivates gambling behavior? Insight into dopamine's role. Front. Behav. Neurosci. **7**, 182 (2013). http://dx.doi.org/10.3389/fnbeh.2013.00182
3. Bartle, R.: Hearts, clubs, diamonds, spades: players who suit MUDs. J. MUD Res. **1**(1), 19 (1996)
4. Bartley, J., Forsyth, J., Pendse, P., Xin, D., Brown, G., Hagseth, P., Agrawal, A., Goldberg, D.W., Hammond, T.: World of workout: a contextual mobile RPG to encourage long term fitness. In: Proceedings of the 2nd ACM SIGSPATIAL International Workshop on the Use of GIS in Public Health, HealthGIS 2013, pp. 60–67. ACM, New York (2013)
5. Berkovsky, S., Bhandari, D., Kimani, S., Colineau, N., Paris, C.: Designing games to motivate physical activity. In: Proceedings of the 4th International Conference on Persuasive Technology, Persuasive 2009, pp. 37:1–37:4. ACM, New York (2009)
6. Bogost, I.: Persuasive Games: The Expressive Power of Videogames. MIT Press, Cambridge (2007)
7. Consolvo, S., Klasnja, P., McDonald, D.W., Landay, J.A.: Goal-setting considerations for persuasive technologies that encourage physical activity. In: Proceedings of the 4th International Conference on Persuasive Technology, Persuasive 2009, pp. 8:1–8:8. ACM, New York (2009)
8. Dorward, L.J., Mittermeier, J.C., Sandbrook, C., Spooner, F.: Pokémon go: benefits, costs, and lessons for the conservation movement. Conserv. Lett. **10**(1), 160–165 (2017). doi:10.1111/conl.12326
9. Gram-Hansen, L.B.: Geocaching in a persuasive perspective. In: Proceedings of the 4th International Conference on Persuasive Technology, Persuasive 2009, pp. 34:1–34:8. ACM, New York (2009)
10. Herrmanny, K., Ziegler, J., Dogangün, A.: Supporting users in setting effective goals in activity tracking. In: Meschtscherjakov, A., Ruyter, B., Fuchsberger, V., Murer, M., Tscheligi, M. (eds.) PERSUASIVE 2016. LNCS, vol. 9638, pp. 15–26. Springer, Cham (2016). doi:10.1007/978-3-319-31510-2_2
11. Khaled, R., Barr, P., Noble, J., Fischer, R., Biddle, R.: Fine tuning the persuasion in persuasive games. In: Kort, Y., IJsselsteijn, W., Midden, C., Eggen, B., Fogg, B.J. (eds.) PERSUASIVE 2007. LNCS, vol. 4744, pp. 36–47. Springer, Heidelberg (2007). doi:10.1007/978-3-540-77006-0_5
12. Orji, R., Mandryk, R.L., Vassileva, J., Gerling, K.M.: Tailoring persuasive health games to gamer type. In: Proceedings of the SIGCHI Conference on Human Factors in Computing Systems, CHI 2013, pp. 2467–2476. ACM, New York (2013)
13. Sedikides, C., Wildschut, T.: Past forward: Nostalgia as a motivational force. Trends Cogn. Sci. **20**(5), 319–321 (2016)
14. Williamson, J.W.: Will the 'Pokémon' be heroes in the battle against physical inactivity? Sports Medicine Open J. **2**(1), 13–14 (2016)

Why Are Persuasive Strategies Effective? Exploring the Strengths and Weaknesses of Socially-Oriented Persuasive Strategies

Rita Orji(✉)

Cheriton School of Computer Science, University of Waterloo, Waterloo, ON, Canada
rita.orji@uwaterloo.ca

Abstract. Socially-oriented persuasive strategies (*competition, social comparison,* and *cooperation*) which leverage the power of social influence have been widely employed in Persuasive Technologies (PTs) designs. However, there have been mixed findings regarding their effectiveness. There is still a dearth of knowledge on the mechanism through which these strategies could motivate or demotivate behaviors. To advance research in this area, we conduct a large-scale qualitative and quantitative study of 1768 participants to investigate the strengths and weaknesses of these strategies and their comparative effectiveness at motivating healthy behaviors. Our results reveal important strengths and weaknesses of individual strategies that could facilitate or hinder their effectiveness such as their tendency to *simplify a behavior and make it fun, challenge people and make them accountable* and their tendency to *jeopardize individual's privacy and relationships, creates unnecessary tension,* and *reduce self-confidence*, respectively. The results, also show that there are significant differences between the strategies with respect to their persuasiveness overall with the social comparison being the most persuasive of the strategies. We contribute to the PT community by revealing the strengths and weaknesses of individual strategies that should be taken into account by designers when employing each of the strategies. Furthermore, we offer design guidelines for operationalizing the strategies to amplify their strengths and overcome their weaknesses.

Keywords: Persuasive technology · Socially influencing systems · Persuasive strategy · Persuasion · Gamified design · Competition · Cooperation · Comparison · Captology

1 Introduction

There is growing evidence that interactive systems can be strategically designed to promote desirable behavior or motivate a change of undesirable behavior using various persuasive strategies [1–6]. Persuasive strategies are techniques that can be employed in Persuasive Technologies (PTs) design to promote desirable behaviors or attitudes. Socially-oriented persuasive strategies such as *competition, comparison,* and *cooperation* are the most widely and frequently employed strategies both in persuasive and gamified systems and other online support systems because of their ability to leverage

P.W. de Vries et al. (Eds.): PERSUASIVE 2017, LNCS 10171, pp. 253–266, 2017.
DOI: 10.1007/978-3-319-55134-0_20

the power of social influence to motivate behavior change. However, despite their wide application across several PT domains due to their perceived effectiveness [1–6], research have also shown that employing the socially-oriented strategies in PT design could be counterproductive [2, 4, 7, 8].

Although, PT researchers have recently begun to investigate the comparative effectiveness of various persuasive strategies (for example, see [2, 4, 9]), there is still a dearth of knowledge on how and why these strategies could influence behavior either positively or negatively and the mechanisms through which the strategies motivate or demotivate behaviors. This is essential for effective operationalization of the strategies in designs.

To advance research in this area, we conducted two large-scale qualitative and quantitative study of 1108 and 660 (a total of 1768) participants to examine the strengths and weaknesses of the three socially-oriented strategies (*competition, comparison, cooperation)* that are widely used in PT designs. We investigated these strategies in the context of PTs for promoting healthy eating behavior (study one) and PTs for motivating change in risky health behavior such as binge drinking (study two). Investigating two different health domains allows us to uncover a wide range of strengths and weaknesses that could be generalized across other domains. As a secondary objective, we validated the persuasiveness of the strategies and showed their comparative persuasiveness with respect to their ability to motivate health behavior change. We used prototype persuasive implementation of the individual strategies that has been validated in other studies [9]. Our results reveal important strengths and weaknesses of individual strategies that could facilitate or hinder their effectiveness such as their tendency to *simplify behaviors and make them fun, challenge people and make them accountable* and their tendency to *jeopardize individual's privacy and relationships, creates unnecessary tension, reduce self-confidence*, and *could backfire,* respectively. The results, also show that there are significant differences between the strategies with respect to their persuasiveness overall with the social comparison being the most persuasive of the strategies. This knowledge of the strengths and weaknesses of the strategies will not only inform the choice of the strategies to employ in a particular PT design but will also inform the manner in which a particular strategy will be operationalized in a PT design. The operationalization of the strategies can amplify their strengths or their weaknesses.

Our work offers three main contributions to the field of persuasive design. First, we reveal the strengths and weaknesses of individual persuasive strategies that should be taken into account by PTs designers when employing each of the strategies. Second, based on our findings, we offer design guidelines on how the strategies could be operationalized/implemented in persuasive design to amplify their strengths and overcome their weaknesses. Third, we validate and compare the perceived effectiveness of individual strategies and show that they differ significantly in their overall persuasiveness.

2 Related Work

In this section, we present a brief overview of socially-oriented persuasive strategies.

2.1 Persuasive Strategies and Their Applications in PTs

Persuasion is often achieved in PTs design using various persuasive strategies. Over the years, a number of persuasive strategies have been developed. For example, Fogg [8] developed seven persuasive strategies, and Oinas-Kukkonen [21] built on Fogg's strategies to develop 28 persuasive system design strategies.

According to the literature, socially-oriented strategies *Competition, Social Comparison, Cooperation* which leverage the power of social influence are among the commonly employed strategies in PTs especially those targeted at health behaviors [9–11]. The **Competition** strategy allows users to compete with each other to perform the desired behavior. **Comparison** strategy provides a means for the user to view and compare their performance with the performance of other user(s). The **Cooperation** strategy requires users to cooperate (work together) to achieve a shared objective and rewards them for achieving their goals collectively. For a detailed discussion of the strategies, see [6].

Studies have shown that these strategies can be effective for motivating desired behavior change [12–15]. As a result, researchers have employed the strategies in PTs for motivating behavior change in many domains including binge drinking prevention [16–18], drug and substance use prevention [19], smoking cessation [20], energy conservation [21], healthy eating [3, 4, 22], and physical activity [23–25]. Among all the domains, the use of PTs for health promotion and disease prevention have received special attention, likely due to the importance of maintaining good health and wellness. Therefore, this study will focus on PTs for health with special emphasis on PTs for promoting healthy eating and PTs for risky health behavior change (risky alcohol behavior change).

Despite the wide application of the socially-oriented strategies across several PT domains due to their perceived effectiveness [1–6], research has also shown that employing these strategies in PT designs could be counterproductive/backfire [2, 4, 7, 8] by demotivating behavior. However, there is yet no research into the mechanism through which the strategies could motivate or demotivate behaviors; why they may work in one context and fail in another. This will shed light on why many PTs record varying degrees of success, mixed findings, and even failures [2].

3 Study Design and Methods

For the purpose of this study, we chose to focus on two common applications of PTs for health to ensure uniformity and generalizability: PTs for encouraging healthy eating behavior and PTs for motivating change of risky healthy behavior (risky alcohol behavior change).

To achieve this, we conducted two separate studies, the first focuses on PTs for motivating healthy eating behavior and the second focuses on PTs for motivating a change of risky alcohol behaviors. To collect data for our studies, we used prototype persuasive implementation of the individual strategies that has been validated in other studies [9, 26]. Specifically, we represented each persuasive strategy in a storyboard about PTs for encouraging healthy eating (study one) and PTs promoting change of risky alcohol behaviors (study two). The three storyboards were drawn by an artist and were based on storyboard design guidelines by Truong et al. [27]. Implementing the strategies in storyboards makes it easier to elicit responses from diverse populations because storyboards provide a common visual language that individuals from diverse backgrounds can read and understand [28]. For study one, the storyboards show a character and their interactions with PTs for motivating healthy eating behavior and in study two, the storyboards show a character and their interactions with PT for promoting change of risky alcohol behavior. We evaluated and iteratively refined the storyboards. Figure 1 shows an example of one of the storyboards illustrating the comparison strategy.

Fig. 1. Storyboard illustrating the persuasive strategies comparison- adapted from [9]

To elicit quantitative feedback on the effectiveness of the strategies, each storyboard was followed by a validated scale for assessing perceived persuasiveness. The scale was adapted from Orji et al. [12] and has been used in many PT studies [4, 9, 14]. Specifically, we asked participants the following questions:

Imagine that you are using the system presented in storyboard above to track your daily eating (or alcohol in study 2), on a scale of 1 to 7 (1-Strongly disagree to 7-Strongly agree), to what extent do you agree with the following statements: (a) The system would influence me. (b) The system would be convincing. (c) The system would be personally relevant for me. (d) The system would make me reconsider my eating (or alcohol drinking) habits.

To obtain qualitative feedback about the strategies, each of the strategy was followed with an open-ended question allowing participants to provide comments to describe the strategy represented in the system, how they would use the system, and justify their ratings of each strategy with respect to the effectiveness – the strengths and weaknesses. Prior to assessing the persuasiveness of the strategies, we ensured that the participants understood the strategy depicted in each storyboard by asking them two comprehension questions – first, to identify the illustrated strategy from a list of 3 different strategies ("*What strategy does this storyboard represent*"); and second, to describe what is happening in the storyboard in their own words ("*In your own words, please describe what is happening in this storyboard*"). We also included questions for assessing the participants' demographic information, and eating and drinking behaviors.

We recruited participants for this study using Amazon's Mechanical Turk (AMT). A total of 1768 responses were included in our analysis (1108 responses from study one and 660 responses from study two), after filtering out incomplete responses and incorrect responses to comprehension and attention-determining questions. In the two studies, our participants were at least 15 years of age at the time of data collection, read and understand English well. In addition to this, for study two, participants consume or have ever consumed alcohol. The participants received a small compensation in appreciation for their time. In general, we have a relatively diverse population in terms of gender, age, education level attained, see Table 1.

Table 1. Participants' demographic information

Total participants = 1,768	
Gender	Females (49%), Males (51%)
Age	15–25 (32%), 26–35 (38%), 36–45 (18%), Over 45 (12%).
Education	Less than high school (1%), High school (31%), College diploma (13%), Bachelor's degree (37%), Master's degree (15%), Doctorate degree (2%), Others (1%).

4 Data Analysis and Results

To ensure that participants understood the strategy depicted in each of the storyboards, we ran chi-square tests on the participants' responses to the multiple-choice questions that required them to identify the represented strategy for each of the storyboards. The results for all the strategies were significant at $p < .01$, which indicates that the storyboards were understood by the participants and that the storyboards successfully depicted the intended persuasive strategies [9]. Second, we determined the consistency of the scale using Cronbach's alpha (α). The α for the strategies were all greater than 0.70 showing that the scales have good internal consistency. Next, to examine and compare the persuasiveness of the strategies, we conducted a Repeated-Measure ANOVA (RM-ANOVA) followed by a pairwise comparison, after validating for the ANOVA assumptions. We also used notched boxplot to visualize the persuasiveness of the strategies.

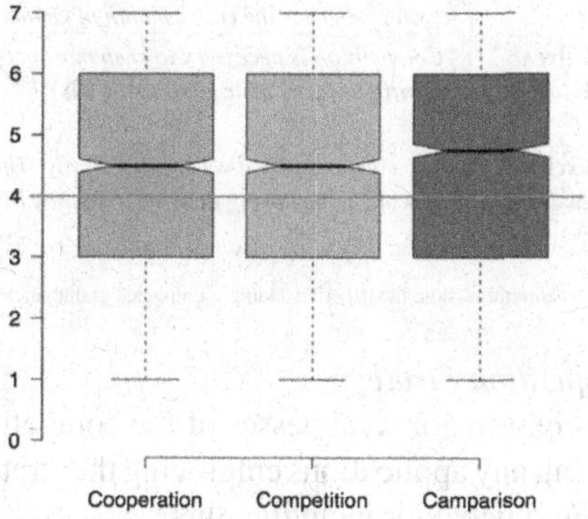

Fig. 2. A boxplot showing the overall persuasiveness (y-axis) of the strategies on a scale ranging from 1 to 7. The horizontal line indicates a neutral rating of 4.

4.1 Comparing the Persuasiveness of the Strategies Overall

The results of the RM-ANOVA show significant main effects of strategy type ($F_{1.93, 3567.635} = 8.807$, $p \approx .000$) on persuasiveness. This means that there are significant differences between the strategies with respect to their persuasiveness overall. In general, *Social Comparison* emerged as the most persuasive of all the strategies (significantly different from *Competition* and *Cooperation* as shown by the Bonferonni-corrected pairwise comparisons). Figure 2 shows the notch boxplots of the strategies. The notch represents the 95% confidence interval of the median. In general, our participants perceive all the strategies as persuasive, with persuasiveness score significantly higher than the neutral median rating of 4 ($p < .0001$), indicated by the horizontal line in Fig. 2.

4.2 Strengths and Weaknesses of Individual Strategies

To tease out the strengths and weaknesses of individual persuasive strategies, we conducted a thematic analysis [29] of 213 pages of qualitative comments about each of the strategies from our participants. The comments were analysed in an iterative manner, identifying the central ideas within them and their relationship and classifying them into strengths and weaknesses until it seemed that no further ideas were emerging from them. The themes reported here are the major ones that transpire from the analysis.

Strengths of the Competition Strategy

There are five major strengths of the competition strategy that make it effective at motivating behavior change:

SN	Strengths of the competition strategy	Sample supporting participants' comments
1	It makes one **committed** to the behavior	*"I am highly competitive, this is a good strategy to **motivate** me and keep me **committed**."*[a] [P51]
2	It gives a sense of **accomplishment**	*"Winning gives a great feeling of **accomplishment!**"* [P80]. *"Competition gives a sense of **accomplishment** that motivates most people to continue to do good in their behavior."* [P115]
3	It **engages** and **challenges** one to be better	*"...I could imagine it being a **challenge** type thing for people like me,...like staying sober for the entire month of October and see who wins."* [P471]
4	It keeps one **focused** and gives them more reason to push towards their goals	*"Competition is necessary to keep most people **interested, focused** and **pushing** towards their goals."* [P1108]
5	It makes behaviors **fun, exciting, interesting**, and appear **easier** to do than usual	*"The competition aspect is **fun**."* P[16] *"The competition aspect makes supposed to be **boring activity** of healthy living **exciting**."* P[957]

[a]Quotes from participants are included verbatim throughout the paper, including spelling and grammatical mistakes.

Weaknesses of the Competition Strategy

Participants highlighted eight major weaknesses of the competition strategy that could either deter them from using any applications employing the strategy or demotivate them from performing the desired behavior including that:

SN	Weaknesses of the competition strategy	Sample supporting participants' comments
1	It reduces **self-esteem** and makes people **lose confidence** in themselves and their ability	*"This could kill someone's **self-esteem** if they are struggling with a diet and see all their friends are doing better."* [P911]
2	It causes unnecessary **stress** and makes people **anxious**	*"This would also **embarrass me**. I'd be **nervous** that my friends would think of me as a fatty if I ate more than them often"* [P89] *"Interesting idea for competitive people, though I think that this would be **stressful** for me."* [P12]
3	It could **jeopardize an** individual's **privacy** and **relationships**	*"People need a supportive, encouraging environment to grow and change within. Competition **drives** people **apart**."* [P1056] *"Living healthy as a competition is **dangerous** for **relationship**."* [P19] *"It seems like my **privacy** would be compromised."* [P6]
4	It could **backfire** and **demotivate** behavior	*"... Points could end up being **rewarded for risky behavior**, just because they were better than their friends. That's wrong. It could **backfire**."* [P1590]
5	It could **annoy** users and **discourage** them from using the PT intervention	*"I do not like the competition aspect of this system, I think it could be very **discouraging** if you are constantly on the **losing** end."* [P519]
6	It **trivializes** the important of healthy behavior	*"This is **infantile**"* [P16] *"It's very **juvenile**"* [P1013] *"... I like competition but I think in healthy living, it might turn to **jealousy** and cause people to **stop altogether**."* P[698]
7	It does not often take individual unique conditions into account and may lead to **falsification** of behavior	*"Living a healthy life is a **personal goal**. One's personal use of alcohol cannot be in competition with another person's. It is a competition against oneself. Competing to improve to be better than my previous self should be the strategy."* [P6] *"I think adding competition to the app would **lower the likelihood of people being truthful** about their behaviour as the attention would be shifted to **winning by all means** even without altering behaviour."* [P999]
8	It encourages **body shaming** and health **disorder** (e.g., eating disorder)	*"This kind of system institutes a **belief that a person is better than another** because of their lifestyle. I personally think it's a **terrible idea**, and instead of people helping one another stay healthy it would turn into a competition that **encourages shaming** and **poor attitudes** about diet and personal value."* [P1001] *"It might make someone have some weird **eating disorder**, whereby they try to beat their friends by eating less and not better."* [P508]

Strengths of the Comparison Strategy

Our participants highlighted three major reasons why social comparison is effective and may be preferred over the competition strategy:

SN	Strengths of the comparison strategy	Sample supporting participants' comments
1	It allows for **subtle** and **empowering** peer pressure	*"I would really **prefer a subtle comparison** with my friends than making it a competition.* "[P235]
2	It offers opportunity to serve as a **positive role model**	*"I like the fact that my friends and I can act as **good role models** for each other by drinking less and that will trigger a **positive response**."* P[600]
3	It makes it **easier** to reach one's **goal** by providing **group reinforcement**	*"...if you have friends who are sharing the same struggles as you are, the motivation provided by the group would, I think, be **very reinforcing**."* P[1115] *"It is easier to meet a goal when you know others are **checking your progress**."* [P215]

Weaknesses of the Comparison Strategy
Our participants highlighted three main weaknesses of the comparison strategies:

SN	Weaknesses of the comparison strategy	Sample supporting participants' comments
1	It could be **invasive** and interfere with individual's **privacy**	*"Seems a bit **invasive** to have your performance displayed for all your friends to see. As a comparison method, I would prefer to compare my average overall daily performance, not my friends - keep it **private**."* [P471]
2	It could reduce **self-confidence, cause depression**, and make one **alienated**	*"I think this application would make me feel not as good as my friends. **I would lose my confidence**."* P[998] *"Due to prior addiction I think **shame** and **depression** would kick in more."* [P68]
3	It could **Backfire by encouraging unhealthy behavior**	*"This type of game could **go both ways**…if I drink less than my friends, it may make me drink more. So it basically comes down to what type of friends one is comparing with and if they promote drinking or not."* [P394]

Strengths of the Cooperation Strategy
Our participants highlighted four main strengths of the cooperation strategy:

SN	Strengths of the cooperation strategy	Sample supporting participants' comments
1	It provides opportunities for **mutual support** and **encouragement**	*"This system is good and will get users to **encourage each other** to meet their goals because they will both benefit."* [P654]
2	It provides opportunities for people to **stay responsible** and **accountable** to others which propels them to meet their behavior goals	*"It's easier to meet goals when you're held **accountable** by others or you know others are checking your progress."* [P9]
3	It **raises users' sensitivity** to disappointment and makes them **work harder** to avoid disappointing others	*"It is good; I won't want others to be **disappointed** by me. So it would work well."* [P1661] *"I wouldn't want to **let the other player down**, so I think this sort of game would work well for me."* [P870]
4	It offers opportunities to **collaborate, make** and **interact** with friends	*"It is very convincing and provides for **healthy collaboration** and opportunity **to interact with friends** more often."* [P918] "…*Instead of inspiring rivalry, this method **inspires companionship**. Awesome!!"* [P908]

Weaknesses of the Cooperation Strategy
Our participants highlighted four main weaknesses of the comparison strategies:

SN	Weaknesses of the cooperation strategy	Sample supporting participants' comments
1	It could create **unnecessarily tension** and **pressure**	*"If I did go in with a friend, I would feel* ***more pressure*** *to perform to not let them down. It creates unnecessary* ***tension.*** *Better to pace myself."* [P57]
2	It could be **tedious** and **stressful**	*"It would be* ***tedious*** *for me to convince someone; I would rather work on my own personal goals."* [P357]
3	It may **jeopardize relationships** and **privacy**	*"It would not work for me; I'd get upset with my friend if he fails! I may not want to* ***risk my relationship with*** *him"* [P19]. *"I like this better, although I may not use it because it seems like an* ***invasion of privacy"*** [P78] *"It can create* ***tension*** *between people and create opportunity for* ***negative criticism."*** [P56]
4	It requires at least **one motivated** and **trusted partner** to be effective to **avoid backfiring**	*"I wouldn't like to have something as* ***personal*** *as this tied to someone else's performance, it could* ***backfire*** *if team members are not motivated."*[P698]

5 Discussion

Our findings show that in general, the socially-oriented strategies are perceived as persuasive with respect to their tendency to motivate the desired health behavior. Although each of the strategies explores the principle and the power of social influence to motivate behavior, they differ in their operationalization of the principle. As a result, there are some common strengths and weaknesses that are shared by the strategies and at the same time, some unique strengths and weaknesses that are peculiar to each of the strategies that designers need to take into consideration when employing the strategies. Therefore, we develop some design recommendations based on our findings to guide designers when employing these strategies in PT designs.

5.1 Design Guidelines for Implementing Socially-Oriented Strategies to Overcome Their Weaknesses and Amplify Their Strengths

This section consists of recommendations for employing socially-oriented persuasive strategies in PT designs that were deduced from the findings of this study.

R1: Preserve people's privacy when employing any socially-oriented persuasive strategies

Our findings reveal that all the three socially-oriented strategies share a common weakness of being perceived as *invasive, less privacy-preserving,* and *has a high potential of harming friendship* that may hinder acceptance and use of any system employing them or their effectiveness in general. Most of the participants noted that they would only use the persuasive applications if it does not require disclosing their personal data. *"This works for me only if you make all participants* ***anonymous****."* [P301] Therefore, we suggest that **to overcome the privacy issues associated with socially-oriented**

strategies, designers should include mechanisms that allow users to hide their identity or use nicknames. Designers should also make it possible for individuals to work with strangers that cannot possibly identify them. Although, anonymizing user's identity may reduce the persuasive effect, we argue that it is a good trade-off to achieve a privacy-preserving system that will be adopted and used by the users. As an alternative to anonymizing user's identity, **designers should abstract user's behavior performance data as a way of preserving privacy**. For example, reporting behaviors as a percentage of an individual's goal may be a good abstraction and less invasive alternative to showing actual data as shown by the comment: "*Learning specific information about another user's calorie intake would be a* ***huge invasion of privacy*** *to me. If it were* ***a simple percentage reported****, then I would enjoy using the system.*" [P1590].

R2: Ensure that employing socially-oriented strategies will not be counterproductive (backfire); there is possibility of positive and negative social influence

Our study also shows that all the three socially-oriented strategies (competition, comparison, and cooperation) have the potential of being counterproductive by demotivating behaviors if not carefully implemented. As highlighted by the comment "*This type of game could go* ***both ways****…if I drink less than my friends, it may make me drink more. So it basically comes down to what type of friends one is comparing with and if they promote drinking or not*" [P394]. Therefore, **care should be taken when employing socially-oriented strategies to reduce the possibility of downward comparison and negative social influence**. Designers can pre-screen and redistribute users across groups to ensure effective social circle for comparison. Designers can also include mechanisms that allow for both within and between-group comparisons.

R3: Implement mechanism to deter cheating and build trust among users

A common weakness of the competition and comparison strategies is the lack of trust among the users that no one is cheating. As highlighted by the comment "*This will work only if I really trust that the people I'm playing with will* ***not cheat****.*" [P957]. "*Competition is the best motivation, there would need to be system so that people* ***couldn't cheat*** …" [P980] Therefore, for competition and comparison-oriented applications to be really effective, **designers should implement mechanisms that deter PT users from cheating and hence win participants' trust in the system**. One way of achieving this is to automatically monitor individual's behaviors using various activity sensing tools and avoid self-reporting of behaviors. We acknowledge that many behaviors to date cannot be automatically monitored, for such behaviors, emphasising the intrinsic benefit of behaviors to the individual's self may help reduce the urge to cheat to outperform others.

R4: Allow for self-competition and self-comparison in PT design

A major drawback of the competition and comparison strategy is their inherent tendency to cause tension and jeopardize privacy and friendship as discussed above. Most of our participants feel that they are inappropriate, especially for motivating healthy behavior. They highlighted that it should be a competition and comparison with oneself. As shown by the comment "*Living a healthy life is a* ***personal goal****. One's personal use of alcohol cannot be in competition with another person's. It is a* *competition against oneself. Competing to improve to be better than my previous self should be the strategy"* [P6]. This is in line with previous research that found that users were uncomfortable with using competition to motivate healthy behavior [30, 31], and may even become demotivated if they lose [7]. Therefore, **designers should provide a mechanism for users to compare their performance and compete to break their own record and reward them accordingly.** This could complement conventional competition strategy and can also substitute it for people who react negatively to competing with others to improve their behaviors.

R5: Reduce the tendency of unhealthy competition by allowing for multiple winners and winning conditions

A major strength of the cooperation strategy as highlighted by our participants is that it provides an opportunity for people to work together and win together as shown by the comment "*This is the best of the systems in my opinion because the more people with the same goal,* ***the more power there is available for achieving*** *that goal in many ways, including* ***support*** *and* ***moral boosting*** *for one another.*" [P559] On the other hand, one of the drawbacks of the competition and comparison strategy is that winning often entails the other party losing – mutually exclusive achievement. This tendency is summarized in a statement by Kohn [32] "*to say that an activity is structurally competitive is to say that it is characterized by what I will call mutually exclusive goal attainment. This means, very simply, that my success requires your failure.*" However, for many behaviors especially healthy behavior, the healthier we are as a community the better. Hence, **designers should design to avoid unhealthy competition and encourage changing together by allowing for multiple winners at a time and winning categories in their systems**. This is supported by participants comment such as "*It is very motivational; it should be possible for* ***more than one person to win*** *by staying healthy.*" [P759].

R6: Ensure fair comparison and competition

Our participants highlighted the need to ensure fair competition and comparison by avoiding matching and comparing dissimilar people and by comparing only realistic behavior measures. A particular participant who was 65 years narrated how it would be unfair to compare her physical activity level with that of a youth in his 20's. This, she said, would obviously discourage her since it is an unfair comparison. Similarly, another participant highlighted why Blood Alcohol Contents (BAC) may not be a reliable behavior measure for comparison since it depends on many characteristics that may vary from one individual to another as can be seen in the comment "*Amount of alcohol doesn't affect BAC in the same way for everyone. Some people would have* ***an inherent advantage*** *in this game, and that would remove the little motivation I'd feel for winning 'points'.*" [P928]. Therefore, designers should ensure fair competition and comparison by not **comparing dissimilar people and by measuring and comparing only realistic behavior measures.**

6 Conclusion

In this study, we investigated the mechanism through which socially-oriented persuasive strategies (competition, social comparison, and cooperation) which are widely employed in persuasive designs could influence behavior either positively or negatively by exploring their strengths and weaknesses via a large-scale study of 1,768 participants. As a secondary objective, we validated the persuasiveness of the strategies and showed their comparative persuasiveness with respect to their ability to motivate healthy behavior change. Our findings show that in general, the socially-oriented strategies are perceived as persuasive with the social comparison strategy being the most persuasive of all. The strengths of the strategies include that: *competition* makes one **committed** and gives a sense of **accomplishment**; *comparison* allows for **subtle** and **empowering peer pressure** and provides an opportunity to serve as a **positive role model;** and *cooperation* provides opportunities for **mutual support** and **encouragement.** On the other hand, some major weaknesses shared by the strategies include their **tendency to jeopardize people's privacy and friendship, be counterproductive** by allowing for **negative social influence, creates unnecessary tension, and reduce self-confidence**. Based on our findings, we offer design recommendations for implementing socially-oriented persuasive strategies in PTs to overcome their weaknesses and amplify their strengths.

Acknowledgments. The authors would like to thank the NSERC Banting Fellowship and the University of Waterloo for supporting Dr. Rita Orji's research. We also thank the anonymous reviewers and study participants for insightful comments. The studies complied with the research ethics guidelines provided by the University.

References

1. Cialdini, R.B.: The science of persuasion. Sci. Am. Mind **284**, 76–84 (2004)
2. Kaptein, et al.: Adaptive persuasive systems. ACM Trans. Interact. Intell. Syst. **2**, 1–25 (2012)
3. Orji, R., Vassileva, J., Mandryk, R.L.: LunchTime: a slow-casual game for long-term dietary behavior change. Pers. Ubiquit. Comput. **17**, 1211–1221 (2012). doi:10.1007/s00779-012-0590-6
4. Orji, R.: Design for behaviour change: a model-driven approach for tailoring persuasive technologies. Ph.D. thesis, pp. 1–257 (2014)
5. Fogg, B.J.: Persuasive technology: using computers to change what we think and do (2003)
6. Oinas-Kukkonen, H., Harjumaa, M.: A systematic framework for designing and evaluating persuasive systems. In: Oinas-Kukkonen, H., Hasle, P., Harjumaa, M., Segerståhl, K., Øhrstrøm, P. (eds.) PERSUASIVE 2008. LNCS, vol. 5033, pp. 164–176. Springer, Heidelberg (2008). doi:10.1007/978-3-540-68504-3_15
7. Bell, M., Chalmers, M., Barkhuus, L. et al.: Interweaving mobile games with everyday life. In: Proceedings of Conference on Human Factors Computing Systems, pp. 417–426. ACM (2006)
8. Stibe, A., Cugelman, B.: Persuasive backfiring: when behavior change interventions trigger unintended negative outcomes. In: International Conference on Persuasive Technology, pp. 65–77 (2016)

9. Orji, R., Vassileva, J., Mandryk, R.L.: Modeling the efficacy of persuasive strategies for different gamer types in serious games for health. User Model User-adapt. Interact. **24**, 453–498 (2014). doi:10.1007/s11257-014-9149-8
10. Lehto, T., Oinas-Kukkonen, H.: Persuasive features in web-based alcohol and smoking interventions: a systematic review of the literature. J. Med. Internet Res. **13**, e46 (2011)
11. Orji, R., Moffatt, K.: Persuasive technology for health and wellness: State-of-the-art and emerging trends. Health Inf. J. **1**, 7–9 (2016). doi:10.1177/1460458216650979
12. Orji, R.: Exploring the persuasiveness of behavior change support strategies and possible gender differences. In: CEUR Workshop Proceedings, pp. 41–57 (2014)
13. Orji, R.: Persuasion and Culture: Individualism–Collectivism and Susceptibility to Influence Strategies. In: Workshop on Personalization in Persuasive Technology (2016)
14. Busch, et al.: More than sex: the role of femininity and masculinity in the design of personalized persuasive games. In: International Conference on Persuasive Technology, pp. 219–229 (2016)
15. Orji, R., Mandryk, R.L., Vassileva, J.: Gender, age, and responsiveness to Cialdini's persuasion strategies. In: Persuasive Technology, pp. 147–159 (2015)
16. Jander, et al.: A web-based computer-tailored game to reduce binge drinking among 16 to 18 year old Dutch adolescents: development and study protocol. Public Health **14**, 1054 (2014)
17. Jander, et al.: Effects of a web-based computer-tailored game to reduce binge drinking among Dutch adolescents: a cluster randomized controlled trial. J. Med. Internet Res. **18**, e29 (2016). doi:10.2196/jmir.4708
18. Klisch, Y., et al.: Teaching the biological consequences of alcohol abuse through an online game: impacts among secondary students. CBE Life Sci. Educ. **11**, 94–102 (2012)
19. Klisch, et al.: The impact of a science education game on students' learning and perception of inhalants as body pollutants. J. Sci. Educ. Technol. **21**, 295–303 (2012)
20. Khaled, et al.: A qualitative study of culture and persuasion in a smoking cessation game. In: Proceedings of Third International Conference on Persuasive Technology, Human Well-Being, pp. 224–236 (2008)
21. Bang, M., Torstensson, C., Katzeff, C.: The PowerHhouse: a persuasive computer game designed to raise awareness of domestic energy consumption. In: IJsselsteijn, W.A., de Kort, Y.A.W., Midden, C., Eggen, B., Hoven, E. (eds.) PERSUASIVE 2006. LNCS, vol. 3962, pp. 123–132. Springer, Heidelberg (2006). doi:10.1007/11755494_18
22. Grimes, A., Kantroo, V., Grinter, R.E.: Let's play!: mobile health games for adults. In: Proceedings of 12th ACM International Conference on Ubiquitous Computing, pp. 241–250 (2010). doi:10.1145/1864349.1864370
23. Berkovsky, S., Coombe, M., Freyne, J. et al.: Physical activity motivating games: virtual rewards for real activity. In: Proceedings of International Conference on Human Factors Computing Systems, pp. 243–252. ACM (2010)
24. Fujiki, Y., Kazakos, K., Puri, C., et al.: NEAT-o-Games: blending physical activity and fun in the daily routine. Comput. Entertaintment **6**, 1–22 (2008). doi:10.1145/1371216.1371224
25. Busch, M., Mattheiss, E., Hochleitner, W. et al.: Using player type models for personalized game design - an empirical investigation. Interact. Des. Architect. J. IxD&A N. 28, 145–163 (2016)
26. Orji, R., Nacke, L.E., DiMarco, C.: Towards personality-driven persuasive health games and gamified systems. In: Proceedings of SIGCHI Conference on Human Factors Computing System (2017)
27. Truong, K.N., Hayes, G.R., Abowd, G.D.: Storyboarding: an empirical determination of best practices and effective guidelines. In: Proceedings of 6th Conference on Designing Interactive Systems, pp. 12–21 (2006)

28. Lelie, C.: The value of storyboards in the product design process. Pers. Ubiquit. Comput. **10**, 159–162 (2005). doi:10.1007/s00779-005-0026-7
29. Braun, V., Clarke, V.: Using thematic analysis in psychology. Qual. Res. Psychol. **3**, 77–101 (2006)
30. Toscos, T., Faber, A., An, S., Gandhi, M.P.: Chick clique: persuasive technology to motivate teenage girls to exercise. In: CHI 2006 Extended Abstracts on Human Factors in Computing Systems, pp. 1873–1878 (2006)
31. Grimes, et al.: Toward technologies that support family reflections on health. In: Proceedings of 2009 International Conference on Supporting Group Work, pp. 311–320 (2009)
32. Kohn, A.: No Contest: The Case Against Competition. Houghton Mifflin, Boston (2006)

Strategies and Design Principles to Minimize Negative Side-Effects of Digital Motivation on Teamwork

Abdullah Algashami(✉), Alimohammad Shahri, John McAlaney, Jacqui Taylor, Keith Phalp, and Raian Ali

Faculty of Science and Technology, Bournemouth University, Poole, UK
{aalgashami,ashahri,jmcalaney,jtaylor,kphalp,rali}@bournemouth.ac.uk

Abstract. Digital Motivation in business refers to the use of technology in order to facilitate a change of attitude, perception and behaviour with regards to adopting policies, achieving goals and executing tasks. It is a broad term to indicate existing and emerging paradigms such as Gamification, Persuasive Technology, Serious Games and Entertainment Computing. Our previous research indicated risks when applying Digital Motivation. One of these main risks is the impact it can have on the interpersonal relationships between colleagues and their individual and collective performance. It may lead to a feeling of unfairness and trigger negative group processes (such as social loafing and unofficial clustering) and adverse work ethics. In this paper, we propose a set of strategies to minimize such risks and then consolidate these strategies through an empirical study involving managers, practitioners and users. The strategies are then analysed for their goal, stage and purpose of use to add further guidance. The strategies and their classification are meant to inform developers and management on how to design, set-up and introduce Digital Motivation to a business environment, maximize its efficiency and minimize its side-effects on teamwork.

Keywords: Digital motivation · Persuasive technology · Gamification · Motivation engineering

1 Introduction

Digital Motivation (hereafter DM) is on the rise and there exist already various established domains which characterize it including Persuasive Technology [1], Gamification [2], Games with Purpose [3] and Entertainment Computing [4]. Central to DM is the use of technology (including games and social computing), to prevent, change, maintain or enhance certain behaviours and attitudes in relation to certain policies, goals, tasks, and social inter-relations. The advances in technology, including mobile and sensing technology, and the increased familiarity of the public with advanced features of Web 2.0, games and social computing have made these techniques possible and acceptable. DM has been used in various application areas including health [5, 6], sport [7], sales [8] and education [9, 10].

P.W. de Vries et al. (Eds.): PERSUASIVE 2017, LNCS 10171, pp. 267–278, 2017.
DOI: 10.1007/978-3-319-55134-0_21

There exist different methods and principles of developing DM. Fogg [11] proposes eight steps of developing and introducing Persuasive Technology. The emphasis in these steps is on the choice of behaviour, the audience and finally, understanding the obstacles. Nicholson [12] proposes a theoretical framework for a 'meaningful gamification' intended to avoid the risk of losing intrinsic motivation when gamifying tasks. Other principles are either focused on single property of DM or coupled with certain application areas. For example, Consolvo et al. [13] focus on goal-setting and explore ways to elicit goals and specify their time frames. Gram-Hansen [14] proposes an approach based on participatory design and constructive ethics to achieve a persuasive design.

We advocate that some DM techniques and methods have potential side effects on teamwork. In [15] we concluded that gamification solutions can cause social and mental well-being problems in the work place and that there is a need to consider ethics and values when adopting such solutions. Nicholson [16] argues that gamification can be seen as exploitation if implemented in certain ways that drive people to do more than their job description would imply. Timmer et al. [17] focus their study on the importance of user-informed consent prior to the use of persuasion. This human aspect in relation to the potential side-effect suggests that we need to take it as an initial requirement when planning and engineering DM. However, while the focus of existing literature is on ways to develop successful DM, there exists little emphasis on how to engineer counter-measures to avoid these side-effects.

Issues that may arise as a result of introducing DM to the work space include reduced collegiality, negative group relations and low group cohesion. For example, introducing a leader-board to a collaborative workplace which is based on measuring individual performance could lead to less collaboration and introduce questions about the measurement of individuals' performances. Social recognition elements, e.g. badges and status, given to groups based on their collective performance may introduce a risk of social loafing [18] and a pressure for social compensation [19].

In our previous work [20] a reference model has been explored and developed, putting together the properties of motives, environment and users which are involved when taking decisions during the development and deployment of DM solutions. In [21], we developed various personas and argued that individual differences need to be catered for DM design and customization to maximize its acceptance and efficiency and also avoid the side-effects discussed in [15]. However, the design principles and tools for preventive and corrective mechanisms to deal with these potential issues of DM have not yet been explored.

In this paper, we build on our previous results presented in [15, 20, 21] and identify strategies that DM development and management can adopt to introduce DM into the workplace with the aim of minimizing the risks it may introduce into teamwork. As a method, we further analyse the results of our previous works and review the literature to come up with an initial set of strategies. This set is then discussed and elaborated in interviews with managers, practitioners and users. A focus group to confirm and categorize the results was then conducted. The results of this paper will be beneficial for developers, management and occupational psychologists to avoid negative experiences DM can facilitate if introduced without careful considerations.

2 Motivating Scenario

We will present two cases to illustrate how an ad-hoc introduction of DM could affect the efficiency of the teamwork environment. In the first case, we highlight workplace intimidation. In the IT department of a company, the front-end development team is responsible for ensuring that the user experience (UX) is kept at a satisfactory level, and also responsible for updating the user interface (UI) when necessary to address customers' requirements. Collaboration of the team members is crucial to the success of the department's work and failure to maintain appropriate communication and collaboration might affect the quality of the final artefact. The UI has great value for the company as they believe this is the client view of the company. Therefore, the company wishes to decrease the chance of failure in the design of the UI as much as possible. Thus, in order to encourage collaboration, the organisation using status as a DM technique to motivate the front-end development team based on its overall performance. For communication and tracking purposes, team members have access to individuals' work performance. This could help them to schedule plans and make changes more easily if needed. However, since team members have access to each other's performance details, there is a risk of negative effect in the group. Team members with better performance may feel closer to each other causing groups to form, and this may pave the way for workplace intimidation, where some high-performance employees bully lower-performance colleagues in the team. This illustrates how using DM might create tension or conflict amongst workers and the need to have strategies to resolve such negative effect.

The second case involves a situation where sabotage could happen within teamwork in the workplace. Two teams are working in an IT department creating a web application. John, Alice and Bob are team A and are working on the design of the UI while Mary, James and Matt are team B and responsible for the back-end development. The manager asks team A to update the design of the UI in a specific time-frame. Bob calls in sick and does not attend work for two weeks. The manager delegates his work to Alice from team B. The department, which uses a leader-board, as a DM technique, to encourage both teams to finish their tasks on time, decides to give points to the team who can finish the task on-time. At the end, the team with most points will receive a reward. Since Alice is from team B and individual efforts are not acknowledged in this setting of DM, there is a risk that she intentionally hinders the job thus causing a delay to enable her team win the reward.

3 Method and Research Settings

This research builds on our previous studies conducted in [15, 20, 21], which include interviews and open-ended surveys with experts, managers, and end-users in the domain of DM. This resulted in identification of various situations where ad-hoc implementation of DM could lead to the creation of negative effect and issues amongst employees. Our analysis resulted in six representative scenarios in which an ad-hoc implementation of DM could create negative impact and issues amongst team members. In order to discover the resolution strategies that could help to resolve the negative effect in such scenarios, a four-stage study shown in Table 1 was designed for this purpose.

Table 1. Research method stages

1st stage	2nd stage	3rd stage	4th stage
Previous studies	Analysis	Interviews	Focus group
The work done in: – DM obstacles and ethical issues identification [15] – DM persona aspect [21] – DM modelling and structuring aspect [20]	– The authors generated six scenarios based on stage 1. – The authors defined resolution strategies based on: – Group dynamics – Group cohesion – Social identity – Conflict theory – Change management – Occupational psychology – Prosocial behaviour – Social norms	The authors refine the strategies through interviews: – Two experts in computing and social informatics – Four experts in psychology and cyber-psychology – Two practitioners – Two managers	The authors refine the results from 1st, 2nd and 3rd stage via a focus group with participants of a multi-disciplinary background (see Table 2)

In the first stage, further analysis of the results from the previous studies was carried out. It was informed by the literature using the main theories in group dynamics [22], group cohesion and development [23], social identity theory [24], group conflict theory [25], change management [26], occupational psychology [27] and prosocial behaviour [28]. Various situations were also investigated where ad-hoc implementations of DM could create negative effect amongst the social actors within the workplace which resulted in six scenarios according to the main theories in conflict resolution. This helped us to generate around seventeen strategies which are intended to help to resolve negative effect in teamwork.

In the next stage of the study, and in order to refine these strategies, we conducted interviews with ten interviewees, including four experts in the domain of psychology; two in computing and social informatics and four from related workplaces of whom two were practitioners and two were managers where DM techniques have been implemented. This helped us to elaborate on our initial set and devise a final set of negative effect management strategies. All of the interviews were recorded and transcribed. The interviews followed a semi-structured style in order to refine with each participant the most appropriate strategies that could help reduce the likelihood of the negative effect, alleviate the adverse effect or resolve it for each scenario. This resulted in 22 strategies which could help in managing teamwork negative impact in relation to DM.

In the final stage, the strategies were classified using a focus group with seven participants with relevant expertise. The participants were familiar with DM and came from diverse domains (see Table 2). Participants were familiarised with the context by means of presentation before the session, the six scenarios were provided as a hard copy, a facilitator explained the scenarios and answered questions during the session, and separate sheets of paper were provided to write down participants' ideas. The session was held in two parts in order to qualify the final results of these strategies. In the first part, the participants were given the scenarios and asked to brainstorm and suggest ideas, strategies and concepts which could help to manage the negative effect in each one. In the second part, they were given a list of possible resolution strategies and the description for each scenario, and then they were asked to provide their perception on these strategies and how they could help to resolve the negative effect on teamwork in relation to DM.

Table 2. 4th stage focus group participants

Participants	Research background
F	Facilitator (one of the authors)
P1, P2	Requirements engineering, computers in human behaviour and cyber psychology
P3, P4	Human factors and user testing
P5	Usability and human computer interaction
P6	Machine intelligence and user modelling
P7	Business management

4 Results

In this section, we report and discuss the results of the study which revolve around two main aspects. In Sect. 4.1 we will describe the first aspect which concerns our proposed strategies to reduce DM side-effects on team work. In Sect. 4.2 we address the second aspect which is about categorising the strategy according to various development and management styles and phases.

4.1 Strategies for Managing DM Negative Effects on Teamwork

- *Commitment:* this strategy is based on the members' agreement and adoption of the choices and actions characterizing how DM is going to operate. This could be achieved by running a negotiation session to discuss views and exchange offers. This would then lead to maximized ownership of DM solution and accepting it.
- *Common ground rules:* this strategy is based on deriving and enforcing rules that articulate the set of acceptable behaviours in relation to DM, in order to facilitate the development of the use of DM within the organisation. Examples of such rules include showing respect for others, appropriate ways in which to express oneself, allowing everyone to 'have a say', openness to different views and confidentiality. This would help to manage and govern the work environment and also reduce the chance of negative effect in the workplace.
- *Facilitator:* this strategy plays an important role in facilitating the design sessions of DM, including running negotiation sessions, helping people to understand the

common objectives of a group and assisting groups to set the common rules of conduct in an effective work environment supported by DM.

- *Anonymity:* the core idea of this strategy is to give opinions or ratings of colleagues or managers in an anonymous way. This could help make the work collaboration environment open. For example, this technique could help with the second case described in Sect. 2 to rate the employees' performances and prevent them from sabotaging the teamwork performances.
- *Voting:* this strategy helps to reach a decision in a facilitated session. When multiple choices are available amongst DM stakeholders in the design process, the facilitator could use a voting technique to try to meet the concerns of team members in a democratic and more acceptable style.
- *Norms:* this technique is based on having a clear understanding of what the organisational culture is, e.g. normal social behaviours. This could help to reduce the likelihood of negative effects within rewarding system environments. For example, an organisation may have a norm of senior managers publicly acknowledging successes of team members in monthly team meetings. A new DM based reward system such as a leader board may aim to serve the same basic function of highlighting success within the team, but the departure from the previously established norm of face to face social approval may cause resentment in team members.
- *Transparency:* this strategy means allowing everyone to see other's performances in DM system. Although some participants involved in this study agreed on the importance of this strategy to resolve DM negative effects, others mentioned that "it should be designed carefully to avoid clustering high performances workers and those of the lower performances".
- *Rotations sensitivity:* this strategy is based on allocating people randomly within DM system so that cliques and rivalries are not created. This could help to eliminate a negative effect caused by workers only supporting their close colleagues to win any reward.
- *Get everyone involved:* this strategy encourage people in different roles to become involved in a discussion to decide behaviours and penalties for their DM system.
- *Story telling:* the core idea of this strategy is to identify negative effect by asking people to present a situation in a story. A manager involved in our study noticed that "when we have a conflict in our company I sometimes go out for walk with some of my staff and ask them to tell the situation in a story, this can help to determine the source of the conflict".
- *Round robin*: this strategy aims to pass the discussion between workers one by one to ensure everyone gives their ideas individually. This would help to ensure the equality amongst workers involved in DM system and ensure everyone gets a chance to speak.
- *External party:* this strategy proposes to use an external authority or expert to check workers' performances and to resolve negative effects which might arise in relation to DM.
- *Non-contentious bargaining:* this strategy encourages team members to control their emotions in a professional way, such as "counting to ten" before taking an action, writing down their concerns carefully in an email or letter with a calm manner [29].

This strategy can be used to reduce negative effects of DM without causing additional side-effects. For example, a group leader may only acknowledge top performing members of a group, via badges and status, despite the remaining group members performing their roles adequately. By expressing their concerns in calm, reasoned (i.e. non-contentious) manner the group members may be able to reach agreement with the group leader on how a DM system can be changed to the mutual benefit of all involved [30].

- *Reward for helping others:* this strategy is related to prosocial theory, in which users can be rewarded for supporting others. This could be used in any DM to encourage collaborative teamwork such as, encouraging workers who always appear in the top list of a leader-board by rewarding them when helping their lower performances colleagues to appear in the leader-board.
- *Acknowledgement of individual efforts:* in some DM situations negative effect on teamwork might arise when individual efforts are not equal. This could arise when some workers rely on others to finish a task and is based on the concept of social loafing, so this strategy could help to inspire individuals to engage in group tasks to completion.
- *Observation strategies:* various strategies using different techniques to help to observe DM teamwork environments include:
  - *Auditing:* means checking individual performances, e.g. giving a quantifiable task and assuming people will also respect quality. Although the *auditing* technique can help to resolve negative effect on teamwork, one practitioner said "it should be used in a very careful style to prevent introducing another conflict".
  - *Random monitoring:* the idea of this technique is to keep workers ready all of the time as their performances might be monitored at any time.
  - *Peer-rating:* this technique means that colleagues can rate each other's efforts and might be checked at any time to avoid a biased evaluation.
  - *Member checking:* this strategy utilises a sample member in order to analyse the eventual DM result after finishing the task.
  - *Managerial level monitoring:* in this strategy managers take the responsibility to check workers' performances in DM workplace.
  - *Self-assessment:* users assess their own performances, and this might be checked by managers at any time.
  - *Regular meetings:* involving teamwork members in regular meetings, e.g. weekly, monthly or annually would help managers to remain updated with the current use of DM system.

4.2 The Categorization of Strategies for DM Management in Teamwork

From the analysis of the interviews and the focus group, it was possible to extract the need for different ways to represent these strategies to resolve negative effect on teamwork within DM workplaces. As a result, three main categorises for better representation of these strategies. These are resolution strategies development aspect, resolution strategies enactment aspect and resolution strategies usage aspect. The concept map for each aspect is represented in Figs. 1, 2 and 3 to illustrates the main characteristics of these aspects and provide examples of strategies which could help to manage the negative impact in teamwork related to DM.

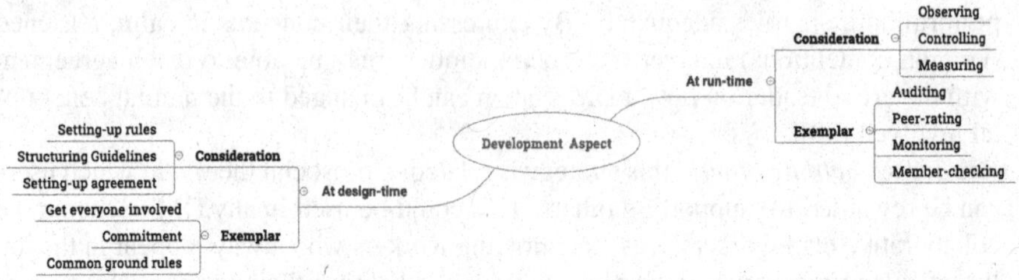

Fig. 1. Concept map of resolution strategies from development stage perspective

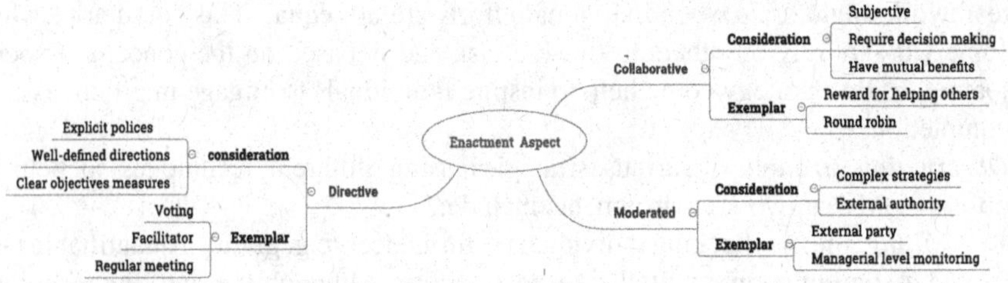

Fig. 2. Concept map for resolution strategies from enactment perspective

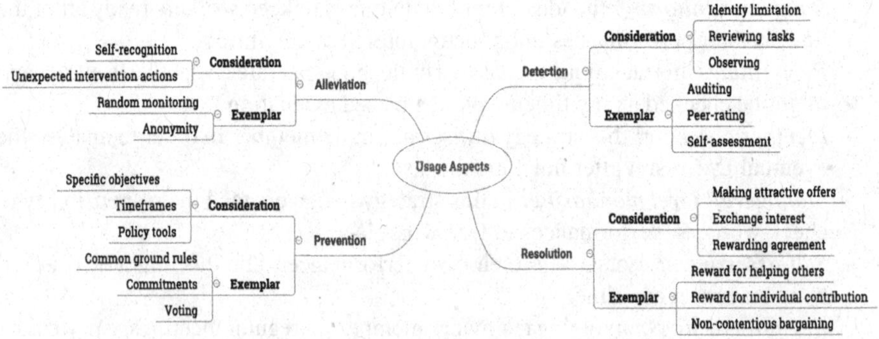

Fig. 3. Concept map for resolution strategies from usage perspective

Resolution Strategies: Development Stage Aspect

The analysis of our empirical studies shows that there are some strategies applicable at the design time of DM, whilst others might be used in real-time when DM is being used in the workplace and finally some may be useful for both.

- *At the design stage:* it seems that strategies would help in setting up agreements, rules and structuring the general guidelines of DM system can be fitted at this stage. For example, some practitioners and psychologists mentioned that we should *get*

everyone involved in a discussion making session at design stage and make them *committed* to the design of their DM. However, others suggested that having a sample of employees could be a help since in large organisations, where the number of employees is very high and it is impossible to engage everyone in the design stage. Moreover, the majority of interviewees agreed on having *common ground rules* at the early stages and asking users to agree on DM rules, which could help reduce the negative effect which might occur during the actual work.
- *At the run-time:* the analysis results suggest that strategies with characteristics like observing and controlling the environments would fit more into this stage. For instance, strategies such as *auditing*, *random monitoring*, *peer-rating* and *member-checking* could help in teamwork to observe the quality of the work and to control and resolve negative effects when they happen.

Resolution Strategies: Enactment Aspects

In order to apply these strategies within a teamwork places there could be different approaches giving more effective implementation results. These have been identified as collaborative, moderated and directive, as explained below. However, there is a difficulty in the suitability of these strategies as some of them might apply for more than one approach and others cannot be easily decided. For example, a strategy like Peer-rating can fit as collaborative amongst workers; however it might have a valuable result when it's moderated by managers or project leaders.

- *Collaborative approach:* our studies suggest that strategies which have mutual benefits e.g. *reward for helping others* is better to be decided collaboratively. Moreover, strategies which are subjective and require decision making would fit in such an approach. For instant, *round robin* strategy where workers can engage in a discussion equally to decide to what extent they should cooperate with each other in the task and how DM could be used to measure their performances would be preferable to be implemented in a collaborative way.
- *Moderated approach*: this approach would help with strategies which are complex and where workers are not able to steer the process to reach the consensus. External authorities or experts work collectively with managers to set up the strategy and moderate the interaction. For example, in the *external party* strategy managers work together with external consultants to decide the effective way to manage the strategy to resolve the negative effect within DM.
- *Directive approach:* the nature of some strategies which are based on explicit polices, with well-defined directions and clear objective measures can be operated effectively. For example, some participants suggested that managers can play the key roles in resolving negative effect within some scenarios through leading the observing or auditing process.

Resolution Strategies: Usage Aspect

The usage aspect is related to the different ways these strategies can be used to manage negative effect on teamwork. As a general principle, the participants used this labelling to categorise characteristics of the effect of these strategies. They assume that some of them would help to alleviate the negative effect when it is impossible to resolve it. Moreover, these strategies are not mutually exclusive strategies they might be used for more than one aspect e.g. detection and resolution at the same time whether at design stage of DM or in real-time at a workplace.

- *Detection strategies:* it seems from the characteristics of some strategies that they would help more to identify where the limitations or weaknesses are in DM more than resolving negative effect. For example, *the observation strategies* have the checking and inspection features which could help more to identify where the negative effect originate.
- *Resolution strategies:* the main mission of these strategies is to help to resolve negative effect on teamwork. Strategies which allow making attractive offers, the exchange of interests and rewarding agreement would fit more as resolution strategies e.g. *rewards for helping others* and *rewards for individual contributions.* Applying such strategies to the second case in Sect. 2 above would help to prevent Alice from sabotaging the teamwork in team *A* and will encourage her to involve in the teamwork.
- *Alleviation strategies:* in some cases, the negative effect cannot be resolved. Thus these types of strategies which support self-recognition and unexpected intervention actions could help to reduce the negative effect. For example, some experts commented that strategies such as *random monitoring* or *anonymity* cannot help to resolve conflict, but it might assist to reduce the negative effect.
- *Prevention strategies:* strategies based on specifying objectives, timeframes and policy tools would play important roles in reducing the likelihood of negative effect from happening in DM work environment. For example, strategies used at the early stages of DM, such as *having common ground rules* and asking users for *commitments* could help to reduce negative effect arising in workplace.

As an example in teamwork negative effect scenario one in Sect. 2, we may apply strategies like *common ground rules* and *commitment* at the design of DM stage and ask the front-end development team to commit on rules such as, everyone should respect all the team members and act with them in similar manner. At the run-time stage we could apply one of the observation strategies e.g. *auditing* or *managerial level monitoring* to check whether workers respect that commitment. In addition, if we detect negative effect happening in this group we might use a strategy like *reward for helping others* to encourage high performance workers to support lower performance which would help to resolve the negative impact in this scenario.

5 Conclusion and Future Work

In spite of the increasing use and success of DM tools to persuade users to be more motivated and engaged, further studies should focus on how to resolve negative issues and side-effects related to its use in the workplace. Amongst various problems which could happen in workplaces, such as a lowering of quality or creation of tension, this paper focused on strategies to manage negative effect on teamwork as one of the significant risks of introducing DM elements into the work environment. We explored the resolution strategies from both psychological and management perspectives, which could help to introduce DM into the work environment in a healthier and coherent way. Our study led to 22 teamwork negative effect management strategies which could help to minimize workplaces negative impact related to DM. We also categorized these strategies into three main aspects based on their goal, stage and purpose of use.

In future work we will further investigate each of the three stages from various stakeholders' perspectives. In particular, we plan to study the use of participatory design [31] in order to engage team members in the development of DM itself as it can incorporate a wide range of the strategies and be by itself a powerful mechanisms for an agreeable and effective DM. We also plan to use negotiation theory as part of the construction of DM solutions so that the rewards can be agreed for tasks involving different stakeholders. A further validation of the suitability and constraints on the proposed strategies will be achieved via practical case study.

Acknowledgment. The research was supported by Bournemouth University through the PGR Development Fund and the FP7 Marie Curie CIG grant (the SOCIAD project). We also thank the participants in our studies for their valuable contributions.

References

1. Fogg, B.J.: Persuasive Technology: Using Computers to Change What We Think and Do (Interactive Technologies) (2002)
2. Deterding, S., Dixon, D., Khaled, R., Nacke, L.: From game design elements to gamefulness - defining "gamification". In: The 15th International Academic MindTrek Conference Envisioning Future Media Environments, Tampere, September (2011)
3. von Ahn, L.: Games with a purpose. IEEE Comput. **39**, 92–94 (2006)
4. Magerkurth, C., Cheok, A.D., Mandryk, R.L., Nilsen, T.: Pervasive games - bringing computer entertainment back to the real world. Comput. Entertainment **3**, 4 (2005)
5. King, D., Greaves, F., Exeter, C., Darzi, A.: "Gamification": influencing health behaviours with games. J. R. Soc. Med. **106**, 76–78 (2013)
6. Spanakis, E.G., Santana, S., Ben-David, B.: Persuasive technology for healthy aging and wellbeing. In: 2014 EAI 4th International Conference on Wireless Mobile Communication and Healthcare (Mobihealth), p. 23. IEEE (2014)
7. Lacroix, J., Saini, P., Goris, A.: Understanding user cognitions to guide the tailoring of persuasive technology-based physical activity interventions. In: 4th International Conference on Persuasive Technology, vol. 350 (2009)
8. Herzig, P., Ameling, M., Schill, A.: A generic platform for enterprise gamification. In: Presented at the 2012 Joint Working IEEE/IFIP Conference on Software Architecture (WICSA) and European Conference on Software Architecture (ECSA)
9. Beenen, G., Ling, K.S., Wang, X., Chang, K., Frankowski, D., Resnick, P., Kraut, R.E.: Using social psychology to motivate contributions to online communities. In: CSCW, pp. 212–221 (2004)
10. Simões, J., Redondo, R.D., Vilas, A.F.: A social gamification framework for a K-6 learning platform. Comput. Hum. Behav. **29**, 345–353 (2013)
11. Fogg, B.J.: Creating persuasive technologies - an eight-step design process. Persuasive **350**, 1 (2009)

12. Nicholson, S.: A user-centered theoretical framework for meaningful gamification. In: Games + Learning + Society (2012)
13. Consolvo, S., Klasnja, P.V., McDonald, D.W., Landay, J.A.: Goal-setting considerations for persuasive technologies that encourage physical activity. Persuasive **350**, 1 (2009)
14. Gram-Hansen, S.B.: The EDIE method – towards an approach to collaboration-based persuasive design. In: Meschtscherjakov, A., Ruyter, B., Fuchsberger, V., Murer, M., Tscheligi, M. (eds.) PERSUASIVE 2016. LNCS, vol. 9638, pp. 53–64. Springer, Heidelberg (2016). doi:10.1007/978-3-319-31510-2_5
15. Shahri, A., Hosseini, M., Phalp, K., Taylor, J., Ali, R.: Towards a code of ethics for gamification at enterprise. PoEM **197**, 235–245 (2014)
16. Nicholson, S.: A user-centered theoretical framework for meaningful gamification. In: Games + Learning + Society (2012)
17. Timmer, J., Kool, L., Est, R.: Ethical challenges in emerging applications of persuasive technology. In: MacTavish, T., Basapur, S. (eds.) PERSUASIVE 2015. LNCS, vol. 9072, pp. 196–201. Springer, Heidelberg (2015). doi:10.1007/978-3-319-20306-5_18
18. Chidambaram, L., Tung, L.L.: Is out of sight, out of mind? an empirical study of social loafing in technology-supported groups. Inf. Syst. Res. **16**, 149–168 (2005)
19. Bajdor, P., Dragolea, L.: The gamification as a tool to improve risk management in the enterprise. Ann. Univ. Apulensis: Ser. Oeconomica **13**(2), 574 (2011)
20. Shahri, A., Hosseini, M., Phalp, K., Taylor, J., Ali, R.: Exploring and conceptualising software-based motivation within enterprise. PoEM **267**, 241–256 (2016)
21. Shahri, A., Hosseini, M., Almaliki, M., Phalp, K., Taylor, J., Ali, R.: Engineering software-based motivation - a persona-based approach. In: RCIS, 1–12 August 2016
22. Forsyth, D.R.: An Introduction to Group Dynamics. Thomson Brooks/Cole, Monterey (1983)
23. Tuckman, B.W., Jensen, M.A.C.: Stages of small-group development revisited. Group Org. Manag. **2**, 419–427 (1977)
24. Ellemers, N., De Gilder, D., Haslam, S.A.: Motivating individuals and groups at work: a social identity perspective on leadership and group performance. Acad. Manag. Rev. **29**, 459–478 (2004)
25. Forsyth, D.: Group Dynamics. Cengage Learning, Belmont (2009)
26. Hayes, J.: The Theory and Practice of Change Management. Palgrave Macmillan, Basingstoke (2014)
27. Ashforth, B.E., Mael, F.: Social identity theory and the organization. Acad. Manag. Rev. **14**, 20–39 (1989)
28. Denham, S.A.: Social cognition, prosocial behavior, and emotion in preschoolers: contextual validation. Child Dev. **57**, 194–202 (1986)
29. McGillicuddy, N.B., Pruitt, D.G., Syna, H.: Perceptions of firmness and strength in negotiation. Pers. Soc. Psychol. Bull. **10**, 402–409 (1984)
30. Forgas, J.P.: On feeling good and getting your way: mood effects on negotiator cognition and bargaining strategies. J. Pers. Soc. Psychol. **74**, 565–577 (1998)
31. Kensing, F., Blomberg, J.: Participatory design: issues and concerns. Computer Support. Coop. Work (CSCW) **7**, 167–185 (1998)

Investigation of Social Predictors of Competitive Behavior in Persuasive Technology

Kiemute Oyibo(✉) and Julita Vassileva

University of Saskatchewan, Saskatoon, Canada
kiemute.oyibo@usask.ca, jiv@cs.usask.ca

Abstract. Research has shown that Competition is one of the most powerful persuasive strategies to intrinsically motivate users in a social context towards performing a target behavior. However, in persuasive technology research, studies showing the predictors of the "persuasiveness of Competition" as a motivational strategy are scarce. Consequently, based on a sample size of 213 Canadians, we tested a model using three socially influential strategies (*Social Learning*, *Social Comparison* and *Reward*) as predictors of *Competition*. Our model accounts for 42% of the variation in *Competition* and reveals that *Reward* is the strongest predictor of *Competition*, followed by *Social Comparison,* but *Social Learning* is not a predictor. Moreover, it reveals that *Social Comparison* mediates the influence of *Reward* on *Social Learning* and *Competition*. Our findings provide designers of persuasive applications with insight into the possibility of implementing Reward, Social Comparison and Competition as effective co-strategies for stimulating user engagement in gamified applications.

Keywords: Persuasive strategies · Gamification · Social influence · Social comparison · Social learning · Reward · Competition · Intrinsic motivation · Path model

1 Introduction

Intrinsic motivation has been identified as an important factor required for behavior change in society at large and persuasive technology in particular [4, 7], especially in the area of gamification, where game elements are employed in a non-game context [11]. It is defined as "*the doing of an activity for its inherent satisfactions rather than for some separable consequence*" (p. 56) [19], for example, performing an activity because it is interesting and enjoyable. As cited in Fogg et al. [6], Malone and Lepper [14] proposed seven basic intrinsic motivators—Challenge, Curiosity, Fantasy, Control, Competition, Cooperation and Recognition—which can be leveraged in educational video games to facilitate learning and assimilation [19]. These intrinsic motivators were adapted by Fogg et al. [6] for microsuasion in video-game-based persuasive technology. They also proposed the inclusion of social influence principles into the video games for microsuasion in order to make them more effective. Similarly, Sundar et al. [28] recommended that persuasive technology should be designed with the intention to increase intrinsic motivation and user participation [26]. In recent years, in gamified persuasive

P.W. de Vries et al. (Eds.): PERSUASIVE 2017, LNCS 10171, pp. 279–291, 2017.
DOI: 10.1007/978-3-319-55134-0_22

applications, group-based intrinsic motivation, such as Competition, Cooperation and Recognition [7], has been put forward to foster more active user participation [26]. However, according to Ryan and Deci [19], "*despite the observable evidence that humans are liberally endowed with intrinsic motivational tendencies, this propensity appears to be expressed only under specifiable conditions*" (p. 58). This makes it important in the persuasive technology domain for researchers to investigate those specific conditions that influence this special type of motivation in order to achieve more active user participation [23]. The Self-Determination Theory [19], for example, is concerned with the social and environmental factors that can facilitate or hinder intrinsic motivation, which, it assumes, is catalyzed when an individual is in certain conditions.

However, in persuasive technology research, limited empirical evidence exists on the social predictors of the interaction-based intrinsic motivators, which can lead to increased user participation [26] and, ultimately, achievement of target behaviors [22]. Consequently, we set out to investigate the persuasiveness of Competition, which is one of the most engaging group-based intrinsic motivators in gamified persuasive technology [7, 11]. We chose this construct because, apart from the central role it plays in human endeavors, such as education, sports, commerce, etc. [3, 8, 27, 29], research in persuasive technology has shown that Competition is a powerful strategy, which can be used to motivate people to act and achieve behavior change [7], given that most people in society desire to win and gain social recognition [3]. As such, using four validated constructs in Busch et al.'s [2] Persuadability Inventory, namely *Social Learning*, *Social Comparison, Competition* and *Reward* and a sample of 213 Canadian participants, we attempt to put forward a model that accounts for the variance of *Competition* as an intrinsic motivator of target behaviors in persuasive technology. Our model accounts for 42% of the variance in *Competition*, with *Reward* being the strongest predictor, followed by *Social Comparison*. *Social Learning* turns out not to be a predictor of *Competition* among our studied population. In addition, we found that: (1) *Reward* influences *Social Learning* for the male group, but not for the female group, when *Social Comparison* is controlled for in the model; and (2) *Social Comparison* fully and partially mediates the influence of *Reward* on *Social Learning* and *Competition* respectively. In summary, our findings provide designers of persuasive technologies with insight into how to predict the persuasiveness of Competition as a socially influential strategy in gamified applications. Specifically, our model reveals that individuals who are more persuadable by Reward and Social Comparison are likely to be more motivated by Competition as a persuasive strategy to foster behavior change.

2 Background

This section provides an overview of Busch et al.'s [2] Persuadability Inventory (PI) constructs (informed by theories in socio-psychology), which are relevant to our study. The social constructs in the inventory refer to their "persuasiveness", i.e., how well they are able to persuade or motivate individuals. Thus, in this paper, we represent all of the socially influential constructs in italics (e.g., *Reward*). However, if we use the same words in their literal sense, they will be represented in regular font (e.g., reward).

Social Comparison. The *Social Comparison* construct is a measure of how well Social Comparison is able to persuade people to perform a target behavior. It is derived from the Social Comparison Theory proposed by Festinger [5]. The theory states that people compare themselves (their opinions and abilities) with other people (the opinions and abilities of others) in order to improve themselves and reduce uncertainties [8], with opinion and abilities interacting to affect human behavior. According to Festinger [5], if people want to improve themselves, they compare themselves with a reference group of superior people (aka upward comparison). On the other hand, if they want to enhance their self-esteem, they compare themselves with an inferior group of people (aka downward comparison). In persuasive technology, Social Comparison can be implemented by providing users with the ability to view and compare their performance with that of others [18]. In empirical studies, it is operationalized as a set of questions. An example question from the PI [2] is "*I like to compare myself with other people.*"

Social Learning. The *Social Learning* construct is a measure of how well Social Learning is able to persuade people to perform a target behavior. It is derived from the Social Learning Theory proposed by Bandura [1]. The theory posits that learning is a cognitive process in which people learn by observing the behaviors of others and their consequences in a social context. According to Bandura [1], it is also a self-regulative process in which learning and regulation of one's behavior occur through the observation of reinforcement and punishment. In persuasive technology, Social Learning is regarded as a persuasive strategy, which is also called Consensus or Social Proof [16, 23]. It is implemented in persuasive applications by informing users that some other users have already performed certain behavior, e.g., bought a certain product [12]. A typical question from the PI [2] is "*I take other people as role models for new behaviors.*"

Reward. The *Reward* construct is a measure of how well Reward is able to persuade people to perform a target behavior. Reward is something which is given to an individual as a result of the accomplishment of a specific task or the achievement of a target behavior. In gamified persuasive technology, it can be implemented as points, badges, etc. A typical question from the PI [2] is "*I put more ambition into something if I know I am going to be rewarded for it.*"

Competition. The *Competition* construct is a measure of how well Competition is able to persuade people to perform a target behavior. Competition is a natural drive (an intrinsic motivation) in humans to outperform one another by adopting certain attitudes or performing certain behaviors [16]. In the persuasive technology domain, people can either compete against one another or against the persuasive system. One way by which Competition (a commonly used persuasive strategy alongside Social Comparison [17]) can be implemented in a gamified persuasive application is the use of leaderboards, which allow users to view and compare their performance (e.g., scores) with the performance of other users of the application [23]. A typical question from the PI [2] is "*It is important to me to be better than other people.*"

3 Related Work

While a number of studies [2, 16, 23, 24] have been carried out on social influence in persuasive technology, few to none have been directed specifically at predicting the persuasiveness of Competition, which is one of the most effective ways of engaging people in performing a target behavior [25]. Busch et al. [2] developed a PI to measure a number of socially influential persuasive strategies. However, they did not investigate the inter-relationships among the constructs that make up the inventory. Stibe [23] proposed a framework for identifying and harnessing the social design principles in feedback-sharing. However, the author focused on predicting perceptions of the effectiveness of sociotechnical systems and user engagement in sharing feedback tweets (and not the persuasiveness of Competition specifically) from Twitter, which may not generalize to other persuasive systems that use other socially influencing media than tweets. Stibe [24] also proposed a similar framework known as Socially Influencing Systems (SIS) framework. However, the framework does not include *Reward* and is not specifically targeted at predicting *Competition* as a persuasive strategy.

4 Method

In this section, we present our research hypotheses, the instruments used to measure the constructs of interest in our proposed model and the demographics of participants.

4.1 Research Hypotheses

Competition has been widely recognized as one of the most powerful ways to intrinsically motivate people in persuasive applications [7]. In this paper, we attempt to address the research question: "*Which socially influential strategies are relevant in the prediction of the persuasiveness of Competition?*" Using the PI, proposed by Busch et al. [2], which was adapted from the Persuasive System Design (PSD) model [16], we formulated six hypotheses (as shown below and in Fig. 1) drawing on knowledge from socio-psychology theories. It is noteworthy that the theories that informed the hypotheses are based on the literal meaning of the constructs. However, our hypothesized relationships are based on the persuasiveness of the constructs. In other words, we assume that if theoretically there is a relationship between two given constructs, then there is possibility of an existence of a corresponding relationship between their persuasiveness.

The first hypothesis (H1: *Reward* positively influences *Competition*) derives from the operant conditioning theory [21], which states that all behaviors are controlled by their consequences. In other words, the performance of certain behaviors is motivated by their rewards or punishments. Based on this theory, we hypothesize that the more persuadable a person is by Reward the more responsive s/he will be to Competition.

H1: *Reward* positively influences *Competition*.
H2: *Social Comparison* positively influences *Competition*.
H3: *Reward* positively influences *Social Comparison*.
H4: *Reward* positively influences *Social Learning*.

H5: *Social Comparison* positively influences *Social Learning*.
H6: *Social Learning* positively influences *Competition*.

The second hypothesis (H2: *Social Comparison* positively influences *Competition*) is based on the work of Garcia et al. [8], who found Social Comparison as an important source of competitive behavior and thus established a direct link between them in their model, which they called the social comparison model of competition (SCMC).

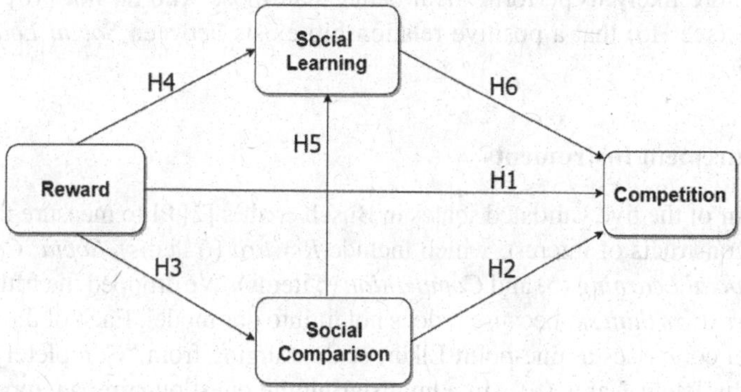

Fig. 1. Hypothesized path model

The third hypothesis (H3: *Reward* positively influences *Social Comparison.*), just like H2, is informed by the SCMC, in which Garcia et al. [8] identified a number of situational factors which influence Social Comparison: incentives, number of competitors, proximity to a standard, uncertainty, etc. They argued that direct incentives, for example, have the potential of influencing Social Comparison and thus Competition. Based on this, with respect to H3, we hypothesize that the more persuadable a person is by Reward the more responsive he will be to Social Comparison.

Further, the fourth hypothesis (H4: *Reward* positively influences *Social Learning*) is informed by the work of Bandura [1] on Social Learning Theory (SLT), which stemmed from the work by Miller and Dollard [15]. Miller and Dollard [15], in their work on social learning and imitation, identified four factors which are essential to learning: drives, cues, responses and rewards. Building on this theory, Bandura [1] in the SLT posited that people learn by observing and imitating the behaviors of others, which can result in two separate outcomes that can either inhibit or facilitate the learned behavior. These outcomes are regarded as punishment and reinforcement. In the former, an observer sees the action of another (the observed) being punished and, as a result, desists from engaging in the act. In the latter, an observer sees the action of another being rewarded (e.g., praised) and, as a result, imitates the act. Based on these two theories, we hypothesize, with respect to H4, that the more persuadable a person is by Reward the more responsive the person will be to Social Learning.

The fifth hypothesis (H5: *Social Comparison* positively influences *Social Learning*) is based on Festinger's [5] theory of upward social comparison for self-improvement. The theory holds that people compare themselves with other people who they think are better than themselves in order to improve themselves. Based on this theory, we argue that when a person compares himself or herself with another better than him or her, there is a tendency for him or her to imitate those behaviors, which s/he thinks make the other person better off. In so doing, the person will be learning new behaviors in a social context. Based on this theory, we hypothesize (see H5) that the more persuadable a person is by Social Comparison the more responsive he will be to Social Learning. Finally, the sixth hypothesis (H6: *Social Learning* positively influences *Competition*) is informed by the finding that people who learn tested behaviors from others are more likely to perform them better than those who do not [16]. Thus, we hypothesize (see H6) that a positive relationship exists between *Social Learning* and *Competition*.

4.2 Measurement Instruments

We used four of the five validated scales in Busch et al.'s [2] PI to measure the socially influential constructs of interest, which include *Reward* (6 items), *Social Comparison* (6 items), *Social Learning* (5) and *Competition* (5 items). We dropped the fifth construct in the PI (*Trustworthiness*) because it does not fit into our model. Each of the constructs in the model comprises a nine-point Likert scale, ranging from "Completely Disagree (1)" to "Completely Agree (9)". In administering the questionnaire, an approximately equal number of items was selected from all four scales, combined and randomized in each webpage of the survey. This prevented participants from knowing which specific construct was being measured at a given time or in a given survey page.

4.3 Participants

Our study was approved by the University of Saskatchewan Research Ethics Board. Respondents were invited to participate in the survey through email, the university's online bulletin, Facebook and Amazon Mechanical Turk (AMT). Those recruited through AMT were compensated with $0.8 each, while those recruited through the other media were given a chance to enter for a draw to win a $50 CAD gift card. 310 subjects participated in the study. After cleaning, we were left with 213 Canadians for analysis. Among them, 31.9% were males, 66.7% were females and 0.01% were unidentified. Moreover, 33.3% were between 18 and 24 years old, 41.8% were between 25 and 34 years old, and 24.9% were above 34 years old. Finally, 25.8% had high school education, 44.6% had bachelor degrees, 11.3% had postgraduate degrees and 18.3% had other qualifications.

5 Results

5.1 Measurement Model

Our model was built using R's *plspm* package [20] and assessed using the indicator reliability, internal consistency reliability, convergent validity and discriminant validity of its constructs [30]. *Indicator Reliability* for all construct indicators had an outer loading greater than 0.5. *Internal Consistency Reliability* for each construct was evaluated using the composite reliability criterion (DG.rho), which was greater than 0.7. *Convergent Validity* for each construct was evaluated using the Average Variance Extracted, which was greater than 0.5. *Discriminant Validity* was based on the cross-loading criterion. No indicator loaded higher on any other construct than the one it measured.

5.2 Data Driven Path Model

We built three models to verify our hypotheses: one global and two subgroup models based on gender (see Fig. 2). Our multigroup analysis shows that there is a significant difference between males and females, but no significant between younger and older participants. All three models are combined in one path model for brevity and easy comparison. The global path coefficients are enclosed in regular brackets, while the subgroup path coefficients for males and females are enclosed in square brackets below the global path coefficients. The bold coefficients indicate the path in which the male and female groups are different. At the global level, our model accounts for 42% of the variance of *Competition*, as indicated by the coefficient of determination (R^2), which measures the predictive accuracy of the model [9]. At the subgroup level, the predictive accuracy is higher for the males (51%) than for the females (39%). Overall, the goodness of fit (GOF) is about 50%. GOF is the predictive performance of the model, which indicates how well the model fits the data [20]. The performance value of 50% is consistent and replicated in the submodels as well.

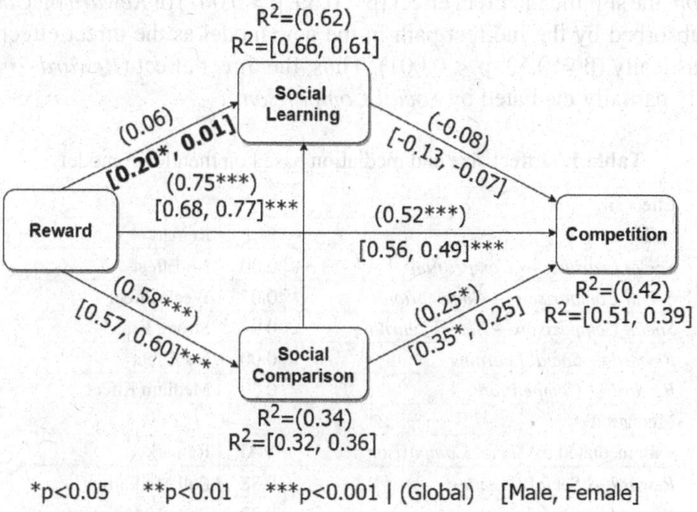

Fig. 2. Global model: GOF = 0.51 and subgroup models: GOF = [0.51, 0.50]

Direct Effects. At the global level (see Fig. 2), four of the six hypotheses (H1, H2, H3 and H5) are supported, with *Reward* positively influencing *Competition* (β = 0.52,

$p < 0.001$) and *Social Comparison* ($\beta = 0.58$, $p < 0.001$), *Social Comparison* positively influencing *Social Learning* ($\beta = 0.75$, $p < 0.001$) and *Competition* ($\beta = 0.25$, $p < 0.05$), respectively. However, *Social Learning* does not influence *Competition* ($\beta = -0.08$, $p > 0.05$). Similarly, *Reward* does not influence *Social Learning* ($\beta = 0.06$, $p > 0.05$) once *Social Comparison* is controlled for (included) in the model. However, based on the gender-based multigroup analysis, we found a significant difference (as indicated in the bold path coefficients) between males and females in the path (*Reward → Social Learning*). It turns out that *Reward* significantly influences *Social Learning* ($\beta = 0.20$, $p < 0.05$) for the male group, but does not for the female group ($\beta = 0.01$, $p > 0.05$).

Effect Size and Mediation. Table 1 shows the effect sizes (f^2) of the exogenous constructs on the endogenous constructs. *Reward* has a medium effect size ($f^2 = 0.29$) on *Competition*, while *Social Comparison* has a strong effect size ($f^2 = 0.97$) on *Social Learning*, but a weak effect size ($f^2 = 0.05$) on *Competition*. Table 1 also shows the mediation effects, computed using the metric, variance accounted for (VAF), i.e., the ratio of the indirect path effect to the total effect [9]. As shown, *Social Comparison* fully and partially mediates the influence of *Reward* on *Social Learning* and *Competition* respectively. Specifically, upon excluding *Social Comparison* from the model, *Reward* has a significant effect ($\beta = 0.50$, $p < 0.001$) on *Social Learning*. However, once *Social Comparison* is included in the model, the significant effect between *Reward* and *Social Learning* is absorbed by the indirect path (*Reward → Social Comparison* and *Social Comparison → Social Learning*), thereby making *Reward → Social Learning* to be non-significant ($\beta = 0.06$, $p > 0.05$). Thus, the direct effect (*Reward → Social Learning*) is fully mediated by *Social Comparison*. On the other hand, upon inclusion of *Social Comparison*, the significant direct effect ($\beta = 0.59$, $p > 0.001$) of *Reward* on *Competition* is hardly absorbed by the indirect path in the new model as the direct effect does not change drastically ($\beta = 0.52$, $p < 0.001$). Thus, the direct effect (*Reward → Competition*) is only partially mediated by *Social Comparison*.

Table 1. Effect size and mediation based on the global model

Effect size		
Path	f^2	Remark
Social Learning → Competition	0.00	No Effect
Social Comparison → Competition	0.05	Weak Effect
Social Comparison → Social Learning	0.97	Strong Effect
Reward → Social Learning	0.00	No Effect
Reward → Competition	0.29	Medium Effect
Mediation		
Path mediated by *Social Comparison*	VAF	Remark
Reward → Social Learning	0.88	Full Mediation
Reward → Competition	0.22	Partial Mediation

6 Discussion

We have proposed a model for predicting the persuasiveness of Competition as a persuasive strategy. The model accounts for over 40% of the variance in the persuasiveness of Competition, which is a powerful group-based intrinsic motivator in persuasive technology. Our path analysis reveals that *Competition* can be predicted by *Reward* and *Social Comparison*, but not by *Social Learning*. Table 2 shows the supported hypotheses at the global and gender-based subgroup levels. At the global level, four of the six hypotheses (H1, H2, H3 and H5) are supported. Moreover, at the subgroup level, H4, which is not supported at the global level, turns out to be supported for the male group.

Table 2. Supported and unsupported hypotheses (relationships)

Hypo	Relationship	Global	Male	Female
H1	*Reward → Competition*	✓	✓	✓
H2	*Social Comparison → Competition*	✓	✓	✓
H3	*Reward → Social Comparison*	✓	✓	✓
H4	*Reward → Social Learning*	×	✓	×
H5	*Social Comparison → Social Learning*	✓	✓	✓
H6	*Social Learning → Competition*	×	×	×

In our model, the first hypothesis (H1: *Reward* positively influences *Competition*) is validated. It suggests that, e.g., in gamified persuasive technology, the more persuadable a person is by Reward as a persuasive strategy the more motivated s/he will be by Competition. Kaptein et al. [13] found that individual differences in persuadability affect the level of compliance of users with persuasive strategies. For example, in the health domain, they found that high persuadables (those with higher degree of persuadability) are more likely to comply with persuasive messages than low persuadables (those with lower degree of persuadability). Drawing on this finding, the validation of the first hypothesis suggests that users who are more persuadable by Reward in gamified context will be more inclined to perform a target behavior based on Competition than those who are less persuadable by Reward. Further, the second hypothesis (H2: *Social Comparison* positively influences *Competition*) is validated. This suggests that the more persuadable a person is by Social Comparison the more motivated s/he will be by Competition. Similarly, the third hypothesis (H3: *Reward* positively influences *Social Comparison*) is validated. This suggests that the more persuadable a person is by Reward the more motivated s/he will be by Social Comparison. Putting it together, based on the validation of these three hypotheses, we can come to the conclusion that the more persuadable a person is by Reward the more motivated s/he will be by Social Comparison and Competition, with *Social Comparison* partially mediating the effect of *Reward* on *Competition*. In design context, this suggests that if a designer of a behavior change application, who wants to motivate its users to engage in a given target behavior, has found out through data gathering (e.g., quantitative surveys, interviews, etc.) that Reward and/or Social Comparison, for example, can motivate his/her target audience, then he can go ahead to implement Competition as well, as a persuasive strategy to strengthen the intrinsic motivation to perform the target behavior. Our model shows that these persuasive strategies can co-exist and are compatible in a given application targeted at a population that is motivated by any of the three persuasive strategies.

Further, the fifth hypothesis (H5: *Social Comparison* positively influences *Social Learning*) is strongly supported. This indicates that individuals who are highly persuadable by Social Comparison are more likely to be responsive to learning from others in a social context through Consensus or Social Proof. This suggests that both Social Comparison and Social Learning could be leveraged in a given persuasive application designed for a target population that is motivated by either persuasive strategy in order to increase the chances of performing the target behavior. This is confirmed by the strong effect size of *Social Comparison* on *Social Learning*. This finding replicates the result of Stibe [23], who found in an experimental setting that there is a strong relationship between *Social Learning* and *Social Comparison.*

Finally, the fourth and sixth hypotheses (H4 and H6) are not supported. Regarding H6, a possible explanation for why *Social Learning* does not influence *Competition* is that the influential people that the study's subjects learn from mostly in real life, which informed their response to the *Social Learning* questions, may not necessarily be those they compete with (e.g., peers, classmates, teammates, etc.). Rather, these influential people may have been more of those they look up to (e.g., those in authority, role models, mentors, teachers, etc.). This is reflected in some of the questions that measure the *Social Learning* construct, such as "*I ask for advice from other people before I make decision*" and "*I take other people as role models for new behaviors.*" Advice, for example, may be more from an older person, a more experienced person, an authority figure, etc., other than the peers they compete with. The same thing applies to role models. Thus, the subjects' responsiveness to Social Learning does not influence their responsiveness to Competition. Regarding H4, the reason why *Reward* does not directly influence *Social Learning* at the global level is that the relationship is mediated by *Social Comparison* once it is included in the model.

6.1 Implication

The implication of our findings is that Reward, Social Comparison and Competition are compatible socially influential strategies that can be implemented in a given persuasive application for a given population that is motivated by any of these three strategies. This is because, as the model reveals, the more motivated a person is by Reward, the more likely s/he will be motivated by Social Comparison and Competition. This suggests that in designing gamified persuasive applications that foster Competition, designers should endeavor to implement motivational affordances [10] such as Reward (e.g. points, badges, etc.) and Social Comparison (e.g., leaderboards) alongside in order to stimulate user participation and engagement towards performing the target behavior.

6.2 Contribution

Our contribution to the body of knowledge is that we have provided a preliminary model for predicting the persuasiveness of Competition, which is a powerful intrinsic motivator used widely in persuasive technology towards user participation and engagement [7]. In addition, we have tested the model and shown empirically that: (1) our model has a moderate predictive accuracy [9, 20], as it is able to predict 42% of the variance of *Competition*; and (2) *Reward* and *Social Comparison* are relevant predictors of *Competition*, with *Social Comparison* playing a mediating role. To the best of our knowledge, this is one of the first empirical studies in the persuasive technology domain, which attempts to uncover the predictors of *Competition* as a persuasive strategy.

6.3 Limitation and Future Work

The limitation of our study is that it is based on self-report and the Canadian population only, which may threaten the generalizability of our findings. However, in future work, we intend to investigate the replication of our findings with participants from other cultures/countries in order to generalize our findings. Another limitation of our study is that, given that we used the Persuadability Inventory by Busch et al. [2], our model of competitive behavior may not have been exhaustive of all of the possible predictors of *Competition*, as there may be other predictors as well (e.g., *Social Recognition, Number of Competitors*, etc.), which future study can consider in order to extend the model.

7 Conclusion

We have presented and tested a proposed model of the persuasiveness of Competition (one of the most powerful intrinsic motivators in video games) based on the Persuadability Inventory developed by Busch et al. [2]. Using data gathered from 213 Canadian participants, our model shows that *Reward* and *Social Comparison* are relevant predictors of *Competition*, but *Social Learning* is not. Our model explains 42% of the variance in *Competition*. It also shows that, when *Social Comparison* is controlled for in our model, *Reward* influences *Social Learning* for the male group, but does not for the female group. Our findings, based on the validated path model, suggest that Reward, Social Comparison and Competition can be implemented together in a given persuasive application where one of these socially influential strategies has been found to be a motivator of the target behavior among a given target audience.

References

1. Bandura, A.: Social learning theory, pp. 1–46. Stanford University (1971)
2. Busch, M., Schrammel, J., Tscheligi, M.: Personalized persuasive technology – development and validation of scales for measuring persuadability. In: Berkovsky, S., Freyne, J. (eds.) PERSUASIVE 2013. LNCS, vol. 7822, pp. 33–38. Springer, Heidelberg (2013). doi: 10.1007/978-3-642-37157-8_6

3. Cheng, R.: Persuasion strategies for computers as persuasive technologies. vol. 10, no. 2010, pp. 1–7. Department of Computer Science University of Saskatchewan (2003)
4. Eyck, A., Geerlings, K., Karimova, D., Meerbeek, B., Wang, L., IJsselsteijn, W., Kort, Y., Roersma, M., Westerink, J.: Effect of a virtual coach on athletes' motivation. In: IJsselsteijn, Wijnand, A., Kort, Yvonne, A.,W., Midden, C., Eggen, B., Hoven, E. (eds.) PERSUASIVE 2006. LNCS, vol. 3962, pp. 158–161. Springer, Heidelberg (2006). doi:10.1007/11755494_22
5. Festinger, L.: A theory of social comparison. Hum. Relat. **7**, 117–140 (1954)
6. Fogg, B.J. et al.: Motivating, influencing, and persuading users. In: Human Computer Interaction Handbook: Fundamentals, Evolving Technologies, and Emerging Applications, pp. 133–146 (2009)
7. Fogg, B.J.: Persuasive Technology: Using Computers to Change What We Think and Do. Morgan Kaufmann, San Francisco (2003)
8. Garcia, S.M., et al.: The psychology of competition: a social comparison perspective. Perspect. Psychol. Sci. **8**(6), 634–650 (2013)
9. Hair, J.F., et al.: A Primer on Partial Least Squares Structural Equation Modeling (PLS-SEM). Sage Publications Inc., Washington DC (2014)
10. Hamari, J., Koivisto, J., Pakkanen, T.: Do persuasive technologies persuade? - a review of empirical studies. In: Spagnolli, A., Chittaro, L., Gamberini, L. (eds.) PERSUASIVE 2014. LNCS, vol. 8462, pp. 118–136. Springer, Heidelberg (2014). doi:10.1007/978-3-319-07127-5_11
11. Huber, M.Z., Hilty, L.M.: Gamification and sustainable consumption: overcoming the limitations of persuasive technologies. Adv. Intell. Syst. Comput. **310**, 367–385 (2015)
12. Kaptein, M.: Adaptive persuasive messages. In: An E-Commerce Setting: The Use of Persuasion Profiles, Ecis 2011, Paper 183 (2011)
13. Kaptein, M., Lacroix, J., Saini, P.: Individual differences in persuadability in the health promotion domain. In: Ploug, T., Hasle, P., Oinas-Kukkonen, H. (eds.) PERSUASIVE 2010. LNCS, vol. 6137, pp. 94–105. Springer, Heidelberg (2010). doi:10.1007/978-3-642-13226-1_11
14. Malone, T.W., Lepper, M.R.: Making learning fun: a taxonomy of intrinsic motivations for learning. Aptit. Learn. Instr. **3**(3), 223–253 (1987)
15. Miller, N.E., Dollard, J.: Social Learning and Imitation. New Haven, CT, US (1941)
16. Oinas-Kukkonen, H.: Persuasive systems design: key issues, process model, and system features. Inf. Syst. **24**, 485–500 (2009). March 2009
17. Orji, R.: Exploring the persuasiveness of behavior change support strategies and possible gender differences. CEUR Workshop Proc. **1153**, 41–57 (2014)
18. Orji, R. et al.: Towards personality-driven persuasive health games and gamified systems. In: Proceedings SIGCHI Conference Human Factors Computer System (2017)
19. Ryan, R.M., Deci, E.L.: Intrinsic and extrinsic motivations: classic definitions and new directions. Contemp. Educ. Psychol. **25**(1), 54–67 (2000)
20. Sanchez, G.: PLS path modeling with R. R News, vol. 235 (2013)
21. Skinner, B.: Science and Human Behavior, p. 480. Simon and Schuster, New York (1965)
22. Spittle, M., et al.: Applying self-determination theory to understand the motivation for becoming a physical education teacher. Teach. Teacher Educ. **25**(1), 190–197 (2009)
23. Stibe, A.: Socially influencing systems: persuading people to engage with publicly displayed twitter-based systems (2014)
24. Stibe, A.: Towards a framework for socially influencing systems: meta-analysis of four PLS-SEM based studies. In: MacTavish, T., Basapur, S. (eds.) PERSUASIVE 2015. LNCS, vol. 9072, pp. 172–183. Springer, Heidelberg (2015). doi:10.1007/978-3-319-20306-5_16

25. Stibe, A., Larson, K.: Persuasive cities for sustainable wellbeing: quantified communities. In: Younas, M., Awan, I., Kryvinska, N., Strauss, C., Thanh, D. (eds.) MobiWIS 2016. LNCS, vol. 9847, pp. 271–282. Springer, Heidelberg (2016). doi:10.1007/978-3-319-44215-0_22
26. Stibe, A., Oinas-Kukkonen, H.: Comparative analysis of recognition and competition as features of social influence using Twitter. In: Bang, M., Ragnemalm, E.L. (eds.) PERSUASIVE 2012. LNCS, vol. 7284, pp. 274–279. Springer, Heidelberg (2012). doi: 10.1007/978-3-642-31037-9_26
27. Stiglitz, J.E.: Theory of Competition, Incentives and Risk. Palgrave Macmillan, London (1986)
28. Sundar, S.S., Bellur, S., Jia, H.: Motivational technologies: a theoretical framework for designing preventive health applications. In: Bang, M., Ragnemalm, E.L. (eds.) PERSUASIVE 2012. LNCS, vol. 7284, pp. 112–122. Springer, Heidelberg (2012). doi: 10.1007/978-3-642-31037-9_10
29. Vansteenkiste, M., Deci, E.L.: Competitively contingent rewards and intrinsic motivation: can losers remain motivated? Motiv. Emot. **27**(4), 273–299 (2003)
30. Wong, K.K.: Partial least squares structural equation modeling (PLS-SEM) techniques using SmartPLS. Mark. Bull. **24**(1), 1–32 (2013)

Perceived Effectiveness, Credibility and Continuance Intention in E-commerce: A Study of Amazon

Ifeoma Adaji(✉) and Julita Vassileva

MADMUC Lab, University of Saskatchewan, Saskatoon, Canada
{ifeoma.adaji,julita.vassileva}@usask.ca

Abstract. The use of persuasive strategies has been identified as one means through which e-businesses can engage their existing clients and make new ones. To contribute to ongoing research in this area, we extend our previous study where Amazon was evaluated as a persuasive system using the Persuasive System's Design (PSD) framework. In the current study, we further investigate the factors that affect the perceived effectiveness, credibility and continuance intention for use of e-commerce systems through the prism of the PSD framework. Using Amazon as a case study and a sample size of 324 Amazon shoppers, we develop and test a research model using partial least-squares structural equation modelling (PLS-SEM) analysis. Our results show that perceived effectiveness of an e-commerce company like Amazon is a great predictor of continuance intention. In addition, social support and primary task support are strong predictors of perceived effectiveness. Furthermore, dialogue support significantly influences perceived product credibility and perceived review credibility and both constructs are strong predictors of system credibility. These findings suggest possible design guidelines in the development of successful e-commerce sites.

Keywords: Perceived effectiveness · Perceived credibility · Continuance intention

1 Introduction

E-commerce is the act of buying and selling or carrying out commercial activities over a network, typically the internet [1]. It takes advantage of the power of digital media to carry out the purchase process where consumers can search for, identify and pay for a product or service [2]. Consumers and businesses are adopting e-commerce as a means of carrying out their transactions, with consumers spending more time online [3]. Because of the success of e-commerce, there has been an increase in the number of companies who do business online. Due to this increase, there is currently a lot of competition among e-businesses to acquire new customers and to maintain existing ones [4]. Hence, simply selling products or conducting e-commerce is no longer enough for businesses to distinguish themselves from their competitors. Companies have to put strategies in place in order to remain attractive to their clients; they need to identify ways and means of engaging their online customers. One strategy that is currently being used

P.W. de Vries et al. (Eds.): PERSUASIVE 2017, LNCS 10171, pp. 293–306, 2017.
DOI: 10.1007/978-3-319-55134-0_23

is persuasive technology [5]. Persuasive technology (PT) is the use of interactive technology to change people's attitude and behavior without coercion or deception [6, 7].

This paper aims to study the perceived effectiveness, credibility and continuance intentions of e-commerce users using a theory driven approach. It examines how the persuasive principles of the Persuasive Systems Design (PSD) framework [8] enhance customers' perceptions of effectiveness and credibility of the Amazon e-commerce platform and their intention to continue using it. This paper is an extension of our previous study [9][1] where we evaluated Amazon as a persuasive system using the PSD framework and further identified the implementation of the persuasive principles of the PSD framework in Amazon.

Using Amazon as a case study and a sample size of 324 Amazon shoppers, in this study, we developed and tested a research model using partial least-squares structural equation modelling (PLS-SEM) analysis. The results of this study can act as a guide to system developers in implementing persuasive strategies that work in e-commerce.

2 Related Work

2.1 Persuasive Technology in E-commerce

The use of persuasive technology in e-commerce is currently an active research area. In their review of theories and models of users' acceptance and use, Alhammad and Gulliver [5] identified variables associated with perceived persuasiveness. These include: informativeness, primary task support, system credibility support, dialog support, social support and design aesthetics. Though the researchers dwelt on some of the principles of the Persuasive Systems Design framework, the authors did not investigate how these principles influence each other.

Chu et al. [10] reviewed a few e-commerce websites in order to identify the persuasive principles that these sites had in common and how they were designed. This study differs from ours because in addition to identifying the persuasive strategies in Amazon, a top e-commerce site, we also developed a model using a validated framework, the PSD framework.

Constantinides [11] carried out a theoretical review on factors affecting e-commerce users' behavior and identified similarities and differences between the persuasiveness of e-commerce customers and traditional brick and mortar customers.

To the best of our knowledge, no other research exists in e-commerce that addresses factors that affect perceived effectiveness, credibility and use continuance using the PSD framework.

2.2 Persuasive Systems Design Framework (PSD)

PSD is a systematic framework used for designing and evaluating persuasive systems. It details the content and design principles that are required in the development and

[1] This paper was part of the 2016 Persuasive Technology conference.

evaluation of persuasive systems. The framework consists of 24 design principles that are categorized into four based on the tasks they aims to accomplish: (1) primary task support – principles in this category support a system's user in achieving their primary objective or goal, (2) dialogue support - design principles in this category support computer-human dialogue which provide feedback to users with the aim of moving users towards their target behavior, (3) system credibility support - principles in this category persuade users through the credibility of the system's design, (4) social support - the principles in this category influence users by leveraging social influence [12]. Despite its strengths, its use has not been extensively explored in the domain of e-commerce.

3 Research Design and Methodology

We developed a hypothetical path model using existing research and the PSD framework. This is because this paper is a continuation of a previous study [9] where the PSD framework was used to evaluate a successful e-commerce site, Amazon.

Our research model is described in Fig. 1. It is made up of twelve hypothesis and eight constructs shown as nodes in Fig. 1. Six of the constructs are derived from the four categories of the PSD framework. For all 8 constructs, we adopted previously validated scales by Lehto and Oinas-Kukkonen [13]. The constructs are described in the following section.

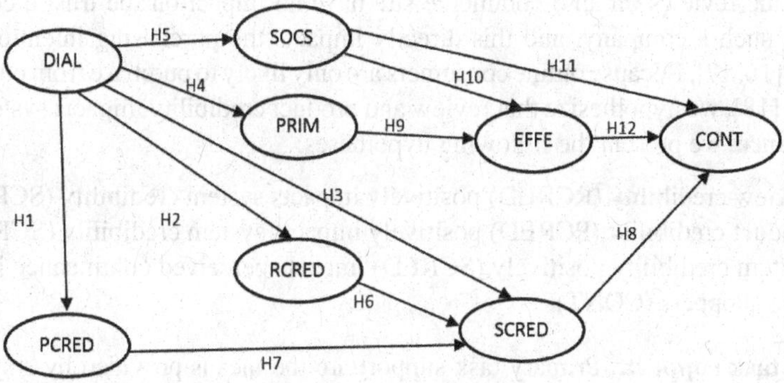

Fig. 1. Research model. All paths assumed positive (DIAL = Dialogue support, SOCS = Social support, PRIM = Primary task support, EFFE = Perceived effectiveness, CONT = Continuance intention, PCRED = Perceived product credibility, RCRED = Perceived review credibility, SCRED = Perceived system credibility).

3.1 Description of Constructs and Definition of Hypothesis

Dialogue task support. According to the PSD framework, dialogue support principles encourage computer-human dialogue which provides feedback to users, with the aim of moving them towards their target behavior [12]. In e-commerce, this feedback is in the form of rating, reviews and communication between buyers and sellers [9]. We hypothesize that dialogue support positively affects the product credibility, because electronic word of mouth, which is a result of communication between customers in e-commerce, has been shown to affect customers' perception of a product [14]. In addition, because the quality and quantity of online reviews have been shown to affect the credibility of the review and consumer's purchase intention, we further hypothesize that dialogue support positively affects the credibility of reviews and the overall system in general. Hence, we propose the following hypotheses:

H1: Dialogue support (DIAL) positively affects product credibility (PCRED) in e-commerce.
H2: Dialogue support (DIAL) positively affects the credibility of reviews (RCRED) in e-commerce
H3: Dialogue support (DIAL) positively affects an e-commerce system's (SCRED) credibility

In e-commerce, feedback in the form of ratings and reviews has been shown to enhance trust between buyer and seller which often leads to an increase in the purchase intention of the buyer [15]. Hence we hypothesize that dialogue positively impacts the purchase intention of e-commerce shoppers. Thus we proffer the following hypotheses:

H4: Dialogue support (DIAL) positively impacts primary task support (PRIM)
H5: Dialogue support (DIAL) positively affects perceived social support (SOCS)

Perceived credibility (review, product and system credibility). The perceived credibility of the reviews on an e-commerce site have an impact on the trust a consumer places on such a company, and this directly impacts the purchasing intention of the customer [16, 17]. Because online consumers are only likely to purchase from companies they trust [18], we hypothesize that review and product credibility impacts system credibility. Hence, we present the following hypotheses:

H6: Review credibility (RCRED) positively impacts system credibility (SCRED)
H7: Product credibility (PCRED) positively impacts system credibility (SCRED)
H8: System credibility positively (SCRED) impacts perceived continuance intention of online shoppers (CONT)

Primary task support. Primary task support are the means provided by a system to aid users in competing their primary tasks [12], while perceived effectiveness, which is related to performance expectancy, describes the extent to which using a technology benefits the users in performing certain activities, for example, online shopping [13]. Primary task support has been found to positively affect perceived effectiveness in other domains [13], hence we hypothesize that in e-commerce, primary task support will also affect perceived effectiveness of the system. We therefore proffer the following hypothesis:

H9: Primary task support (PRIM) positively affects perceived effectiveness (EFFE)

Social support. Social support principles influence users by leveraging social influence [12]. Social support benefits online consumers when making purchase decisions and has been found to critically affect users' future shopping intentions [19]. Perceived effectiveness is a measure of how users perceive the effectiveness of using technology in achieving their goal compared to other methods [13]. We hypothesize that social support will positively impact the perceived effectiveness of an e-business and the future purchase intentions of its customers. Thus, we present the following hypotheses.

H10: Social support (SOCS) positively impacts perceived effectiveness (EFFE)
H11: Social support (SOCS) positively impact continuance intention of online shoppers (CONT)

Perceived effectiveness. Perceived effectiveness is a measure of how users perceive the effectiveness of using technology in achieving their goal compared to other methods [13]. Based on this definition, we define perceived effectiveness in e-commerce, as a measure of how users perceive the effectiveness of the buying process from an e-commerce platform compared to the traditional brick and mortar store. Since the effectiveness of an e-business will likely lead to its customers' continued patronage [20], we hypothesize that the perceived effectiveness of an e-business will lead to continuance shopping intention of its clients. Therefore, we propose the following hypothesis:

H12: Perceived effectiveness of an e-commerce platform positively affects perceived continuance intention of online shoppers

3.2 Measurement Model

To measure the eight constructs of interest described above, we adopted previously validated scales of Lehto and Kukkonen [13]. Dialogue support was measured with four items while primary task support, social support and perceived effectiveness were each measured with three items. Continuance was measured with two items while product credibility, review credibility and system credibility were measured with four items each. All items were measured on a five-point Likert scale (1 = strongly disagree, 5 = strongly agree).

3.3 Participants

For this study, we used Amazon shoppers as our test subjects. A total of 324 participants took part through an online survey. 150 participants were recruited through Amazon's Mechanical Turk, while 174 were recruited through various social media platforms and news boards. Participation was voluntary and the study was approved by the ethics board of the University of Saskatchewan.

Table 1 shows the demographics of the participants. Overall, 54% of the participants were female while 42% were between the ages of 25 and 34. In addition, 45% have used Amazon for between 2 and 5 years, while 38% of the participants shop on Amazon at least once in three months. Finally, most of the participants are registered on at least one social media platform.

Table 1. Demographics of participants

Demographics	Value	Frequency (%)
Age	18–24	28
	25–34	41
	35–44	18
	45–54	7
	>54	4
	Not disclosed	2
Gender	Male	45
	Female	54
	Not disclosed	1
Duration of Amazon use	<1 year	10
	2–5 years	45
	>5 years	42
	Not disclosed	3
Shopping frequency on Amazon	At least once a week	10
	At least once a month	38
	At least once in 3 months	30
	At least once a year	19
	Not disclosed	3
Social media use	At least one social media platform	91
	None	9

4 Data Analysis and Results

We analyzed our data using Partial Least Squares Structural Equation Modelling (PLS-SEM) with the SmartPLS tool. We present the result of our analysis in the following section along with validation of the global measurement and structural models used.

4.1 Evaluation of Global Measurements

According to Wong [21], it is important to determine the reliability and validity of the latent variables in order to complete the examination of the structural model using various reliability and validity items like indicator reliability, internal consistency reliability, convergent validity and discriminant validity.

Indicator Reliability. All indicators in our measurement model had outer loading greater than 0.7, the minimum acceptable level [22]. Hence, reliability criteria was met. Due to space constraints, we do not include the table with the indicator reliability values of the indicators

Internal Consistency Reliability. Composite reliability has been suggested as a replacement to Cronbach's alpha in measuring internal consistency reliability because Cronbach's alpha tends to provide a conservative measurement in PLS-SEM [22], hence, we used composite reliability in this study. The composite reliability values for all latent variables were higher than the acceptable threshold of 0.7 [21], hence high levels of internal consistency reliability were established among all latent variables. Table 2 shows the composite reliability values for all latent variables.

Table 2. Latent variable coefficients and correlations

Latent variables	COR	AVE	VIF
CONT	0.921	0.854	2.005
DIAL	0.792	0.656	1.109
PRIM	0.870	0.692	1.906
SOCS	0.827	0.615	1.356
EFFE	0.904	0.825	1.734
PCRED	0.902	0.696	1.973
RCRED	0.910	0.716	2.150
SCRED	0.917	0.733	2.298

COR = Composite Reliability, AVE = Average Variance Extracted and VIF = Variance Inflation Factor

Convergent Validity. We used the Average Variance Extracted (AVE) to compute the convergent validity of each latent variable. As shown in Table 2, all AVE values are greater than 0.5, the acceptable threshold [22]. Hence, convergent validity is confirmed.

Discriminant Validity. According to Fornell and Larcker [23], the square root of AVE for each latent variable can be used to determine discriminant validity, if it is greater than the other correlation values among the latent variables. Table 3 shows the latent variable correlations with the square roots of AVE shown on the bold diagonal. For each latent variable, the square root of AVE is greater than the other correlation variables (along the row and column). Hence, discriminant validity is well established.

Table 3. Latent variable correlations

	CONT	DIAL	PRIM	SOCS	EFFE	PCRED	RCRED	SCRED
CONT	**0.924**							
DIAL	0.282	**0.810**						
PRIM	0.429	0.657	**0.832**					
SOCS	0.450	0.390	0.387	**0.784**				
EFFE	0.693	0.403	0.532	0.573	**0.908**			
PCRED	0.396	0.401	0.493	0.352	0.459	**0.834**		
RCRED	0.359	0.438	0.469	0.429	0.423	0.694	**0.846**	
SCRED	0.483	0.472	0.565	0.392	0.547	0.766	0.730	**0.856**

Note: Square roots of AVE are shown on the bold diagonal

4.2 Structural Model

Figure 2 shows the partial least square path model (PLS-SM) for our research model. It shows the path coefficients, β, (between the various constructs) that explain how strong the effect of the exogenous variables are on the endogenous variables. In addition, the model indicates how much the variance of the endogenous variables are explained by the exogenous variables. Finally, the number of asteriks indicates the significance of each direct effect. The number of asteriks ranges from 1 to 4, and this corresponds with the p-value of <0.05, <0.01, <0.001 and <0.0001 respectively.

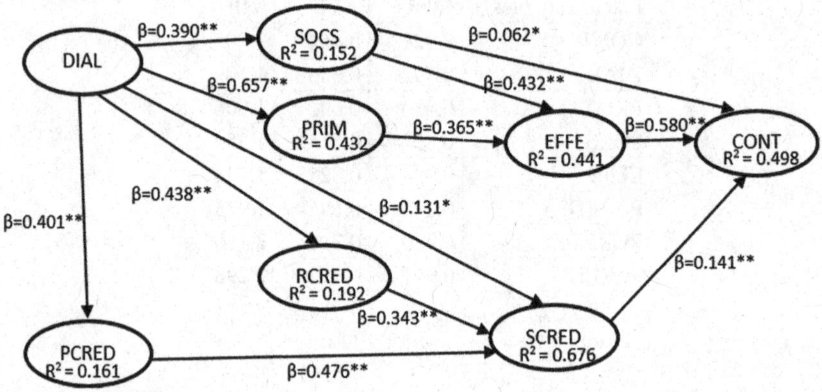

Fig. 2. Structural model with results of PLS-SEM analysis (DIAL = Dialogue support, SOCS = Social support, PRIM = Primary task support, EFFE = Perceived effectiveness, CONT = Continuance intention, PCRED = Perceived product credibility, RCRED = Perceived review credibility, SCRED = Perceived system credibility)

Perceived effectiveness has the strongest effect (β value) on continuance intention, while social support has the least effect on continuance intention. Hence, perceived effectiveness is a strong predictor of continuance intention compared to social support. In addition, perceived product credibility and perceived review credibility have stronger effect on system credibility compared to dialogue support. Therefore, perceived product credibility and perceived review credibility are stronger predictors of system credibility compared to dialogue support. Social support and primary task support both have significant effect on perceived effectiveness, hence they are both good predictors of perceived effectiveness. Because the acceptable threshold for path coefficient is 0.20 [21], we conclude that dialogue support does not have a significant effect on perceived system credibility. Similarly, perceived system credibility and social support do not have significant effect on continuance intention.

According to the PLS-PM standard [24], a coefficient of determination (R^2) value of <0.30 is considered low, while 0.30 < R2 < 0.60 is moderate, and R2 > 0.60 is regarded as high. Since R2 is 0.676 for perceived system credibility, this means that review credibility, product credibility and dialogue support together substantially explain 67.6% of the variance of system credibility. In addition, 44% of the variance in perceived effectiveness is explained by social support and primary task support.

Furthermore, social support, perceived effectiveness and system credibility explain 49.8% of the variance in the continuance intention of the users.

4.3 Path Significance

In order to test the path significance between the constructs as described in Fig. 2, we computed the T-value for significance testing of the structural paths between the constructs as recommended by [21]. Our results as shown in Table 4 indicate that all paths between constructs are significant except SOCS -> CONT which did not meet the minimum threshold of 1.96 [21]. Hence we can conclude that our model loadings as described in Fig. 2 are highly significant.

Table 4. Path significance testing

Path	T-statistics	Path	T-statistics
SOCS -> CONT	1.179	PRIM -> EFFE	6.423
DIAL -> SCRED	3.681	DIAL -> SOCS	7.797
SCRED -> CONT	3.732	DIAL -> PCRED	7.119
RCRED -> SCRED	6.254	SOCS -> EFFE	8.549

4.4 Total Effects and Effect Sizes

To ascertain how much an exodenous latent variable contributes to an endogenous latent variable, for example how much perceived effectiveness contributes to continuance intention, we computed our model's total effects and the corresponding effect sizes. The effect sizes describe the strength or degree of the relationship between the various

Table 5. Total effects and effect sizes

	CONT	DIAL	PRIM	SOCS	EFFE	PCRED	RCRED	SCRED
DIAL			0.657** (0.760)	0.390** (0.179)		0.401** (0.192)	0.438** (0.238)	0.472* (0.042)
PRIM					0.365** (0.202)			
SOCS	0.312* (0.005)				0.432 (0.284)			
EFFE	0.580** (0.368)							
PCRED								0.476** (0.354)
RCRED								0.343** (0.177)
SCRED	0.141** (0.027)							

*p < 0.05; ** <0.01; *** <0.001; **** <0.0001. Effect sizes are shown in parentheses

constructs [21]. This is in line with the findings of Chin et al. [25], who concluded that path significance is not enough to determine the strength of the relationship between two constructs. Based on the recommendation of [21], we adopted effect sizes of <0.02, <0.15 and <0.35 to indicate small, medium and large effect respectively. Table 5 describes the total effects and effects sizes between the 8 constructs of our model. It is evident that the effect size between the constructs range from medium to large except for SCRED -> CONT, SOCS -> CONT and DIAL -> SCRED, hence we can conclude that relationships between these constructs are not very strong.

4.5 Verification of Hypotheses

In this section, we verify or discredit the hypotheses defined in Sect. 3 based on the results of analysing our model.

Because of the significant path coefficient (Fig. 2), the strength of the path significance (Table 4) and the significant effect size (Table 5) between dialogue support and perceived product credibility, dialogue support and perceived review credibility, Dialogue support and primary task support, dialogue support and social support, perceived review credibility and system credibility, perceived product credibility and perceived system credibliity, primary task support and perceived effectiveness, social support and perceived effectiveness, perceived effectiveness and continuance intention, we concluded that our hypothesis H1, H2, H4, H5, H6, H7, H9 and H10 were validated.

On the other hand, because of the low path coefficient (Fig. 2), the low strength of the path significance (Table 4) and the insignificant effect size (Table 5) between dialogue support and perceived system credibility, perceived system credibility and continuance intention, social support and continuance intention, we concluded that our hypothesis were invalid for H3, H8 and H11. For example, for H11, we hypothesized that social support positively impacts continuance intention of online shoppers. However, from Table 4, it is clear that the path coefficient between social support and continuance is not statistically significant ($\beta = 0.062$ whereas the minimum acceptable value for $\beta = 0.2$). In addition, the path significance between social support and continuance intention as described in Table 4 is lower than the minimum acceptable threshold of 1.96 (SOCS -> CONT = 1.179). Furthermore, the effect size between social support and continuance intention, in Table 5, is less than the least acceptable limit of 0.02 (SOCS -> CONT = 0.005). We hypothesized for H3 that dialogue support positively affects an e-commerce system's credibility. Although the path significance between dialogue support and perceived system credibility is above the threshold of 1.96 (DIAL -> SCRED = 3.681) as shown in Table 4, the effect size is significantly low (DIAL -> SCRED = 0.042), and so is the path coefficient ($\beta = 0.131$). Hence H3 was invalid.

5 Discussion

In this study, we created a theoretical model to explore the factors that affect the perceived effectiveness, credibility and use continuance of e-commerce shoppers using Amazon as a case study. It was remarkable to discover that perceived review credibility, perceived product credibility and dialogue support can explain almost 70% of the variance in perceived system credibility. This shows the importance of reviews in e-commerce systems and why e-businesses should incorporate product reviews on their websites. This finding is in line with previous studies that have found that the reviews on an e-commerce site have an impact on the trust a consumer places on such a company, and this directly impacts the purchasing intention of the customer [16, 17].

It was surprising to discover that although perceived review credibility and perceived product credibility significantly influence perceived system credibility and are strong predictors of perceived system credibility (with almost 70% variance), perceived system credibility does not significantly influence continuance intention of shoppers in Amazon. We attributed this to the fact that Amazon is a well-known company, hence people will shop with them irrespective of how credible they feel the system is. This result might not be the same with a new e-commerce company.

Also worthy of mention is that social support and primary task support can explain about 44% of the variance in perceived effectiveness of the system, which alongside perceived system credibility and social support can further predict almost half of the variance in continuance intention of participants to shop with Amazon in the future. Of the three constructs that predict continuance intention: perceived effectiveness, social support and perceived system support, perceived effectiveness had the most significant impact; $\beta = 0.580$. The implication of this is that the effectiveness of an e-business (a measure of how users perceive the effectiveness of the buying process from an e-commerce platform compared to the traditional brick and mortar store) in providing products and services to its clients affect shoppers' behavior not only when they are online, but also affects the future shopping behavior of the customers. This conclusion is in line with previous studies that found that effectiveness of an e-business will likely lead to its customers' continued patronage [20].

Another significant finding from this study is the influence dialogue support has on perceived credibility. In this study, we split perceived credibility into three: perceived product credibility, perceived review credibility and perceived system credibility. While dialogue support significantly influences perceived product credibility and perceived review credibility, it does not influence perceived system credibility. This is not in line with the findings of [13]. Our result indicates that being able to receive feedback from Amazon does not influence the perceived credibility of the system by its's users. Instead, the credibility of the products purchased by customers influences their perceived credibility of the system.

This study extends our previous study [9] where we evaluated Amazon as a persuasive system using the PSD framework of Oinas-Kukkonen and Harjumaa [12], and identified the implementation of the various principles of the framework in Amazon.

Our study contributes significantly to ongoing research in e-commerce by demonstrating the substantial role the various persuasive principles of the PSD framework play in e-commerce.

The implication for web developers and e-business owners includes the following: (1) implementing web applications that encourage dialogue and social support between customers is important. Strategies that could be used to encourage dialogue include feedback, reminders and rewards [9]. System developers should also include platforms where users can share their experience about a product, not only through reviews but using question & answer platforms and comment sections [9]. (2) The primary task of users in e-commerce (reviewing and shopping for products) should be as seamless as possible in order to ensure a user continues with the e-business. There should also be enough online resources available on the website to ensure ease of use of the e-commerce site. E-businesses should also have a means through which customers can review their previous orders in order to keep track of their progress.

There are some limitations to this study. First, we did not verify that the participants were actual Amazon shoppers. However, because Amazon is very popular and is widely used, we are confident that the participants do use Amazon. This is also because we were able to recruit a significant number of participants (174 out of 324) through various social media platform without offering any reward. Second, the number of participants we used in the study (324) might not reflect a significant fraction of the total number of Amazon users. In the future, we plan to carry out this experiment on a larger scale to further validate our findings. In addition, we will evaluate the effect of perceived credibility (system, review and product) on the perceived effectiveness of the system. The main strength of our study is that, to the best of our knowledge, no other research exists in e-commerce that addresses factors that affect perceived effectiveness, credibility and use continuance using the PSD framework.

6 Conclusion

E-businesses have to do more than simply selling goods and services in order to maintain their advantage in a highly competitive market. The use of persuasive strategies has been identified as one means through which e-commerce companies can engage their clients. In this paper, we extended a previous study [9] where Amazon was evaluated as a persuasive system using the PSD framework. In the current study, we explored the factors that affect the perceived effectiveness, credibility and use continuance of e-commerce using the PSD framework. Using Amazon as a case study and a sample size of 324 Amazon shoppers, we developed and tested a research model using partial least-squares structural equation modelling (PLS-SEM) analysis.

Our results showed that perceived effectiveness of an e-commerce company like Amazon is a great predictor of continuance intention. In addition, social support and primary task support are strong predictors of perceived effectiveness. Furthermore, dialogue support significantly influences perceived product credibility and perceived review credibility and both constructs are strong predictors of system credibility.

This research can act as a guide to help e-commerce developers implement effective persuasive strategies. In the future, we intend to conduct a sub-group analysis by splitting the current data we have into two groups based on how participants were recruited. One group will include those recruited through social media without any reward or financial benefit while the other group will include participants that were recruited through Amazon's Mechanical Turk.

References

1. Asokan, N.: Fairness in electronic commerce (1998)
2. Loshin, P., Vacca, J.: Electronic Commerce, 4th edn. Charles River Media, Rockland (2004)
3. Adam, Nabil, R., Yesha, Y.: Electronic commerce: an overview. In: Adam, Nabil, R., Yesha, Y. (eds.) EC 1994. LNCS, vol. 1028, pp. 5–12. Springer, Heidelberg (1996). doi: 10.1007/3-540-60738-2_10
4. How to win online: Advanced personalization in e-commerce: An Oracle white paper (2011)
5. Alhammad, M.M., Gulliver, S.R.: Persuasive technology and users acceptance of e-commerce: exploring perceived persuasiveness. In: Zhang, R., Zhang, Z., Liu, K., Zhang, J. (eds.) LISS 2013, pp. 1099–1103. Springer, Heidelberg (2015)
6. Fogg, B.J.: Persuasive technology: using computers to change what we think and do. Ubiquity **2002**, 5 (2002)
7. Oinas-Kukkonen, H., Harjumaa, M.: Towards deeper understanding of persuasion in software and information systems In: 2008 First International Conference on Advances in Computer-Human Interaction, pp. 200–205 (2008)
8. Torning, K., Oinas-Kukkonen, H.: Persuasive system design: state of the art and future directions. In: Proceedings of the 4th International Conference on Persuasive Technology, p. 30 (2009)
9. Adaji, I., Vassileva, J.: Evaluating personalization and persuasion in E-commerce. In: Proceedings of International Workshop on Personalization in Persuasive Technology (2016)
10. Chu, H., Deng, Y., Chuang, M.: Persuasive web design in e-Commerce. In: Intenational Conference on HCI (2014)
11. Constantinides, E.: Influencing the online consumer's behavior: the web experience. Internet Res. **14**(2), 111–126 (2004)
12. Oinas-Kukkonen, H., Harjumaa, M.: A systematic framework for designing and evaluating persuasive systems. In: Oinas-Kukkonen, H., Hasle, P., Harjumaa, M., Segerståhl, K., Øhrstrøm, P. (eds.) PERSUASIVE 2008. LNCS, vol. 5033, pp. 164–176. Springer, Heidelberg (2008). doi:10.1007/978-3-540-68504-3_15
13. Lehto, T., Oinas-Kukkonen, H.: Explaining and predicting perceived effectiveness and use continuance intention of a behaviour change support system for weight loss. Behav. Inf. Technol. **34**(2), 176–189 (2015)
14. Cheung, C.M.K., Thadani, D.R.: The impact of electronic word-of-mouth communication: a literature analysis and integrative model. Decis. Support Syst. **54**(1), 461–470 (2012)
15. Patton, M., Jøsang, A.: Technologies for trust in electronic commerce. Electron. Commer. Res. **4**(1), 9–21 (2004)
16. Park, D.-H., Lee, J., Han, I.: The effect of on-line consumer reviews on consumer purchasing intention: the moderating role of involvement. Int. J. Electron. Commer. **11**(4), 125–148 (2007)
17. Chevalier, J.A., Mayzlin, D.: The Effect of word of mouth on sales: online book reviews. J. Mark. Res. **43**(3), 345–354 (2006)
18. Gefen, D., Karahanna, E., Straub, D.: Trust and TAM in online shopping: an integrated model. MIS Q. **27**(1), 51–90 (2003)
19. Liang, T.P., Ho, Y.T., Li, Y.W., Turban, E.: What drives social commerce: the role of social support and relationship quality. Int. J. Electron. Commer. **16**(2), 69–90 (2011)
20. Bhattacherjee, A.: An empirical analysis of the antecedents of electronic commerce service continuance. Decis. Support Syst. **32**(2), 201–214 (2001)
21. Wong, K.: Partial least squares structural equation modeling (PLS-SEM) techniques using SmartPLS. Mark. Bull. **24**(1), 1–32 (2013). Technical note 1

22. Hair Jr., J., Hult, T., Ringle, C., Sarstedt, M.: A Primer on Partial Least Squares Structural Equation Modeling (PLS-SEM). Sage Publications, Thousand Oaks (2016)
23. Fornell, C., Larcker, D.: Evaluating structural equation models with unobservable variables and measurement error. J. Mark. Res. **18**(1), 39–50 (1981)
24. Sanchez, G.: PLS Path Modeling with R. Trowchez Editions, Berkeley (2013)
25. Chin, W., Marcolin, B.: A partial least squares latent variable modeling approach for measuring interaction effects: results from a Monte Carlo simulation study and an electronic-mail. Inf. Syst. **14**(2), 189–217 (2003)

Author Index

Printed in the United States
By Bookmasters